Handbook of Turtles

Comstock Classic Handbooks

Handbook of Salamanders: The Salamanders of the United States, of Canada, and of Lower California, by Sherman C. Bishop

Handbook of Turtles: The Turtles of the United States, Canada, and Baja California, by Archie Carr

Handbook of Lizards: Lizards of the United States and Canada, by Hobart M. Smith

Handbook of Snakes of the United States and Canada, by Albert Hazen Wright and Anna Allen Wright

Handbook of

TURTLES

The Turtles of the United States,
Canada, and Baja California

By Archie Carr

Comstock Publishing Associates

A DIVISION OF

Cornell University Press

ITHACA AND LONDON

To my parents, whose affection withstood

the uneasy adolescence of a herpetologist

First published 1952 by Cornell University Press.
Ninth printing 1983
First printing, Cornell Paperbacks, 1995.

Library of Congress Cataloging-in-Publication Data
Carr, Archie Fairly, 1909–
    Handbook of turtles : the turtles of the United States, Canada, and Baja California / by Archie Carr.
        p.   cm. — (Comstock classic handbooks)
    Includes bibliographical references (p.   ) and index.
    ISBN 0-8014-0064-3 (cloth). — ISBN 0-8014-8254-2 (paper)
    1. Turtles—North America.   I. Title.   II. Title: Turtles.
III. Series.
QL666.C5C34   1995   94-44414
597.92′097—dc20

Printed in the United States of America

♾ The paper in this book meets the minimum requirements of the American National Standard for Information Sciences—Permanence of Paper for Printed Library Materials, ANSI Z39.48-1984.

CONTENTS

PART I. INTRODUCTION

PART II. ACCOUNTS OF SPECIES

DISTRIBUTION MAPS

Foreword to the 1995 Printing

J. Whitfield Gibbons

Archie Carr's *Handbook of Turtles* remains a classic. Following its publication in 1952, the book set the foundation and framework for the study of all aspects of the biology of turtles north of Mexico. No biologist interested in turtles was without a copy close at hand—and no turtle enthusiast should be without a copy today.

The *Handbook of Turtles* provides information on everything Archie Carr knew about the geographic range, distinguishing physical features, habitat, habits, reproductive traits, feeding, and perceived economic importance of all recognized species from the United States and Canada. From his own observations and those of others, Carr identified many pressing systematic and ecological problems.

Some biological mysteries, such as the increase of melanism in adult male *Trachemys scripta*, have yet to be explained fully (Lovich, McCoy, and Gartska, 1990). Carr's recognition that the genetic and ecological relationships within the *Pseudemys floridana (concinna)* complex offered "uncommon stimulation for the inquiring mind" has led to a spate of phylogenetic investigations (e.g., Crenshaw, 1965; Fahey, 1980; Ward, 1984; Seidel, 1994). And his history of diamondback terrapin exploitation remains one of the best chronicles of this still-beleaguered species.

When this book was first published, 38 species of turtles were known from North America outside of Mexico. Today, the most recent accounts recognize 55 native species, of which 34 have the same nomenclature as that used by Carr. Nomenclatural changes include eight elevations to the species level of variants recognized by Carr as subspecies, four changes in generic names based on phylogenetic interpretations, and eight species described since publication of the *Handbook of Turtles*. One addition results from a range extension of a primarily Mexican species, *Kinosternon hirtipes*, that Carr was unaware occurred in the United States. Two additional U.S. species, *Trionyx sinensis* and *T. steindachneri*, are now recognized as introductions from Asia that became naturalized in Hawaii following World War II. Because Hawaii was not a state until 1959, Carr did not include either species as part of the U.S. turtle fauna.

Yet, despite the biological knowledge gained from the many turtle studies conducted since 1952 and the resulting taxonomic changes, Carr's timeless perceptions and personal insights retain their value for the turtle biologist, as do his discussions of the basic biology of turtles and the still-to-be-solved problems. And the *Handbook of Turtles* has another enduring feature. It showcases Archie Carr's talent as a writer and storyteller. Most of the species accounts are filled with absorbing personal observations and anecdotes that have direct application to the issue at hand. Carr frequently draws on excerpts from his field notes to provide a complete picture of the turtles he is describing. This narrative approach of the field naturalist has disappeared from most of today's scientific literature. In reading the *Handbook of Turtles,* one wishes that style would reappear, if we still have writer-naturalists like Archie Carr.

REFERENCES

Crenshaw, J. W. 1965. Serum protein variation in an inter-species hybrid swarm of turtles of the genus *Pseudemys.* Evolution 19: 1-15.

Ernst, C. H., J. E. Lovich, and R. W. Barbour. 1994. Turtles of the United States and Canada. Washington, D.C., Smithsonian Institution Press.

Fahey, K. M. 1980. A taxonomic study of the cooter turtles, *Pseudemys floridana* (Le Conte) and *Pseudemys concinna* (Le Conte), in the lower Red River, Atchafalaya River, and Mississippi River basins. Tulane Stud. Zool. 22: 49-66.

Lovich, J. E., C. J. McCoy, and W. R. Gartska. 1990. The development and significance of melanism in the slider turtle. *In* J. W. Gibbons, ed., Life history and ecology of the slider turtle. Washington, D.C., Smithsonian Institution Press. Pp. 233-254.

Seidel, M. E. 1994. Morphometric analysis and taxonomy of cooter and red-bellied turtles in the North American genus *Pseudemys* (Emydidae). Chelonian Conserv. Biol. 1: 117-130.

Ward, J. P. 1984. Relationships of chrysemyd turtles of North America (Testudines: Emydidae). Spec. Publ. Texas Tech. Mus. 21: 1-50.

The following table compares the currently recognized nomenclature (based on Ernst, Lovich, and Barbour, 1994) of turtle species of the United States (excluding Hawaii) and Canada with Carr's taxonomy. Describer and year are given for taxa described since 1952. "Undescribed in 1952" indicates that the species was not recognized as a distinctive taxonomic entity in the *Handbook of Turtles.*

Current use (different from Carr)	Current use (same as Carr)	Use by Carr
	Chelydridae	
	Chelydra serpentina	
	Macroclemys temminckii	

Current use (different from Carr)	Current use (same as Carr)	Use by Carr
	Kinosternidae	
	Sternotherus carinatus	
	Sternotherus odoratus	
Sternotherus depressus Tinkle and Webb, 1955		Undescribed in 1952
Sternotherus minor		*Sternotherus carinatus minor*
	Kinosternon baurii	
	Kinosternon flavescens	
	Kinosternon sonoriense	
	Kinosternon subrubrum	
Kinosternon hirtipes		Not known from USA
	Emydidae	
	Clemmys guttata	
	Clemmys insculpta	
	Clemmys marmorata	
	Clemmys muhlenbergii	
Emydoidea blandingii		*Emys blandingii*
	Terrapene carolina	
	Terrapene ornata	
	Malaclemys terrapin	
	Graptemys barbouri	
	Graptemys geographica	
	Graptemys oculifera	
	Graptemys pseudo-geographica	
	Graptemys pulchra	
Graptemys ernsti Lovich and McCoy, 1992		Undescribed in 1952
Graptemys gibbonsi Lovich and McCoy, 1992		Undescribed in 1952
Graptemys caglei Haynes and McKown, 1974		Undescribed in 1952
Graptemys ouachitensis Cagle, 1953		Undescribed in 1952
Graptemys kohnii		*Graptemys pseudo-geographica kohnii*

Current use (different from Carr)	Current use (same as Carr)	Use by Carr
Graptemys versa		*Graptemys pseudogeographica versa*
Graptemys nigrinoda Cagle, 1954		Undescribed in 1952
Graptemys flavimaculata Cagle, 1954		Undescribed in 1952
	Chrysemys picta	
Trachemys scripta		*Pseudemys scripta*
Trachemys gaigeae		*Pseudemys scripta gaigeae*
	Pseudemys rubriventris	
	Pseudemys nelsoni	
	Pseudemys floridana	
Pseudemys concinna		*Pseudemys floridana concinna*
Pseudemys gorzugi Ward, 1984		Undescribed in 1952
Pseudemys alabamensis		*Pseudemys floridana mobiliensis*
Pseudemys texana		*Pseudemys floridana texana*
	Deirochelys reticularia	

Testudinidae
Gopherus agassizii
Gopherus berlandieri
Gopherus polyphemus

Cheloniidae
Chelonia mydas
Eretmochelys imbricata
Caretta caretta
Lepidochelys kempii
Lepidochelys olivacea

Trionychidae

Trionyx ferox		*Amyda ferox*
Trionyx muticus		*Amyda mutica*
Trionyx spiniferus		*Amyda ferox spinifera*

Dermochelyidae
Dermochelys coriacea

PREFACE

THIS is a book about the 79 species and subspecies of turtles that inhabit the United States, Canada, and Baja California. I included the Mexican state of Baja California because of its intimate geographic relation to California and because its small turtle fauna comprises only representatives of groups conspicuous in the fauna of the United States. The inclusion of more of Mexico would have complicated matters unduly by introducing tropical groups about which next to nothing is known.

The only other book on American turtles is Clifford Pope's *Turtles of the United States and Canada*, published in 1939. The existence of this work, which created a new criterion of excellence in the field of semipopular natural history, has greatly simplified for me the task of combing the literature for information on turtle habits. The mass of new data that has appeared during the thirteen years that Pope's book has been in print demonstrates the extent to which it stimulated the study of turtles in the field and in the laboratory.

The turtle nomenclature used here shows repeated, though for the most part trivial, divergence from that of *A Check List of North American Amphibians and Reptiles* (Stejneger and Barbour, 1943). Additions to, omissions from, and changes in that list are shown below. Where pairs of equivalent names are given, that used in the present book appears to the right of the form employed by the check list.

Sternotherus carinatus (Gray) = *Sternotherus carinatus carinatus* (Gray)
Sternotherus minor (Agassiz) = *Sternotherus carinatus minor* (Agassiz)
Sternotherus carinatus peltifer Smith and Glass added
Kinosternon flavescens (Agassiz) = *Kinosternon flavescens flavescens* (Agassiz)
Kinosternon steindachneri Siebenrock = *Kinosternon subrubrum steindachneri* Siebenrock

Clemmys marmorata (Baird and Girard) = *Clemmys marmorata marmorata* (Baird and Girard) and *Clemmys marmorata pallida* Seeliger

Terrapene carolina (Linné) = *Terrapene carolina carolina* (Linné)

Terrapene bauri Taylor = *Terrapene carolina bauri* Taylor

Terrapene major (Agassiz) = *Terrapene carolina major* (Agassiz)

Terrapene triunguis (Agassiz) = *Terrapene carolina triunguis* (Agassiz)

Malaclemys centrata centrata (Latreille) = *Malaclemys terrapin centrata* (Latreille)

Malaclemys centrata concentrica (Shaw) = *Malaclemys terrapin terrapin* (Schoepff)

Malaclemys pileata pileata (Wied) = *Malaclemys terrapin pileata* (Wied)

Malaclemys pileata macrospilota (W. P. Hay) = *Malaclemys terrapin macrospilota* W. P. Hay

Malaclemys pileata littoralis (W. P. Hay) = *Malaclemys terrapin littoralis* W. P. Hay

Malaclemys terrapin rhizophorarum Fowler added

Graptemys pulchra Baur added

Chrysemys bellii bellii (Gray) = *Chrysemys picta bellii* (Gray)

Chrysemys bellii marginata (Agassiz) = *Chrysemys picta marginata* Agassiz

Pseudemys alabamensis (Baur) omitted

Pseudemys bangsi Babcock = *Pseudemys rubriventris bangsi* Babcock

Pseudemys concinna concinna (Le Conte) = *Pseudemys floridana concinna* (Le Conte)

Pseudemys concinna hieroglyphica (Holbrook) = *Pseudemys floridana hieroglyphica* (Holbrook)

Pseudemys concinna hoyi (Agassiz) = *Pseudemys floridana hoyi* (Agassiz)

Pseudemys concinna mobiliensis (Holbrook) = *Pseudemys floridana mobiliensis* (Holbrook)

Pseudemys concinna suwanniensis (Carr) = *Pseudemys floridana suwanniensis* Carr

Pseudemys gaigeae (Hartweg) = *Pseudemys scripta gaigeae* Hartweg

Pseudemys nebulosa (Van Denburgh) = *Pseudemys scripta nebulosa* (Van Denburgh)

Pseudemys rubriventris (Le Conte) = *Pseudemys rubriventris rubriventris* (Le Conte)

Pseudemys scripta (Schoepff) = *Pseudemys scripta scripta* (Schoepff)

Pseudemys texana Baur = *Pseudemys floridana texana* Baur

Pseudemys troostii troostii (Holbrook) = *Pseudemys scripta troostii* (Holbrook)

Pseudemys troostii elegans (Wied) = *Pseudemys scripta elegans* (Wied)

Chelonia mydas (Linné) = *Chelonia mydas mydas* (Linné)

Chelonia agassizii Bocourt = *Chelonia mydas agassizii* Bocourt

Eretmochelys imbricata (Linné) = *Eretmochelys imbricata imbricata* (Linné)

Eretmochelys squamata Agassiz = *Eretmochelys imbricata squamata* Agassiz

Caretta caretta (Linné) = *Caretta caretta caretta* (Linné)

Caretta caretta gigas Deraniyagala added

Dermochelys coriacea (Linné) = *Dermochelys coriacea coriacea* (Linné)

Dermochelys schlegelii (Garman) = *Dermochelys coriacea schlegelii* (Garman)

Amyda ferox hartwegi Conant and Goin added

Amyda agassizii (Baur) = *Amyda ferox agassizii* (Baur)

Amyda emoryi (Agassiz) = *Amyda ferox emoryi* (Agassiz)

Amyda ferox (Schneider) = *Amyda ferox ferox* (Schneider)

Amyda spinifera spinifera (Le Sueur) = *Amyda ferox spinifera* (Le Sueur)

Amyda spinifera aspera (Agassiz) = *Amyda ferox aspera* (Agassiz)

A four-and-a-half-year interlude in Honduras separated the initial and final stages in the preparation of the present manuscript, and it is probably foolhardy for me to attempt to list from memory the many people who have helped me with it in one way or another. Despite the inevitability of omissions, however, I must acknowledge the kindness of the following individuals, who supplied specimens, information, or illustrations, who called my attention to references or collaborated in searching the literature, or who read and criticized the manuscript: Lewis Berner, Sherman C. Bishop, Fred Cagle, Thomas D. Carr, Alphonse Chable, Roger and Isabelle Hunt Conant, John W. Crenshaw, Coleman and Olive Goin, Arnold Grobman, Clara Harris, Norman Hartweg, Bessie Hecht, Robert Hellman, Margaret Hogaboom, Arthur Loveridge, Lewis Marchand, Wilfred Neill, James Oliver, Amado Pelen, J. Speed Rogers, Karl P. Schmidt, H. B. Sherman, Kirk Strawn, Frank and Frances Young, and Eric Waering.

During the years of my association with the United Fruit Company I accumulated indebtedness to a number of its officials in the tropics. Of those who helped me further my acquaintance with sea turtles I owe special thanks to Wilson Popenoe, Director of Escuela Agrícola Panamericana, Honduras, and to Henry Hogaboom of La Lima, Virgil Scott of Tegucigalpa, and Paul Shank of El Zamorano.

The co-operation of my friends Charles M. Bogert, Leonard Giovannoli, and J. C. and Lucy Dickinson and the constant aid and encouragement of my wife, Marjorie Harris Carr, have done much to lighten the task of compilation, while Ross Allen, with customary generosity, placed at my disposal the facilities of his establishment as a source of pertinent material. For the line drawings I am indebted to Esther Coogle, Staff Artist, Department of Biology, University of Florida, and to Doris Cochran, United States National Museum, who contributed also data and photographs from among the notes of the late Leonhard Stejneger. Marine Studios, Marineland, Florida, have been most generous in furnishing data on sea turtles from their files and access to their collections and exhibits, both before and since the tragic death of Arthur McBride.

The majority of the photographic illustrations used here (all those not otherwise acknowledged except a few made by me) are the work of Albert Hazen Wright and Anna Allen Wright, and to them I am grateful for many other favors as well.

For typing the main body of the text, I wish to thank the staff of the Personnel Office, University of Florida, and for additional typing and bibliographic work I am indebted to Irene Ruibal, The American Museum of Natural History. Without the wholehearted co-operation of the library staffs of The American Museum of Natural History and the Museum of Comparative Zoology, the bibliography could not have been assembled in its present form.

PART I

INTRODUCTION

TWO hundred million years ago the reptiles, newly arisen from an uncommonly doughty set of amphibians, were on the verge of great adventures. They bore the mark of destiny in the shape of impervious scales and the new cunning to lay shelled eggs, and these devices insured them against the age-old disaster of drying out, both before birth and after, and let them gratify their growing curiosity about the vast and almost empty land. Along with the new equipment they had imagination and no end of notions for novel body designs. Today we call these old beasts cotylosaurs, or stem reptiles, because all the lines of vertebrate life above the amphibian level lead back to them as branches converge in the trunk of a tree.

The first of the innovations made by the stem reptiles was in a way the most extraordinary and ambitious of all—the most drastic departure from the basic reptile plan ever attempted before or since. By a cryptic series of changes, few of which are illustrated in the fossil record, there evolved a curious and improbable creature which, though it retained the old cotylosaur skull (with no opening in the temporal region), had a horny, toothless beak and a bent and twisted body encased in a bony box the like of which had never been seen. And more than this, within the box the girdles connecting the legs with the rest of the skeleton had by some legerdemain been uprooted and hauled down to an awkward position underneath the ribs.

The new animal was a turtle. Having once performed the spectacular feat of getting its girdles inside its ribs, it lapsed into a state of complacent conservatism that has been the chief mark of the breed ever since.

Of course, the turtle was not completed overnight. Just when he was finished is not clear. We know that there were no turtles in the Carboniferous, that in the Triassic there were very satisfactory ones, and that by Cretaceous times most of the modern trends had been established, and a lot besides that were later cut off. But no one knows with certainty what

turtles looked like during the several million years between the Carboniferous and the early Mesozoic. Almost the only glimpse permitted us is the
skeleton of a reptile from the Permian of South Africa, which has teeth
in the jaws and a wide, flat body and conspicuously broadened ribs that
seem to forecast the bony turtle carapace.

This animal, called *Eunotosaurus,* is intriguing to turtle genealogists
because it appears to substantiate their postulated derivation of turtles from
cotylosaurs. It comes from the proper time, it is evidently a modified
cotylosaur, and it appears to be on the verge of developing a shell. It has,
however, one serious drawback as a hypothetical turtle ancestor. The
carapace of modern turtles does not, as is often claimed, develop just from
wide ribs but from independent plates of dermal bone which expand
markedly and fuse with one another and with the underlying ribs and
vertebrae. There is no such carapace as this in *Eunotosaurus,* nor is there
any bottom shell or plastron, which turtles seem to have pieced together
from the abdominal ribs of the primitive reptiles, finished off in front by
parts of the old dermal shoulder girdle (clavicle and interclavicle).

The unique chelonian armor and the contortions the skeleton had to
undergo to fit into it, combined with the toothless beak, place the modern
turtles in a position apart from all living reptiles. It has even been proposed
(by Jaekel, 1910) to separate them entirely from other reptiles and place
them, along with the duck-billed platypus, the echidna, and a few other
misfits, in a new class of vertebrates (Paratheria). Most modern students,
however, agree with Williston and Gregory in their belief that the turtles
are direct descendants of the stem reptiles.

But what forces in the environment wheedled or coerced the plastic
cotylosaurs along this particular line of specialization? To this question all
sorts of explanations have been proposed. It has been suggested that the
turtle ancestor learned to curl into a ball like an armadillo to conceal itself
within a flexibly segmented armor now stiffened to form the shell. This
idea was introduced as a possible explanation of the lack of coincidence
between the laminae and the plates of the inner bony shell. One recent
author envisioned the primitive turtle as a haunter of dense brush where
a heavy shell permitted a bulldozerlike locomotion through the restraining
vegetation, but most paleontologists will probably agree with Schmidt
(1939) that this "seems to be sheer nonsense." Hay (1898) believed the extra
scale rows on the shell of the alligator snapper to be all that was needed to
establish the homology of the shell laminae of ordinary turtles and the
twelve longitudinal keels of the widely divergent leatherback turtle, and

accordingly concluded that *Macrochelys* is the most primitive living turtle and most like the ancestral form.

The consensus of opinion sees the first turtles as marsh reptiles like their cotylosaur forebears. They were almost certainly scaly and similar to lizards in form, and the series of changes that they underwent may have attended the development of a defense reaction by which the shoulders were hunched and the back bowed to draw the head back out of harm's way.

In many cases some inkling of the historical origin of an anatomical feature may be gained by studying its development in a growing embryo. The occurrence of lateral folds in turtle embryos, for instance, probably means that the ancestral form had these structures, which are today found only in lizards. It might accordingly have been hoped that the evolution of the relationship between the shell, ribs, and girdles during embryology would shed some light on the original history of these events, but such is not the case. Ruckes (1929a) presented the first adequate account of the embryonic changes that result in the unorthodox orientation of the girdles, but as one writer (Walker, 1947) puts it, "it is felt that phylogeny could not have followed the same course." In other words, the internal gymnastics required to shift the position of the girdles are so strenuous that it is hard to believe that the intermediate stages represent the design of ancestral forms.

Ruckes showed that the girdles do not, as formerly supposed, change their relative positions by migrating downward. They actually undergo no absolute motion, the whole change in the relative positions of skeletal parts being due to the manner of growth of the carapace and plastron. According to Ruckes, the shell rudiment, as seen in an embryo of the limb-bud stage, consists of a dermal band surrounding the embryo between the girdles, and readily evident on the middorsal line as a thickened strip. This middorsal area is the site of maximum growth, and since growth proceeds in all directions, the active, outer edges of the dermal area soon come to overlap the pectoral and pelvic regions. This radial spreading of the shell rudiment takes place while the ribs are still mere spurs projecting at right angles to the vertebrae. As the ribs begin their development, they become intimately associated with the spreading shell elements above them and these so influence the course of their growth that they begin to curve downward only after their ends have passed laterally beyond the girdles.

However this skeletal rearrangement may have come about historically, it satisfied the early turtles and allowed full expression of their philosophy of meditation and passive resistance. Down through the faraway Permian they sat in their shells and meditated as great events took shape. The coal forests withered, and with the coming of a new climate and flora the archaic animal types began to drop out. Winged insects arose and the reptiles grew in vigor and restlessness. The synapsidan clan moved aside to begin the experiments which, epochs later, produced the mammals.

The Permian ended, and the turtles watched as the main reptile stock found its evolutionary stride and through a hundred million years staged the most dramatic show the world has ever seen—the rise and spread and the incomprehensible decline of the incredible archosaurs. The turtles remained conservative through it all and though some of them took to the sea—sacrificing parts of the beloved shell for greater bouyancy and producing such multiton monsters as *Archelon,* with a twelve-foot flipper spread, and *Meiolania,* with a horned skull two feet wide—they always clung to their basic structural plan, as other lines tested and exploited and abandoned a thousand specious schemes.

They remained unimpressed as *Pteranodon* cruised the skies and another strain of slim and athletic archosaurs devised Archaeopteryx and the birds, and as the Squamata dabbled in mososaurs and snakes. They remained turtles; they even began to prosper as never before, while the dinosaurs bellowed and pounded down through the Jura toward their utter and senseless doom in the Cretaceous, when the last *Brachiosaurus* laid down his fifty tons to rest and the final tyrannosaur gasped out the anticlimax to nature's greatest venture in mayhem. (Among those present when Triceratops fell was a genus of turtles, now known as *Podocnemis,* which to this day populate in placid abundance the streams of Brazil and Madagascar!)

The Cenozoic came, and with it progressive drought, and the turtles joined the great hegira of swamp and forest animals to steppe and prairie, and watched again as the mammals rose to heights of evolutionary frenzy reminiscent of the dinosaurs in their day, and swept across the grasslands in an endless cavalcade of restless, warm-blooded types. Turtles went with them, as tortoises now, with high shells and columnar, elephantine feet, but always making as few compromises as possible with the new environment, for by now their architecture and their philosophy had been proved by the eons; and there is no wonder that they just kept on watching as *Eohippus* begat Man o' War and a mob of irresponsible and shifty-eyed little shrews swarmed down out of the trees to chip at stones, and fidget around fires, and build atom bombs.

Turtle Functions and Capacities
RESPIRATION

The respiratory apparatus of turtles includes a glottis, which opens to receive air that has entered the pharynx from the internal nares and to admit it into the larynx, supported by the hyoid apparatus and connected with the trachea or windpipe. The trachea divides within the thorax into two bronchi, one of which enters each lung. The lungs are complex sacs with a large number of interior folds and subdivisions.

It is evident that the breathing process of a turtle cannot be wholly similar to that of an animal with a distensible thorax. Just how air is drawn in and expelled from the lungs has been much debated. Despite the fact that several early writers suggested the real character of the breathing mechanism, it has been repeatedly stated that the movements of the throat of a turtle represent impulses by which the hyoid apparatus pumps air into the larynx, as in frogs. It now seems more probable, however, that the hyoid movements are involved only in the olfactory sense and that breathing is accomplished more as in mammals than as in amphibia. According to Hansen (1941) and McCutcheon (1941), inhalation is effected by muscles located at each leg pocket and beneath the viscera. These operate like the mammalian diaphragm, contracting, enlarging the coelom and reducing pressure within it, and thus drawing air into the lungs. During inspiration the shoulder girdle rotates forward, but this is thought to be merely a passive movement. Expiration is produced by the contraction of two pairs of ventral muscles which push the viscera against the lungs and compress and deflate them.

To augment their oxygen supply, many aquatic turtles use the highly vascular pharyngeal cavity as a sort of gill, sucking in and expelling water and obtaining by this means sufficient oxygen to increase materially their capacity for remaining submerged. In a similar way, additional underwater respiration is effected by some species that augment the work of the pharynx by filling and emptying, through the anus, two thin-walled sacs that communicate with the cloaca. The currents set up by these pumping movements may be easily demonstrated if a small amount of dye or suspended silt is placed near the anal or nasal openings of a live turtle.

The possibility that soft-shelled turtles may be more dependent upon their aquatic respiration than has been suspected may be indicated by the observations of Dr. O. Lloyd Meehean, who during his fisheries studies in Florida lakes used large amounts of derris in poisoning lakes for fish population analyses. Dr. Meehean found that in nearly all the lakes which he poisoned, *Amyda ferox ferox* was killed in numbers, while no other turtle was affected. Since rotenone, the active principle of derris, apparently works through the respiratory system to kill fishes, one is tempted to conclude that *Amyda* is susceptible because of its cloacal or pharyngeal respiration. Lumsden (1924) stated that the respiratory needs of turtles are so slight that one inspiration may suffice for as long as two hours. There is a record of a turtle surviving for twenty-four hours in an atmosphere of pure nitrogen, and the difficulty of drowning turtles at ordinary temperatures is well known to collectors. A writer reported that one of his turtles lived under water for eight days, presumably satisfying its respiratory needs by anal and pharyngeal breathing.

Shaw and Baldwin (1935) suggested that the ability of turtles to go for long periods without breathing when on land is due to an ability to ventilate the lungs thoroughly, combined with a low "unloading tension" of chelonian hemoglobin as well as to unusually low oxygen requirements.

CIRCULATION

The heart of the turtle is three-chambered, with two auricles and a single ventricle. Tendency toward the four-chambered condition is seen in a rudimentary partition that incompletely divides the ventricle. Venous blood returns to the heart from the body over a large postcaval and two precaval veins and passes into the right auricle by way of the *sinus venosus*. From here it enters the right side of the ventricle, and when this contracts the blood is pushed out either through the pulmonary artery to the lung or through the left aorta, whence it may go directly to the viscera or enter the dorsal aorta. The pulmonary circuit is completed when oxygenated blood from the lungs returns to the left auricle through the pulmonary veins. This blood passes into the left side of the ventricle and with the next contraction it is pumped out through the right aorta, which joins the dorsal aorta. Owing to the incomplete nature of the septum in the ventricle, the blood is partially mixed there, and that which enters the aorta includes oxygenated blood from the left auricle and venous blood from the right auricle.

There are a renal portal system and a well-developed hepatic portal system.

EXCRETION

The excretory organs are two kidneys, each of which communicates by way of a ureter with the cloaca. The excretory product is stored in a bladder which in some species may be markedly distensible. The nitrogenous part of the urine of some turtles, at least, is insoluble uric acid instead of soluble urea, and the importance of this as a device permitting water conservation is mentioned in the discussion of *Gopherus agassizii.* It is of interest that the urine of the wholly aquatic *Chelonia mydas,* which is obviously never faced with the problem of conserving water, was found by Khalil (1947) to contain ammonia as the principal nitrogen end product. The bladder communicates with the outside by way of the cloaca and the anus.

DIGESTION

The digestive system of the turtle shows no distinctive features. In the mouth there is a broad tongue, fixed immovably to the floor of the cavity. Dorsally the mouth receives the internal nostrils (choanae of the skull), and behind the tongue a longitudinal slit or valve, the glottis, marks the divergence of the respiratory tube from the alimentary canal. A thin-walled pharynx leads into the narrow, thick-walled esophagus, and this communicates with the stomach. The narrow passage from the stomach into the small intestine is controlled by the pyloric valve, and the ileocecal valve separates the small intestine from the large intestine. Bile secreted by the liver reaches the small intestine by way of the bile duct, and the pancreas contributes its enzymes to the intestine by way of several ducts.

REPRODUCTION

Turtles reproduce sexually and fertilization is internal. The male copulatory organ is a distensible, grooved, unforked penis, attached to the fore wall of the cloaca and receiving the ends of a pair of *vasa deferentia* that bring spermatozoa from the two testes. The female organs include a pair of ovaries, each of which communicates with the cloaca by an oviduct.

All female turtles lay eggs; these are spherical or elliptical, usually white, and with a shell that varies from soft and leathery to calcareous and brittle. In some cases the shell is soft when first laid but hardens during incubation. Soft-shelled eggs usually have a dent in one side which normally disappears after a variable period of time in the nest.

Burger (1937) found that the sexual cycle of *Pseudemys s. elegans* could be initiated by gradually lengthened periods of illumination and suggested that the sexual behavior of turtles might be largely controlled by light, as in birds.

Courtship varies from the simple pursuit and mounting of the female by the male to a fairly complicated program of prenuptial tilting during which the male may bite the neck and shell of the female or butt and push her about, in some cases accompanying his activities with vocal noises. The most divergent courtship behavior appears to be that of *Chrysemys* and the *scripta* section of *Pseudemys,* in both of which the male swims backward before the female as he strokes or taps her face with his greatly elongated fingernails. Copulation occurs in the water in aquatic forms and either in the water or on land in terrestrial species. In some cases, notably in the sea turtles, the male grasps the anterior rim of the carapace of the female during coitus, while in others contact is established only posteriorly, where the hind legs of the male usually hook under the posterior edge of the shell of the female. There are almost no data on duration of copulation in turtles. There is one printed mention of a five-minute coital period, but the average is certainly much longer and in some cases, at least, may possibly continue for hours. Among the many bizarre amatory feats that popular legend ascribes to sea turtles, I find one that I believe will escape editorial censorship. While it is of course apocryphal, it is entertainingly so, and it also serves to emphasize the remark that we really know almost nothing about the subject:

All the turtles from the Charibbeas to the Bay of Mexico, repair in Summer to the Cayman Islands. . . . They coot for 14 days together, then lay in one Night three Hundred Eggs, with White and Yolk but no shells. Then they coot again, and lay in the Sand; and so thrice; when the Male is reduced to a kind of Gelly within, and blind; and is so carry'd home by the Female [Old-mixon, *The British Empire in America,* 1708, Vol. ii, p. 338; as quoted by Lewis, 1940].

Darwin long ago observed that Galápagos tortoises roar during coitus, and Evans and Quaranta (1949) and Evans (1949) stated that the roars are emitted throughout the duration of the act, by captive specimens, at intervals of about five seconds.

Spermatozoa may be stored in the genital tract of the female and continue to fertilize eggs there for periods of at least four years after copulation, although the percentage of fertile eggs in clutches produced after the first season decreases progressively.

Sexual dimorphism usually includes a longer, thicker tail in the male, since the penis is housed in the base of the tail and is extruded from the anus near the tip. Besides being longer and with more terminally located vent, the tail of the male is more strongly prehensile than that of the female, being in some species markedly so, and is sometimes equipped with an enlarged conical terminal scale that serves as an instrument to use in grappling for copulatory contact. In some species the plastron of the male is concave, to accommodate the convex carapace of the female during coitus. As was mentioned above, in a few emydid turtles, and most strikingly in the *floridana* section of *Pseudemys,* the nails of the fore feet are markedly elongate. In the marine species, a single nail on each flipper of the male is enlarged and curved inward for grasping the shell of the female.

Sexual differences in coloration are most marked in the *scripta* section of *Pseudemys.* The development of a secondary melanistic coloration by some old males is described under the account of that group. In species in which the male is much smaller than the female, the former usually tends to retain the vivid juvenile color pattern, while this fades in the larger females.

The sexes may be of similar size, or either may be to varying degrees the larger. The most extraordinary disparity in size is found in *Graptemys* and *Malaclemys.* In *Graptemys barbouri,* for instance, males mature at less than 100 mm. (3.94 inches) in shell length, while most of the mature females have shells about 200 mm. (7.88 inches) long and some may exceed 250 mm. (9.85 inches).

In some turtles (*Graptemys, Malaclemys,* some *Pseudemys*) the heads of the older females may become astonishingly enlarged; in other forms (Kinosternidae) enlarged heads may be characteristic of old males.

It has been observed that the female box turtle is more timid than the male; in aquatic species the reverse is frequently true.

Sex is determined in turtles (in one form at least) by a difference in the egg cells, part of which have one more chromosome than the rest and produce ma'es when fertilized, while those lacking the extra chromosome make females. In many species, at least, females outnumber males. Forbes (1940) pointed out that most reptiles deviate from the usual 1:1 ratio. Risely (1933) found 2.3 females for each male of *Sternotherus odoratus,* and for diamondback terrapins (*M. t. centrata*) Hildebrand (1933) derived a ratio of 5.9 females per male. Marchand (1942, MS) counted 3.6 females for every male of Tennessee *Chrysemys p. dorsalis.* Of 281 desert tortoises examined by Woodbury and Hardy (1949), 151 were females and 101 were males (sex not determined in 29).

The eggs of turtles are laid in holes in soil, sand, or decaying vegetation

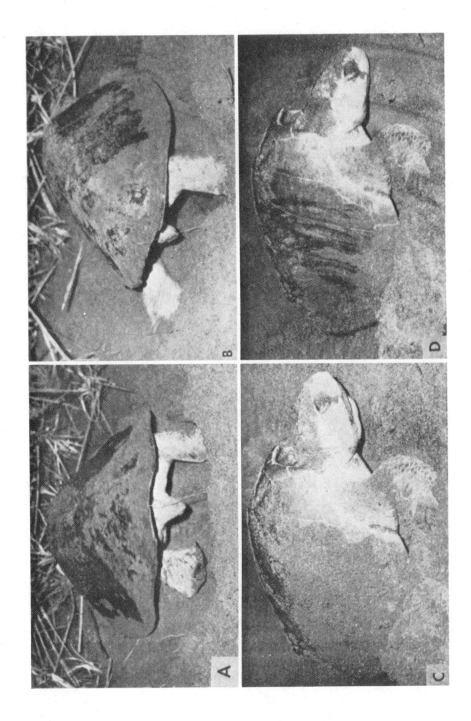

Plate 1. (*Facing page.*) A series of flashlight photographs of the nesting activities of a Pacific ridley (*Lepidochelys olivacea*) on Isla de Ratones, Honduras. *A:* Digging the nest. Rear view of a turtle with a flipperful of sand poised before being thrown to the side. This feat of delicately curling the edges of a highly modified swimming appendage and meticulously removing sand with it from a narrow-mouthed hole 18 inches deep never fails to astonish the observer. *B:* Sand from the deepening hole is thrown to one side. *C:* Head up. A bunch of eggs will fall into the nest momentarily. Note the closed and tearful eyes and the protruding hyoid apparatus. *D:* Head down. At this moment the eggs are falling into the nest.

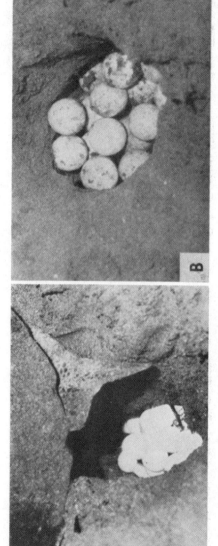

Plate 2. A continuation of the series of photographs showing the nesting activities of *Lepidochelys olivacea*. *A:* A clutch half laid. Removal of the back wall of the nest appeared to cause the turtle no anxiety. She merely pressed a flipper against the side of the cavity to prevent caving and continued to lay. The egg with the blurred upper outline is falling. *B:* One hundred and twenty eggs ready for covering. This complement was laid by a turtle that was turned as she finished laying so that the nest with the eggs in place but still uncovered might be photographed.

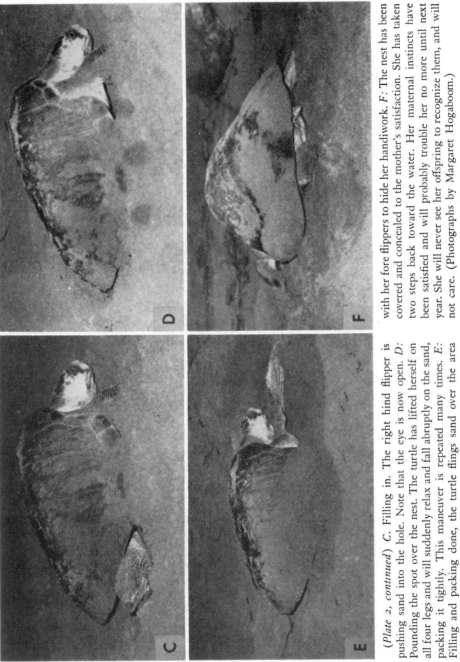

(*Plate 2, continued*) *C.* Filling in. The right hind flipper is pushing sand into the hole. Note that the eye is now open. *D*: Pounding the spot over the nest. The turtle has lifted herself on all four legs and will suddenly relax and fall abruptly on the sand, packing it tightly. This maneuver is repeated many times. *E*: Filling and packing done, the turtle flings sand over the area with her fore flippers to hide her handiwork. *F*: The nest has been covered and concealed to the mother's satisfaction. She has taken two steps back toward the water. Her maternal instincts have been satisfied and will probably trouble her no more until next year. She will never see her offspring or recognize them, and will not care. (Photographs by Margaret Hogaboom.)

such as leaf mold or rotten wood. The nests are dug by the female. She may excavate a preliminary basin by sweeping motions of the fore legs; the nest proper, however, is dug by the hind feet, which work alternately throughout the operation to form a narrow-mouthed, more or less flask-shaped hole of a depth determined by the length of the stretched hind leg. The process is sometimes, but not always, attended by the wetting of the soil by cloacal bladder water which softens it and makes digging easier. It is often stated that the chief purpose of this wetting down of the site of the cavity is to provide a moist environment for the eggs. Cunningham (1923), however, showed that the eggs do not take on water during the first few days after deposition, and it appears likely that the water from the cloaca is designed merely to improve the mechanical condition of the soil.

As the eggs are laid, the female usually makes some attempt to arrange them in the bottom of the nest with a hind foot. The covering process, likewise, is performed by the hind feet, which rake in soil from behind and from the sides. When the hole has been filled the site is pressed and trodden by the hind feet and may be pounded by the plastron. Some effort at concealment, such as the flinging about of sand with the fore feet or the kicking of sand by the hind feet, may be made. Females of some species crawl back and forth across the site to obliterate signs of their work.

Among the mud turtles of the genus *Kinosternon* the nesting instinct is poorly developed, some of them often making no nest at all, but planting their eggs singly in shallow holes and sometimes only half covering them.

The number of eggs in a clutch varies from one (the single large egg of the African *Testudo tornieri*) to a maximum obscured by the numerous unfounded guesses and estimates that have appeared in print, but which probably lies between 200 and 300. The largest complement recorded for any North American fresh-water turtle was one of eighty eggs deposited by a Canadian snapper. There is some correlation between the size of the species and the size of the clutch, and considerable correlation between the size of the individual and that of the egg complement. A large mature turtle may lay as many as twice the number of eggs deposited by a newly matured and smaller individual. Many turtles lay more than once a year, seven visits to the nesting beaches (by a green turtle) being the maximum number that has been definitely demonstrated. There is evidence that some fresh-water turtles may lay at least three times during a season, although others almost certainly lay all their eggs for the season

Plate 3. A captive peninsular turtle (*Pseudemys f. peninsularis*) laying its eggs. (Courtesy Ross Allen.)

at one time. There is a conspicuous need for further investigation of this subject.

Incubation periods normally range between two and three months, but are so strongly affected by humidity and temperature that no given species adheres very closely to a definite schedule. Overwintering is common, both by unhatched eggs and by hatchlings that may await either the more pro-

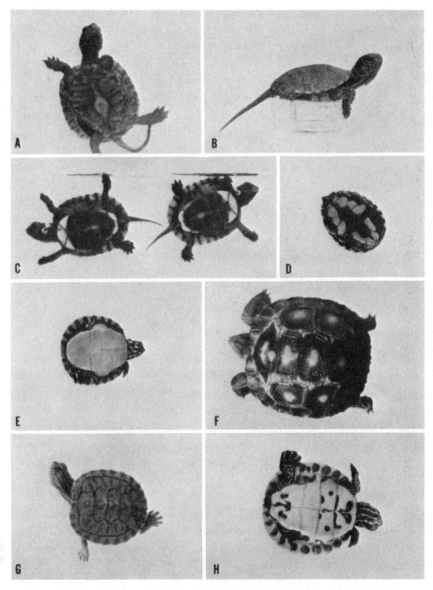

Plate 4. Young turtles. *A: Clemmys insculpta;* Ithaca, New York. *B: Clemmys marmorata pallida;* Perris, California. *C: Clemmys muhlenbergii;* New York. *D: Kinosternon subrubrum steindachneri;* Gainesville, Florida. *E: Chrysemys picta dorsalis;* Plaquemine, Louisiana. *F: Gopherus berlandieri;* Edinburg, Texas. *G: Pseudemys scripta elegans;* Plaquemine, Louisiana. *H: Pseudemys floridana mobiliensis;* Plaquemine, Louisiana.

pitious temperatures of spring or the spring rains that soften the roof of the nest and free them. Thus the time actually required to hatch the eggs and that which elapses prior to the emergence of the young from the nest may be very different, and for this reason there is available little reliable information on natural incubation periods.

Interspecific hybridization among turtles apparently is not common. Hildebrand (1933) described reciprocal crosses between the Texas and northern diamondback terrapins, but these forms have subsequently proved to be intergrading races of one species. Shaw found an aberrant sea turtle that he suspected might be a hybrid between *Caretta caretta* and *Lepidochelys olivacea,* and in many places fishermen believe that the loggerhead crosses regularly with both the green turtle and the hawksbill, but this has never been demonstrated. There are indications that *Pseudemys floridana peninsularis* and *P. nelsoni* may sometimes interbreed in the extreme southern part of peninsular Florida. Risely (1941) suggested that the occurrence of hermaphroditic turtles may be the result of hybridization, but the cases which he cited involved intergrade specimens between subspecies and not the progeny of interspecific crosses.

SENSES, BEHAVIOR, AND INTELLIGENCE

The chelonian nervous system has no marked peculiarities. It is considerably in advance of that of amphibians, the cerebral hemispheres being much more prominent and a gray cerebral cortex being distinguishable from a white medulla. The cerebellum is also larger, indicating a greater capacity for co-ordination and correlation of movements. There are twelve pairs of cranial nerves. Some notes on senses and reactions are given below.

Hearing and voice: Anyone who has watched a dozen turtles slide off a log in concerted response to a slight noise a half mile down-river will be loath to accept the pronouncement that turtles do not hear well, but such appears to be the case. Although the auditory apparatus is complete, there being a tympanic membrane connected with a well-developed inner ear by a slender columellar bone and a Eustachian tube leading from the middle ear into the pharynx, atmospheric vibrations are apparently not picked up readily and what appears to be hearing is often actually the "feeling" of vibrations of the water or of the substratum. It is thus curious to note that a diverse scattering of species is able to emit sounds that have appeared to numerous writers to warrant the name "voice."

Some instances of turtle vocalization are surely illusory, like that of an early New England naturalist who wrote of *Chrysemys:* "The shrill, piping note of this species is frequently heard in May and June—especially during intervals between showers on hot, sultry days."

Cope (1865) quoted Berendt's assertion that the Mexican *Staurotypus* has "two very distinct voices" and that *Geomyda* makes "a soft, melancholy piping that is rather touching when killed." Both a mercy cry and a roar or grunt of anger are attributed to sea turtles, the leatherback and green turtle being most frequently mentioned in this respect. An old generic name of the leatherback (*Sphargis*) is derived from a Greek word signifying "to make a noise."

In many cases the "voice" credited to turtles is merely sound incidental to the exhalation of the breath or to frictional contact between parts of the body. I have several times heard from mud turtles squeaks or yelplike noises which I am sure were caused by the "gnashing" of their jaw surfaces, and similar sounds are occasionally made by *Terrapene* in closing its shell. On the other hand, there is no question but that some species bark or grunt as a regular and perhaps functional feature of their mating program, and in such cases the word "voice" is no misnomer even though the sounds are not made by orthodox vocal cords.

The enlarged scales on the inner surface of the hind leg of kinosternid turtles have often been called "stridulating organs," one writer even professing to have heard them in use. It is now believed, however, that the big scales merely help the male hold the shell of the female during copulation.

In the Florida backwoods there are still plenty of people who will tell you that the baby-chick cheep of oak toads in chorus is made by scorpions (by which they mean lizards of the genus *Eumeces*), and in the Nicaraguan rain forest I was repeatedly assured by otherwise astute machetemen that the baritone trill of the big marine toads was the voice of the *terciopelo* or fer-de-lance. When such intrenched misconceptions as these, and all instances of merely incidental mechanical noise, have been discounted in the case for the voice of the turtle, there still remain a few examples of genuine vocalization that would well repay further investigation.

Smell and taste: It has been repeatedly stated that the sense of smell is acute in turtles, and while this probably is true, there are few experimental data to support the assertion. The pulsating movements of the throat are thought to be connected in some way with the olfactory sense. Allard (1949) found no evidence that box turtles are capable of long-range smelling, and performed a simple experiment to determine whether odors emanating from nearby objects are perceived. He wrapped a fish in burlap, made a similar package containing only a stone, and placed both on the ground near a number of box turtles. The bundle containing the stone was not noticed, but one turtle made a considerable effort to open that containing the fish.

The presence in turtles of axillary and inguinal (and other) scent glands is common but the odors secreted by them may be purely protective, and the extent to which the animals may use them in keeping track of each other is not known.

The nature and relative acuity of the sense of taste in turtles has not been determined. Its existence is probably indicated by the fact that turtles often take into their mouths bites of substances that are subsequently spat out, presumably owing to their unpalatability. The distinction between taste and olfaction in aquatic forms is probably academic.

Touch: The sense of touch is well developed, although so intimately involved with the vibratory sense that the two are at times difficult to distinguish. Even a hulking horny-skinned tortoise is able to feel the contact of a straw tip dragged along its flank. A delicate tactile sense is demonstrated by a nesting female arranging her eggs with a hind foot, and the sensitivity of the partly everted cloaca of the nesting ridley to variations in the texture of the soil is extraordinary.

Behavior, intelligence, and visual perception: The magnetism of turtle personality stems more from good-humored quaintness and elfin drollery than from intellectuality. Turtles, it must be said, are not very intelligent. If this remark engenders in the bosom of the cheloniophile the urge to disprove my statement by citing the genius of a pet turtle that comes to the table to lift an appealing foot and gaze wistfully up at the diners, or even at some one especially beloved diner, my rebuttal is ready. I have seen a possum and a salamander do the very same thing, and surely these must represent the nadir of cerebral evolution.

Tinklepaugh (1932) averred that the wood turtle equaled the "expected accomplishment of a rat" in learning the intricacies of an experimental maze, but one must conclude that Tinklepaugh had known only feeble-minded rats.

Yerkes (1901) found that terrestrial turtles hesitated to walk off platforms a foot in height, and that when placed in positions six feet in elevation they were obviously apprehensive. Aquatic turtles, on the contrary, showed little concern at either elevation, an attitude that Yerkes charitably attributed to the impunity that attends their falling from logs into the water.

A hundred and twenty years before Yerkes made his inquiries into the subject of chelonian space perception, Gilbert White had written of a pet tortoise:

Because we call this creature an abject reptile, we are too apt to undervalue his abilities and depreciate his powers of instinct. Yet he is, as Mr. Pope says of his lord,

> "—Much too wise to walk into a well:"

and has so much discernment as not to fall down an haha, but to stop and withdraw from the brink with the readiest precaution.

The mental superiority of terrestrial turtles as compared with aquatic species has been mentioned by numerous writers.

The most penetrating investigations of turtle mentality were those of Casteel (1911), who tested discriminatory ability by decorating with different patterns of lines the walls of the tunnels leading to two boxes, in which, respectively, palatable food and an electric shock were administered. In this device, painted turtles learned to distinguish between black and white, between vertical and horizontal lines, and between lines of varying widths, in one case even recognizing the difference between lines two and three millimeters wide.

Quaranta and Evans (1949) found the principle of "reward training" to be effective in experiments with Galápagos tortoises, and Quaranta (1949) devised tests to measure color discrimination in these animals. He found that they readily learned to distinguish between orange and blue-green, between blue and green of the same intensity, and even between blue and gray.

A tendency toward elementary social organization has been noted in the order of precedence that is soon established when tortoises or box turtles are grouped in captivity.

Kitchin (1949) was convinced that the rather complex behavior of nesting diamondback terrapins is probably merely a series of mechanical and reflexive responses to contact stimuli. Allard, however, reached a somewhat different conclusion after observing the changes in the reactions of a female box turtle that he took from a nest she was in the process of covering and placed first on bare soil and then on an artificial nest:

These versatile changes in behavior, involving a return to earlier phases of a train of reactions, would indicate something either remotely approaching reason, or a perceptive sensitivity that involves a very appropriate adaptiveness of behavior, step by step. The animal seemed to behave like an automatic machine so long as a normal course of action was allowed, but a change in conditions readily broke up the covering process at any point, and reestablished an earlier phase wherever this was demanded. Would not most men follow a similar chain of reactions and behaviors under such conditions?

Disposition: In general, turtles are an inoffensive race. Having specialized in defense rather than in equipment for aggression, they usually retire into their shells when molested. It is interesting to note that those forms with the most inadequate shells are generally the most irascible; Eigenmann (1918) offered the following succinct comments in this connection:

Correlated with the defective armature in the soft-shelled turtle we find the extreme of pugnacity. A soft-shelled turtle will snap and bite on suspicion from the time it is half way out of the shell. The disposition of the snapping turtle, with exposed ventral surface, is proverbial. The musk turtle will bite, as anyone who has collected their eggs can testify. On the other hand, the well-protected painted, geographic and Blanding's turtles, and above all the terrestrial and perfectly armored box turtle are the gentlest of creatures which no amount of provocation will induce to bite.

It should be pointed out that Eigenmann's remarks refer merely to the average temperament of a given species, since various writers have noted the marked variety of personality and disposition that is revealed when an observer becomes intimately acquainted with the individuals in a population. Allard showed that even the confirmedly meek and pacific box turtle may have periods of aggressive ill temper.

SIZE, GROWTH, AND AGE

If there is one thing that the casually interested person knows about turtles, it is that they live to be very old. Their enviable longevity has been extolled in classic mythology, in literature, and in folk belief, and so dominates man's thinking about turtles that when one is found the remark "I wonder how old *he* is" is automatic, whether the subject be a truly antique tortoise or a sprightly virgin cooter of two short summers. Unlike most animal myths, this one is so well grounded in fact that no real debunking is necessary, although the popular concept will bear some qualification.

The whole subject of turtle growth and longevity can be embraced by two generalizations, as follows: (1) turtles do attain great age in some cases, perhaps the greatest of any living vertebrate animals; and (2) they grow and reach maturity far more rapidly than has until recently been supposed. The box turtle engraved with the legend "G. Washington, 1751" beyond doubt came from the hands of charlatans, although even such uncertain evidence as this permits gullible souls like me to daydream. Sadly, though, the authentic minority of these inscribed tortoises are indistinguishable from the fakes, and the dates can be accepted only in a very few exceptionally clear cases.

The only reliable means of gaining an idea of how long turtles live is to live with them, and this course can become monotonous in the extreme. Gilbert White kept his famous tortoise fifteen years after it had reached an age known to be at least forty years, and on numerous occasions captive tortoises have officially passed the century mark. A giant Aldabran tortoise was supposed to have lived on St. Helena for more than 120 years and to have been well known to the exiled Napoleon. However, my friend Arthur Loveridge has found evidence that this animal was actually a composite of two individuals whose periods of residence on the island overlapped. Another tortoise from the Seychelles was said to have lived for more than one hundred and fifty years in Mauritius.

That these huge tortoises mature much more rapidly than might be suspected is indicated by the fact that a specimen of *Testudo vicina*, taken in the Galápagos Islands when it weighed 29 pounds, attained a weight of 450 pounds by the time of its death fifteen years later. Ninety other young Galápagos tortoises brought to the United States at an average weight of 18.5 pounds weighed 44.3 pounds each at the end of two years. Six others taken to Hawaii in 1929, when they weighed 26.5 pounds each, reached average weights of 63 pounds by the end of the following year. Another individual, probably three years old and weighing 29 pounds, reached a weight of 360 pounds during seven years' residence in California. (What would it have done in Florida!) During the succeeding seven years, however, this tortoise gained only 65 pounds more.

A captive loggerhead grew from a hatchling to a weight of 80 pounds in four and one-half years. Hawksbills, which are smaller than loggerheads but still large as turtles go, may attain sexual maturity at the surprisingly early age of three years, at which time they may weigh 30 pounds.

Smaller turtles may require as much time to reach maturity as the most ponderous ones, and their expected life span is not by any means proportionately shorter. In the northern parts of their range, box turtles mature sexually in four or five years and at shell lengths of from 3½ to 5 inches. Nichols estimated that at least 20 years may be required for a New York box turtle to complete its growth, and that the life expectancy is about 40 or 50 years, with good evidence that individuals may attain ages of between 80 and 123 years.

Cagle's work with *Pseudemys scripta elegans* (1944b, 1948a) revealed that females of this form may become sexually mature when only about three years old and with shells about 6 inches long, while the males may reach their mature length of about 3½ inches within "slightly more than one growing season."

The very largest turtle is the leatherback (*Dermochelys*), and while considerable vagueness surrounds its upper weight limits, it probably reaches a weight of 1,500 pounds and possibly even a ton. The biggest green turtles and loggerheads weigh between 500 and 1,000 pounds, although here also the most imposing records are enshrouded in doubt. Runners-up to the marine turtles in point of size are the giant Testudos, which, pushed by the occult force that makes gigantism common among island animals of many sorts, have reached measured weights of as much as 560 pounds. If a recently published record of a 403-pound alligator snapper is accepted, then I suppose *Macrochelys* must be regarded as the largest fresh-water turtle in the world. That its longevity may match its size is indicated by the fact that Conant and Hudson referred to two individuals that had in 1948 resided in the Philadelphia Zoo for 57 and 47 years, respectively.

A limited amount of very interesting information concerning growth rates in turtles has been obtained from a study of rings in the scales of the carapace and plastron. The laminae of the turtle shell adjust to the growth of the bones beneath them by eccentric growth around the granular infantile scale or "areola." During periods of rapid growth the scale is enlarged by accretion of new scale substance, which is generally believed to be applied over the entire under surface of the previous scale, around the edges of which it projects as a marginal ring. During periods of inactivity growth slows down or stops and a peripheral wrinkle appears and often is impressed on the underlying bone. Thus, as in the case of tree rings, clay varves, and fish scales, a count of the accumulated rings should afford some idea of the number of good seasons and lean seasons that alternated throughout the period of ring formation. Moreover, since the normal annual winter-summer cycle in a temperate region produces alternating periods of high and low activity in turtles, it is reasonable to conclude that the striae on turtle scales are seasonal growth rings, or "annuli," which reflect, individually, the height of the sun and, in the aggregate, the age of the turtle in years.

Unfortunately, however, for the student of turtle age and growth, several factors tend to disrupt the sequence of annuli. The more important of these are as follows:

(1) The fact that the more infantile (older) parts of each lamina wear away, destroying all but the more recently formed annuli.

(2) The fact that many turtles shed the laminae sporadically or periodically, which eventually results in the smoothing out of the topography of the scale, as well as that of the bony plates.

(3) The fact that any marked fluctuation in the physiologic level of the animal may cause a break in the rate of scale deposition. Minor breaks tend to confuse the chronology of major ones. Either too few or too many rings for the elapsed years may result from such factors.

(4) The fact that in southern latitudes without marked annual temperature cycles, the growth of turtle scales, like that of the wood of rainforest trees or of the scales of Florida bass, is often even and without major interruptions. In areas with a monsoon climate the alternation of a wet and dry season may substitute as an annulus builder, but in many places even this effect is lacking and laminal rings are merely reflections of small or sporadic physiologic changes.

In spite of the limitations imposed by these factors, growth rings are a useful tool for investigations of the natural history of turtles. Thus, though in *Pseudemys* the first annulus is rarely visible after three years, Cagle has used growth rings extensively in his valuable studies of this group. As he points out, even though loss of the earliest rings may make it impossible to determine the exact age of the specimen, the fact remains that "(a) in most individuals [in Illinois] not more than one is formed annually, (b) the zone between any two annuli represents one season of growth, (c) the approximate period of time represented by any area between rings is determinable."

The equation used by Cagle in computing rates of growth for periods covered by discernible annuli is one in wide use among ichthyologists, having been used first with turtles by Sergeev (1937). It is as follows:

$\dfrac{L_1}{L_2} = \dfrac{C_1}{C_2}$, with C_1 representing the length of the annulus, C_2 the length of the whole plastral lamina, L_2 the length of the turtle in terms of the sums of the lengths of the plastral laminae, and L_1 the unknown, or length of the plastron of the turtle at the time the annulus was formed. This method was found to have high validity in furnishing data for previous seasons of growth.

The obliteration of laminae by such seasonal shedding as is undergone by *Chrysemys* and *Pseudemys* is not abrupt but is gradual and progressive. Nichols believed that in *Terrapene,* which apparently does not molt, growth rings give a reliable accounting of age through the first five years of life and a fair degree of accuracy for ages between seven and fifteen years. After fifteen years, however, they are of little value.

Woodbury and Hardy (1949) agreed with Miller (1932) that the age of the desert tortoise cannot be determined from growth rings.

Mattox showed a correlation between the number of rings in sections of the long bones of *Chrysemys picta marginata* and the size of the individual. His correlation could probably be extended to include age as well, although this has not been demonstrated.

TURTLE ADAPTATIONS

It seems a curious anomaly that the more noteworthy adaptations to be found in modern turtles are mostly reversals of the primary chelonian tendency toward skeletal hypertrophy. The heavy shell of the primitive turtle, though a valuable asset to a paludal or semiaquatic animal more concerned with turning the teeth of its rapacious contemporaries than with improving its own locomotor ability, has been repeatedly modified as one stock after another has spread into new environments. Three principal adaptational trends are discernible: one toward more effective aquatic locomotion, as seen in sea turtles; another along lines demanded by terrestrial life; and a third toward the type of bottom-lurking that has molded the soft-shelled trionychids in their extraordinary pattern. Each of these courses has, for different reasons, brought about a reduction of the bony shell.

The most highly aquatic of all reptiles, with the possible exception of the sea snakes, is *Dermochelys;* in it the most extensive loss of shell bones may be seen. In the other marine turtles the plastron is incomplete and embryonic gaps in the bony carapace persist toward the lower ends of the ribs for varying periods after hatching, and even into maturity.

Along with their jettisoning of shell bone, the sea turtles have abandoned the typical chelonian swimming stroke, in which diagonal fore and hind legs kick alternately, to adopt a new and much more effective method whereby the highly modified fore flippers are raised and lowered like the wings of a bird in flight. This system, which necessitates some remodeling of the pectoral girdle, enables the sea turtles to attain the greatest speeds of any modern reptiles. Deraniyagala (1939a) called *Dermochelys* "the swiftest living tetrapod," and mentioned 100 meters in 10 seconds as the maximum speed of some thecophorans. Anyone who has watched a green turtle (which I think is the best swimmer among Thecophora) in full flight, as, for instance, after being surprised by a glancing blow from an iron, will probably have little difficulty in accepting Deraniyagala's figures, even though they place leatherbacks on a par with the best human runners of the hundred-yard dash.

Land turtles also have reduced their shells, although for a somewhat different reason. An emydine turtle grown to the size and shape of an Aldabran tortoise, and with proportionately massive shell bones, would be so ponderous that locomotion would be out of the question. To avoid awkward increase in the weight-to-volume ratio, tortoises, and especially some of the giant island forms, have decreased the thickness of their shells to a striking degree. The carapace is often much elevated, apparently as a measure to increase total volume and thereby heighten resistance to desiccation, and perhaps also as an inconvenience to gnawing predators that might easily crack a flat shell. There may be extra spaces for water storage in such tortoises, and water conservation is also furthered by extended retention of urine, a course made possible by the insolubility of the excreted uric acid. Tortoises walk on the tips of their toes, and their feet are columnar and elephantine except in the case of the few burrowers, in which the fore feet may be vertically flattened, laterally sweeping shovels.

As an additional protective device some land turtles (as well as a few kinosternids) have in their shells transverse hinges that allow them to be closed, either partially or completely. The movable parts are usually sections of the plastron, but in the African genus *Cinixis* there is a hinge across the carapace.

Of all the terrestrial turtles the most drastically modified is another African form, the soft-shelled tortoise (*Testudo tornieri*). In this small species, which is an inhabitant of rocky terrain, reduction of the thickness and distribution of the bones of the carapace have rendered it so flexible that the animal can squeeze itself into narrow rock crevices or beneath or between boulders, inflate its body with air like a toad caught by a snake, and resist the most determined efforts to extricate it.

Although the soft-shelled turtles of the family Trionychidae appear at first glance to owe their unique features to the exigencies of aquatic locomotion, a closer perusal shows them to be more probably adaptations to bottom dwelling. That the ambushing of prey is an ancient turtle custom is probably indicated by the concealing conformation, adornment, and coloration of the primitive Chelydridae (in one of which a lure for the attraction of potential victims is added to the equipment for ambuscade) and of such luridly inanimate-looking creatures as the pleurodiran *Chelys fimbriata* of South America. While it has yet to be demonstrated that soft-shells hide themselves for the purpose of waylaying the animals that they eat, instead of merely to escape notice, there seems little doubt that the greater part of their time is spent in concealment. The thin, soft edges of the pancake shell conform readily to bottom contours and make it easy for the turtles to shuffle themselves into sand or silt. They usually lie in water so shallow that the long neck can carry the schnorkellike nostrils to the surface for air, from time to time, without the necessity of the turtle's dislodging itself. Between times, the head may be withdrawn beneath the mud with only the tip of the tubular snout protruding to prolong the period of submergence by admitting water for pharyngeal respiration.

This appears to be the world-wide trionychid way of life, and it seems likely that the flatness of the shell and the flexibility imparted to its margin by the loss of the peripheral bones may be primarily adaptations favoring concealment beneath bottom deposits, and that the elongate neck and proboscis, the capacity for effective pharyngeal respiration, and the fast strike and steel-trap grip of the jaws may likewise all be modifications toward this end.

Turtles and Men

ECONOMIC USES

The practical importance of turtles to man lies mostly in their contribution, actual and potential, to his diet. There can be little doubt that people have been eating turtles pretty steadily for as long as they have had the wits to get them out of their shells; and they probably had been eating turtle eggs long before that. While for some groups turtles have served merely as an occasional delicacy, other people, more fortunately located, have leaned heavily upon them as a dietary staple. In a few cases the role of the turtle as gastronomic accessory to man's machinations has been spectacular. For instance, the very reliable cheloniologist Georg Baur, who visited the Galápagos Islands in the 1880's, gathered data on which he based an estimate that early whaling ships had carried away from the teeming tortoise population of this tiny archipelago no fewer than ten million giant testudos. Exploitation of the giant tortoises of the Indian Ocean was probably little less extensive, and the part the tea turtles played in supporting early exploration and colonization, though hard to appraise accurately, was unquestionably a vital one.

Perhaps no reptile has attained the gastronomic glamour of the little diamondback terrapin, which during the gilded first quarter of this century brought prices as high as $120 a dozen. No less profound reverence, however, has been held by the English for the green turtle and by the Japanese for their succulent *Suppon* (soft-shell). These are the recognized aristocrats among table turtles, but the whole group is edible, and I suppose that all the species are eaten by someone, except, perhaps, for a few which are sporadically poisonous. I never heard of anyone in the United States eating a mud or musk turtle, but *Staurotypus* is relished in parts of Mexico and the Mosquito people of Nicaragua roast *Kinosternon* in its shell and eat it with gusto.

The recent development of improved refrigeration methods will almost certainly impose heavy drains on turtle populations. Markets have shifted because of disruptions brought by the two wars. The diamondback has declined in favor, and British aldermen eat fewer green turtles than formerly; but on the other hand I know a number of people who within the past few months have made their first enthusiastic acquaintance with green-turtle meat through a local establishment that features it on a popular carry-away lunch. The deep-fried steaks here are a far cry from Key West chicken turtle with chines and calipees, or from an authentic curry, or green-turtle soup in New Orleans; or for that matter from Carr's broiled ridley filets with lime butter; but people like them, and this bodes no good for the green turtle.

Of the fresh-water turtles of the United States, the snapper (*Chelydra,* and to a lesser extent *Macrochelys*) is easily the most important. Besides being a turtle, and accordingly good to eat, the snapper is big and widely distributed. It has made wide gains in popularity in recent times, although Philadelphia has of course been aware of its virtues for many decades. The soft-shells are just as savory as the snapper, although less well known; and it seems probable that these two could be made the basis for a much-expanded industry. Better than either of them (in my private judgment) are the gopher and the "Suwannee chicken," but like most of their near relatives these are too locally distributed to support a commercial market.

It seems to me that in any plan to extend human food resources by more intelligent exploitation of the sea, the green turtle should receive careful consideration. It not only furnishes meat of unsurpassed quality but, being herbivorous, it is able to utilize huge volumes of forage provided by the submarine pastures of turtle grass and manatee grass that cover immense circumtropical areas. While at the present time the green turtle is not abundant, we have good evidence that it once grazed these pastures in numbers incomparably greater than now, and that the depletion was a result of short-sighted exploitation by man. To me the fact that green turtles have survived at all indicates that they are an uncommonly tough breed.

These points suggest that in this animal important economic potentialities await only an intelligent plan of management. The chief obstacle would be our almost complete ignorance of the bionomics of the present populations. When painstaking investigations have furnished a sound basis for practical schemes of protection and control, and international agreements have implemented these, it seems highly probable that the green turtle hordes could be restored to their three-fathom meadows to harvest for us the almost inexhaustible stores of energy held there.

People do other things with turtles besides eat them. The tortoise-shell industry is as old as civilization. Oils extracted from turtle fat are used around the world for various purposes. There is no way of estimating the monetary value of the swarms of baby turtles sold as pets, but millions of hatchlings are caught annually for this market. Great numbers of turtles are bought by biological supply houses for laboratory dissection, and in parts of the Mississippi Valley large quantities of turtle eggs are sold as fish bait.

Certain species of fresh-water turtles are in slightly bad odor with one school of conservationists because of alleged predation on or competition with other useful aquatic organisms—mainly waterfowl and game fishes. On the whole, however, this aspect of their relation to man is probably of far less significance than their value as a source of food, and there would appear to be much need for conservation measures based on this concept. The work of Lagler in Michigan suggests that as more is learned of the detailed feeding activity of turtles the popular notion that they are implacable enemies of the fish culturist will be discarded.

METHODS OF COLLECTING TURTLES

Fresh-water turtles are most frequently taken in traps, and the variety of devices used for ensnaring them is imposing. Besides the conventional baited barrel net or fyke net rigs, which are of limited value for catching mature individuals of herbivorous species, various traps that exploit the sunning habit of turtles are used. In some of these a trap door is sprung when the turtle crawls onto a platform just above water level; but a simpler variation is a wire basket nailed to the side of a favorite sunning station, with its open top flush with the surface of the water. The efficacy of this is much enhanced if the collector can make occasional abrupt appearances from the off side of the log and thus force the sunning turtles to dive into the basket. This is one of the favorite methods of catching the famous "Suwannee chickens," and as many as twenty or thirty are sometimes taken at one time in this way. Sunning stations are also sometimes strewn with steel traps or with treble fishhooks fastened by short lengths of line.

In some places turtles may be caught in a seine, but mature turtles are skillful at eluding this type of attack and it is useful mainly where large numbers of young are concentrated in suitable waters.

In northern regions, where turtles sometimes congregate to hibernate, large catches are occasionally made when a group is located by probing with a pole in holes under logs, in old muskrat houses, and in similar favored sites. Soft-shells lurking in the sand of river bars and diamondback terrapins in the mud of coastal marshes are also often located by the probing bar, although the latter are best taken in stop nets, where such are still legal.

In clear water up to twenty feet in depth, a bullen, or net rigged on a hoop at right angles to the end of a long shaft, is often useful. If the water is no more than three feet or so in depth and is choked with aquatic vegetation, no method surpasses the simple procedure of wading about at night with a flashlight. In the numerous Florida lakes and ponds that afford such conditions this technique has been found very effective, and as many as seventy-five turtles have been taken by three people in one evening in this way.

Another direct-approach method, more adaptable than the above, is that developed by Mr. E. Ross Allen and Dr. Louis Marchand whereby a swimmer equipped with diving mask is pushed along, face under water for reconnaissance, by a slowly moving boat equipped with an outboard motor. When a turtle is sighted it is chased down by the swimmer, seized, and dumped into the boat.

Most carnivorous turtles and young individuals of many species bite baited hooks with some enthusiasm. The horny jaws make them adroit bait stealers, but some species, notably soft-shells and kinosternids, are easily caught by hook-and-line fishing.

Soft-shelled turtles have been found susceptible to suffocation by the concentrations of rotenone in standard use in poisoning fishes.

All turtles wander more or less widely on land, either during laying season or during sporadic migrations, and at such times they are of course easily taken.

Sea turtles are nowadays caught commercially in large-meshed (six-to-twelve-inch) nets tied of twine about 33 threads, and usually from 100 to 200 feet long and from 20 to 40 feet deep. These are buoyed by floats and either anchored or left to drift. They are often used in conjunction with painted wooden decoys, which appear to increase their efficacy considerably.

Where sea turtles are still abundant, more direct methods may be used. A bullen or hoopnet can be used to drop over a cornered or surprised turtle, or where the catch is to be butchered for local consumption, grains or harpoons are often used, but the injuries caused by these almost preclude their use for taking green turtles for export or hawksbills from which carey is to be stripped. What to me seems an incomprehensibly refined skill is that sometimes attained by veteran turtle hunters—Carib, Mosquito, or "Conch," who with almost incredible timing are able to intercept the swift and erratic course of a fleeing green turtle with the long curving trajectory of a heavy grains pole thrown from a moving boat. In former days much use was made of the "peg," or spear with a loosely socketed, cylindrical, naillike point that caused much less damage than barbed points.

If sea turtles can be found while sleeping they often may be approached, caught, and held at the surface by a swimmer who prevents their sounding by tilting the fore edge of the shell upward. In this operation the possibility of a wound from the raking claw of a flipper is a real hazard.

Far and away the most colorful method by which men catch turtles is that which employs the remora or shark sucker (*Echeneis naucrates* and other species). Comparable to falconry in its appeal to the imagination, this ancient custom was apparently prevalent among all the sea Caribs, and it still persists among some of their descendants, as well as among various peoples in the Pacific and Indian oceans. Gudger (1919) wrote an exhaustive account of the practice, and DeSola (1932) told of witnessing the capture of a hawksbill in this way off the southern coast of Cuba in the Bight of Manzanillo, where Columbus had observed the same thing on his second voyage in 1494.

The following description of the use of the remora in catching turtles in Mozambique waters appeared in the *Histoire naturelle des poissons* (1829;

vol. III, p. 490) of Lacépède, who obtained his information from the naturalist Commerson:

There is attached to the tail of the living Naucrates a ring of diameter sufficiently large not to incommode the fish, and small enough to be retained by the caudal fin. A very long cord is attached to this ring. When the Echeneis has been thus prepared, it is placed in a vessel full of salt water and the fishermen place this in their boats. They then sail toward those regions frequented by marine turtles. These animals have the habit of sleeping at the surface of the water on which they float and their sleep is so light that the least noise of an approaching fishing boat is sufficient to wake them and cause them to flee to great distances or to plunge to great depths. But behold the snare which is set from afar for the first turtle which the fishermen perceive asleep. They put into the sea the Naucrates, furnished with its long cord. The animal, delivered in part from its captivity, seeks to escape by swimming in all directions. There is paid out to it a length of cord equal to the distance which separates the sea turtle from the boat of the fishermen. The Naucrates, restrained by the line, makes at first new efforts to get away from the hand that masters it. Soon, however, perceiving that its efforts are in vain, and that it cannot free itself, it travels around a circle of which its line is in some fashion a radius, in order to meet with some point of adhesion, and consequently to find rest. It finds this asylum under the plastron of the fleeing turtle, to which it attaches itself easily by means of its sucker, and gives thus to the fishermen to whom it serves as a fulcrum, the means of drawing to them the turtle by pulling the cord.

Although I have never had the privilege of seeing a turtle taken by remora fishing, that the adhesive force which one might furnish would be more than adequate was demonstrated to me when the bow of my small boat was once pulled around through half a circle by a big remora that took my hook when attached to a good-sized shark. The pull on the hand line was sufficient to cause it to slip through my fingers.

THE INSCRUTABLE TURTLE

In the mystic philosophy of the ancient Hindus the earth was a hemisphere, resting flat side down on the backs of four elephants; and these mighty beasts stood on the back of a single ponderous tortoise. American Indians believed that before there was anything else there was a great turtle floating in a primal sea, and all the animals lived upon his carapace. The earth as man knows it was built on this foundation by the crayfish, which vanquished the beaver and the muskrat in a diving contest to bring up mud from the bottom of the sea and build the dry land.

Although no other folk system appears to have assigned to the turtle such a fundamental role as that on which Hindu and Amerind so strikingly agree, nearly all races that have inhabited turtle country have shown preoccupation with these engaging animals. Among many Asiatic peoples the turtle is still sacred. Burmese Buddhists adorn the shells of Batagurs with gold leaf and release them at the river's edge with propitiatory ceremony. Smith (1931) mentioned a temple in Bangkok in which *Hieremys* and *Geomyda* are placed in great numbers by people who hope to gain merit in the next world by thus saving a life. With the inconsistency that is so often a part of religious practice, those who leave the turtles there lose immediately all interest in them and allow them to starve.

In Chinese religious belief, too, the turtle has always held a high station, from time immemorial having been revered for its wisdom and benevolence. The ancient Chinese consulted the complex fissures traversing burned turtle shells to learn of the will of the Supreme Ruler. Pope identified fragments used in such rites as belonging to the now extinct Chinese turtle *Pseudocadia anyangensis.*

The Greeks believed that even the gods held the turtle to be sacred. The lute was said to have been invented in about the year of the World 2000 by Mercury, who made the first model from the shell of a turtle found on the bank of the Nile. On Mount Parthenon in Arcadia the gods killed the venerated trunkback (alleged to have been the original lute turtle from the seven longitudinal ridges on the back which suggested the strings of a lute) only when pressed by need of a new lute. Even then they probably had trouble wheedling a specimen out of the god Pan, who was custodian of turtles.

In his *Natural History,* Pliny the Elder attributed to turtles strongly developed medicinal properties and described sixty remedies derived from them. A sample nostrum from his list follows:

> The flesh of the sea tortoise, mixed with that of frogs, is an excellent remedy for injuries caused by the salamander; indeed there is nothing that is a better neutralizer of the secretions of the salamander than the sea tortoise [*Natural History of Pliny,* Bohn's Classic Library, London, 1885, vol. 6, p. 16].

Nearly all the tribes of North American Indians were much interested in turtles, and one species or another was regarded as sacred throughout most of the continent. According to Speck (1943), the tortoise was everywhere seen as "endlessly wise and shrewd and a kindly advisor"; and the occurrence of the charmed numbers 12 and 13 in the laminae of the shell demanded added veneration. Turtle shells were used as ceremonial rattles by many tribes, and the Seminoles sometimes tied them in clusters to the knees of dancers. The benign virtue of the Indian folk turtle was locally impugned by the Creeks, who thought box turtles caused droughts and floods and accordingly dashed them to pieces on sight.

In the rural United States today, turtle tales lack the character and imagination of those about snakes, falling short of them both in variety and in dramatic impact. In this country, other things being equal, the richness of a given folklore is in direct proportion to the violence or morbidity of the subject, and by these criteria snakes, of course, have it all over turtles. You can still hear it said that a turtle won't let go till it thunders, or that a beheaded turtle won't die until sundown, or that tortoises live five hundred years, or that snappers bite broomsticks in two. But on the whole, turtle stories are merely exaggerations of attributes that turtles to some degree actually possess, and they seem pale and anemic compared with snake stories.

This generalization is true except in the case of the sea turtles. There is nothing pale about sea-turtle stories. They make as virile and dynamic a body of legend as I know of; and they are alive today. I have heard the same gripping yarns in Spanish and in English of every Caribbean shade, and they are told every day in Carib, Mosquito, French, and Danish, and who knows in what other recondite tongues elsewhere. The stories are of all sorts, and I have referred to a few of them in the accounts of the species to follow. But the best of them scrupulously eschew all subjects except sex in bizarre forms and I have had to bow to convention and exclude them. As Audubon put it, "The loves of the [sea] turtle are conducted in a most extraordinary manner, but as the recital of them here must prove out of place I shall pass them over." I might add, "with real regret," and point to the relatively mild quotation on page 8 of this book as an innocuous sample from a robust branch of ethnozoology.

Terminology of Turtle Structures

In order to prepare adequate descriptions of turtles it is necessary to have for the taxonomically significant features a terminology that is as concise as it can be made. There must be available a set of anatomical terms each of which clearly applies to one single structure and to no other. While such a nomenclature has been devised for the skull and plastron of turtles, the loose ambiguity of the names in vogue for the epidermal and bony elements of the carapace is astonishing.

This confusion has existed since the time of the earliest writers on the classification of the turtles, and even in the works of such important turtle students as Gray, Bell, Agassiz, LeConte, and Baur we find the epidermal segments referred to more or less indiscriminately as "scales," "plates," "scutes," "scutella," and "shields," and in some cases most of these words are used synonymously in the description of a single specimen. Moreover,

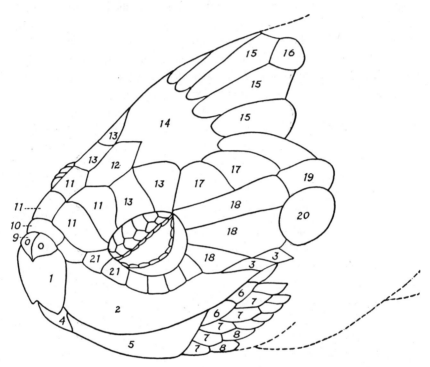

Figure 1. Generalized scalation of the head of a turtle. *1*, rostral; *2*, first supralabial; *3*, supralabials; *4*, mental; *5*, first infralabial; *6*, infralabials; *7–8*, postmandibulars; *9*, nasal; *10*, frontonasal; *11*, prefrontals; *12*, frontal; *13*, supraoculars; *14*, frontoparietal; *15*, parietals; *16*, interparietal; *17*, temporals; *18*, postoculars; *19*, supratympanic; *20*, tympanum; *21*, preoculars. (After Deraniyagala.)

at one time or another most of these synonymous terms have been transferred temporarily to the corresponding section of the underlying bony shell. It is easy to see that the chances for confusion in such a situation would be great; and they have been fully realized in many cases.

Boulenger, to whom we owe a debt for stabilizing the descriptive terminology of reptiles, employed the words "nuchal" and "costal" for both horny segments and bones but was consistent in distinguishing the shell-scales from the bony elements by the respective terms "shield" and "plate."

Thus, since Boulenger usually associated these dual-purpose terms nuchal and costal with the qualifying words plate or shield, it is nearly always possible to understand from his context how the term has been applied.

However, this course sometimes demands close attention if the association is to be kept in mind. Moreover, later turtle students have been less careful than Boulenger, often reverting to the older, indiscriminate designation of scales and bones alike as either plates or shields, as their moods have dictated. As one example among many possible, Freiberg in his "Catálogo sistemático y descriptivo de las tortugas argentinas" (1938) uses the words *nucal, costales,* and *marginales* for both bones and scales.

This is a melancholy situation, and there is no real reason for its existence. The late Dr. Leonhard Stejneger was well aware of this awkward weakness in terminology and many years ago adopted in his own writings a more rational scheme, by which the horny and bony parts of the carapace are designated by two wholly distinct sets of names. His system was a consistent adherence to the policy of some early writers to use Greek names for the bony pieces and Latin names for the scales of the carapace. Unfortunately, Dr. Stejneger published no full explanation of his revised nomenclature, and consequently it was not extensively adopted by other herpetologists.

Like all well-devised nomenclature, Dr. Stejneger's series of names offers the dual advantage of maximum clarity and brevity. I have adopted it with slight modifications in this book and commend it to the consideration of turtle students in general.

REVISED TERMINOLOGY

Because of the impracticability of salvaging any precision with the popular old terms such as plate, shield, and scute, they are abandoned and the scales of the carapace are referred to as "laminae," singular "lamina."

For the so-called "vertebral" laminae the term "centrals" is selected, while the corresponding bones retain the name "neurals."

Similarly, the bony plates covering the ribs become "pleurals" and the horny laminae above them, "laterals."

For the outer series of laminae the word "marginals" is retained, while the bones there are called "peripherals."

For the bone commonly known as "nuchal" the name "proneural" is proposed, and the corresponding lamina is designated "precentral."

The term "pygal" is restricted to the bony plate, and the last pair of marginal laminae are called "postcentrals" or, when fused, "postcentral."

The ambiguous word "supracaudals" refers more correctly to epidermal structures on the upper side of the tail itself, as for instance the supracaudal tubercles of the snapping turtles.

The terms indicating the boundaries between the bones of the shell and those between the laminae must also be revised. Thus, that between two laminae becomes a "seam" and the impression of the seam on the surface of the bone beneath a "sulcus," while the articulations between the bones retain the name "sutures."

There has been considerable confusion likewise from the indiscriminate use of the words "jaws" and "mandibles" for both the bony structures and their external bony coverings. It thus seems convenient to employ the terms "mandible" and "maxilla" for the bony jaws alone and to adopt the word "beak" for the horny sheath and "tomium" for its cutting edge.

The following brief accounts of some of the structural features commonly used in the classification of turtles have been largely abstracted from Hay's *Fossil Turtles of North America*, with a few abbreviations and modifications. The accounts refer primarily to the generalized *Pseudemys s. scripta*, and the modifications to be seen in such groups as the soft-shelled and marine turtles will be pointed out later under the appropriate headings. For more detailed discussions of turtle osteology the reader is referred to Boulenger's *Catalogue of the Turtles of the British Museum* and Smith's section on reptiles and amphibians in the *Fauna of British India*.

PARTS OF THE BONY SHELL

The shell of the turtle consists of two parts, an upper, the *carapace,* and a lower, the *plastron*. On each side between the fore and the hind leg, the two parts are joined by the *bridge*.

The carapace is composed of a large number of bones, each of which is articulated with the adjoining bones by jagged sutures. In front of and behind the bridge the outer ring of bones (the peripherals) projects freely like the eaves of a roof.

Along the mid-line twelve of the bones of the carapace are arranged in a row. In front is the *proneural* bone (usually known as the nuchal). Behind this comes a series of eight *neurals*. The last of these is followed by two *epipygals* and the hindermost of these is succeeded by the *pygal,* as the last of the median series. The neurals are connected with the neural arches of the dorsal vertebrae, of which they appear to be mere expansions. The proneural, the epipygals, and the pygal are not connected with the vertebrae.

On each side of the series of neurals, and articulating with them, are eight *pleurals*. Outside of the pleurals, extending around either side of the body from the proneural to the pygal is a series of bones (usually eleven in number), the *peripherals*.

The plastron comprises a median bone, the *entoplastron*, and four paired bones: the *epiplastra*, the *hyoplastra*, the *hypoplastra*, and the *xiphiplastra*. On each side, between the fore and hind legs, the hyoplastron and the hypoplastron articulate with peripherals three to seven inclusive to form the bridge. The notch from which the front leg protrudes is the *axillary notch* and that from which the hind leg emerges is the *inguinal notch*. The part of the plastron in front of the axillary notches is the *anterior lobe*, and that behind the inguinal notches is the *posterior lobe*.

Just behind the axillary notch the hyoplastron sends upward a strong process, the *axillary buttress*, which fuses immovably (ankyloses) with the inside of the lower end of the first pleural. In front of the inguinal notch the hypoplastron sends up a similar process, the *inguinal buttress*, which fuses with the inside of the lower end of the fifth pleural.

EPIDERMAL LAMINAE

The bones of the carapace and plastron are covered by a number of horny laminae. The meeting edges of these are called *seams* and the impression of a seam on the underlying bones appears as a furrow, or so-called *sulcus*. The laminae coincide neither in number nor position with the underlying bones, and it is seldom that the seams coincide with the sutures.

Carapace. At the fore end of the carapace, on the mid-line, there is a small lamina, the *precentral*. Then comes a row of five large laminae, the *centrals*, each extending laterally beyond the neural bones. The seams separating these laminae cross, respectively, the first, third, fifth, and eighth neurals. On each side of these centrals is a row of four large laminae, the *laterals*. The seams between these descend respectively on the second, fourth, sixth, and eighth pleural bones.

The borders of the carapace are invested by a series of marginal laminae, eleven on each side. The seams dividing these from the lateral laminae run along near the upper border of the peripheral bones. At the free borders of these peripherals the laminae turn downward and appear on the under side of the bony shelf. The last pair of marginals, often fused into one lamina, are called the *postcentrals*, or the *postcentral*.

A short series of small laminae is found in *Macrochelys* between the posterior laterals and the marginals; these are called *supramarginals*.

Plastron. A median longitudinal seam runs from the front to the rear

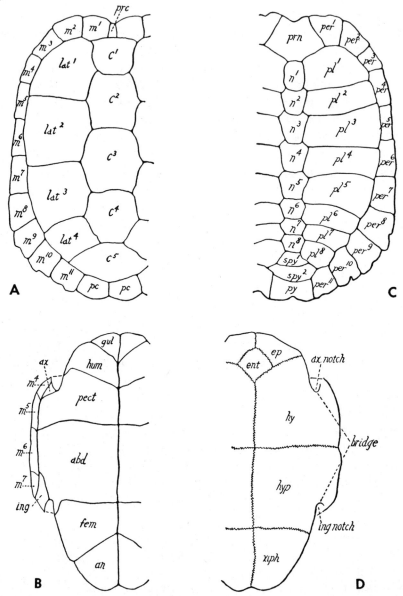

Figure 2. Epidermal shell (*A and B*) and bony shell (*C and D*) of an emydid turtle (*Pseudemys s. scripta*). Epidermal laminae: *prc*, precentral; *m*, marginals; *lat*, laterals; *c*, centrals; *pc*, postcentrals; *gul*, gular; *hum*, humeral; *pect*, pectoral; *abd*, abdominal; *fem*, femoral; *an*, anal; *ax*, axillary; *ing*, inguinal. Shell bones: *prn*, proneural; *per*, peripherals; *pl*, pleurals; *n*, neurals; *spy*, suprapygals; *py*, pygal; *ep*, epiplastron; *ent*, entoplastron; *hy*, hyoplastron; *hyp*, hypoplastron; *xiph*, xiphiplastron; *ax notch*, axillary notch; *ing notch*, inguinal notch.

of the plastron, separating the laminae of each pair. In front there is a pair of *gular* laminae; then a pair of *humerals,* followed by the *pectorals,* the *abdominals,* the *femorals,* and finally by the *anals.* At the back edge of each axillary notch there is an axillary lamina, while at the front edge of each inguinal notch is found an *inguinal* lamina.

In some turtles there is a single lamina between the gulars which is known as the *intergular.* Similarly in others (Cheloniidae) there is a lamina in the posterior angle between the anals, the so-called *interanal.*

The axillary and inguinal laminae belong to a series known as *inframarginals,* which form a complete row in the Cheloniidae and in others.

THE BONES OF THE SKULL

The skull [1] seen from above shows three pairs of bones which join at the median line. In front are the *prefrontals,* extending backward to the middle of the orbits and with their anterior ends forming the roof of the nasal cavity. A strong process descends from each to join the vomer and the palatine, and to form the front wall of the orbit. Behind the prefrontals are placed the larger *frontals.* The *parietals* form the roof and much of the lateral walls of the brain case, and the exterior end of each sends downward a strong process that joins the *pterygoid.* Besides the pterygoid, the lower border of the parietal articulates with the *pro-otic* (otosphenoid) and the *supraoccipital.* The latter bone is greatly prolonged backward and it forms a small part of the boundary of the *foramen magnum.*

The *maxilla* bounds the nasal cavity on the side and the orbit below, and its lower border forms an acute cutting edge. Posteriorly the maxilla articulates with the *jugal.* The hinder part of the rim of the orbit is formed of the jugal below and of the *postfrontal* above. These two bones form a postorbital bar of moderate width.

A large *tympanic cavity* is excavated in the *quadrate* and below this cavity the quadrate descends to form a movable articulation with the lower jaw. In the hinder border of the quadrate there is a small, deep notch for the passage of a long, rodlike bone, the *columella.* Interposed between the anterior border of the quadrate and the jugal bone is the *quadratojugal* (paraquadratum). The jugal and quadratojugal form the zygomatic arch. Above and behind the quadrate is the *squamosal.*

The sides of the hind part of the skull are each occupied by a long ex-

---

[1] For a detailed account of the bones of the skull, see Siebenrock, Das Kopfskelet der Schildkröten, Sitz. Ber. Akad. Wiss. Wien, Math.-Nat. C., vol. 106, pt. 1, July, 1897, 84 pp., 6 pls.

cavation, the *temporal fossa.* The floor of this is formed of the parietal on the inside, of the pro-otic and *paroccipital* in the middle, and of the quadrate and squamosal on the outside. The pro-otic and the quadrate enclose the *foramen carotico-temporale.*

As seen from below, the anteriormost bones of the skull are the *premaxillae,* bounding the nasal cavity below and forming part of the roof of the mouth. On each side and behind the premaxilla is the *maxilla.* It presents outwardly the cutting edge already mentioned. Its inner border joins, in

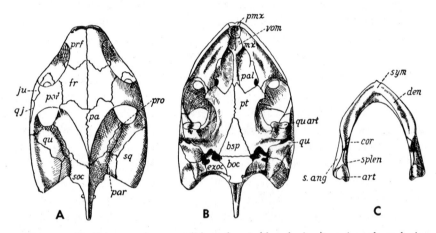

Figure 3. Skull bones of an emydid turtle. *A* (dorsal view): *prf,* prefrontal; *fr,* frontal; *pa,* parietal; *soc,* supraoccipital; *pof,* postfrontal; *ju,* jugal; *qj,* quadratojugal; *qu,* quadrate; *sq,* squamosal; *par,* paroccipital. *B* (ventral view): *pmx,* premaxillary; *mx,* maxillary; *vom,* vomer; *pal,* palatine; *pt,* pterygoid; *bsp,* basisphenoid; *boc,* basioccipital; *exoc,* exoccipital; *qu art,* articulating process of the quadrate. *C* (lower jaw, dorsal view): *sym,* symphysis; *den,* dentary; *cor,* coronoid; *splen,* splenial; *art,* articular; *s ang,* supra-angular.

front, the *vomer* and, posteriorly, the *palatine;* while the middle portion is mostly a free edge, forming the lateral boundary of the *choana.* Between the two borders is a broad *alveolar,* or *triturating,* surface.

The mid-line behind the premaxillae is occupied by the single *vomer.* Anteriorly it divides the nasal passages from each other; laterally it articulates with the *palatines,* and posteriorly, with the *pterygoids.* The free lateral projection of the latter is the *ectopterygoid* process. The palatines assist in roofing the nasal passages and in forming the alveolar surface already mentioned. Between each palatine and the maxilla of its side there is an opening, the *posterior palatine foramen.*

The pterygoids meet each other at the mid-line anteriorly, but posteriorly are separated by the *basisphenoid*. They extend backward so far as to exclude the latter bone from contact with the quadrates. The lateral border of each pterygoid is mostly a sharp free edge. Behind the basisphenoid comes the *basioccipital*. It is met on each side by the *exoccipital,* and all three of these bones join in forming the *occipital condyle*. From the ventral view is seen also a portion of the paroccipital and squamosal. On each side of the basicranial axis, foramina for the passage of various nerves and blood vessels are evident.

Each ramus of the lower jaw is composed of six bones. In front is the *dentary,* forming the alveolar surface of the jaw, and completely co-ossified with its fellow of the opposite side at the *symphysis*. On the lower border of the jaw this bone extends backward nearly to the articulation with the quadrate. The upper border of the jaw, behind the alveolar surface, is formed in front by the *coronoid* bone, posteriorly by the *supra-angular*. These two bones are to be seen both from the outside and from the inside of the jaw. Behind the supra-angular is a nodular bone, the *articular,* that articulates with the quadrate. On the inner surface of the jaw, near the posterior end, are two bones, the lower called the *angular* or splenial, the upper the *prearticular*.

Shell Abnormalities and Their Meaning

The horny laminae of the turtle shell are subject to extensive variation. Although some irregularities appear to be random changes, others may involve the consistent recurrence of a lamina in a particular position, either in one or in several species. Moreover, laminae that occur as supernumerary scales in one turtle may form an integral part of the shell in another. Although such regular or correlated variations are today not especially surprising, similar trends having been observed in most groups of animals that have received attention, they were avidly seized upon by the less genetically sophisticated students of the early 1900's as of possible significance in the vexing problem of the ancestry of turtles, and a voluminous literature on the subject has been built up.

One of the first students to make an inventory and suggest an explanation for laminal anomalies was the erudite naturalist Gadow, who, however, unwittingly based his analysis of 1899 on material that included at least two, and perhaps as many as four, different forms of sea turtles, which of course vitiates the significance of his data. Moreover, Gadow became hopelessly entangled in a mystical, orthogenetic interpretation of the apparent decrease in the adult laminal count from that of the juvenile. He

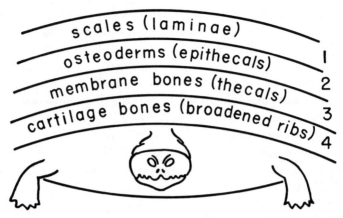

Figure 4. Diagram showing layers in the carapace of the theoretical turtle ancestor. Among living forms, *Dermochelys* has layers 2 and 4, and ordinary turtles have 1, 3, and 4.

saw in this a (presumably guided) progression of the individual toward a goal of perfection—the same goal, evidently, toward which the whole turtle line had been striving for millenniums! Although this vaporous interpretation of scale variations was in itself of little value, it at least served to focus the interest of other turtle students on the subject.

The most important descriptive papers in this field were those of Newman (1906b) and Coker (1910). The former gave the results of an examination of 476 specimens of *Graptemys* and 183 of *Chrysemys*, while Coker took his data from 243 specimens of *Malaclemys*, of which 45 per cent showed some laminal abnormality and 20 per cent had more or fewer than the normal number.

Besides an addition or loss of laminae, these writers found numerous cases of fusion or incomplete division of normal or abnormal scales, and variation in the size or form of a scale was common. Among the diamond-back terrapins showing abnormalities, asymmetry was as frequent as symmetry, and the addition of laterals occurred with or without disturbance of the central series. No significant difference in the incidence of variations in young and old individuals could be shown, but females showed more disorder than males. The most frequent variation was the occurrence of an inguinal lamina in 21 per cent of all specimens. The plastron was much less variable than the carapace, but in it asymmetry or partial fusion was noted occasionally, as was the presence of extra laminae in the plastral series.

Various students have noticed that a reduction in the number of laminae is of much less frequent occurrence than an increase, and that variations tend to occur more often at the posterior end of the carapace than at the anterior end.

The most marked variation is found in the shells of sea turtles, in some of which it is difficult to determine what the characteristic scale number actually is on a basis of the limited series of specimens available. Asymmetry in *Lepidochelys olivacea* appears to be almost as common as symmetry. *Caretta* is also extremely unstable in this respect, and one specimen in which the entire carapace was covered by only three laminae has been noted. It is at least possible that the occasional hawksbills that turn up with the shell covered by a solid, unsegmented sheet of horn may be variants of this type, although they are more generally believed to be individuals from which the carey has been stripped by heat.

Coker suggested that supernumerary scales might in some cases be due to injuries to the embryo produced by crushing or pressure in the nest, since he found a relatively high incidence of irregularity in loggerhead hatchlings from transplanted eggs. Lynn (1937), however, was inclined to attribute them to chance "fusion or division" during embryology.

The majority of students have regarded variation in chelonian laminae as of some phylogenetic significance. An increase in number has been interpreted as reminiscent of the many-scaled armor from which the modern shell has been derived through reduction in laminal number.

Several writers have noted a coincident change in the number of laminae and the number of bones beneath them, and this has been taken as evidence for the origin of the bony shell in an old carapace of osteoderms, or dermal platelets of bone imbedded in the skin. However, both Newman and Coker called attention to the fact that while coincidence of variation in marginals and underlying peripherals was common, little or no correlation between centrals and neurals or between laterals and costals could be shown. Lynn suggested that the correspondence between marginals and peripherals might be the result of the establishment of a new center for laminal development when a new bone appeared; while he saw in the more infrequent coincidence of bones and scales elsewhere merely a mechanical necessity for the filling of gaps in the horny covering when extra bones lengthen the shell.

Interpretation of laminal instability, and especially of correlated changes in the scales and bones of the shell, is necessarily shaded by one's concept of the shell of the turtle ancestor and thus of the taxonomic position of

Dermochelys. Some earlier writers regarded the leatherback as directly ancestral to all modern turtles. Baur (1887b) commented as follows:

"That the carapace of the Thecophora has developed from the carapace of the Athecae has been proved by a specimen of *Eretmochelys imbricata* in which I found small, polygonal plates of the same shape as those of *Dermatochelys* suturally connected with the third, fourth, fifth and sixth costal plates."

If such a position for the leatherback were accepted, the variation in laminae might be interpreted as harking back to a time when each scale covered an ossicle of the primitive armor.

Today, however, few students regard *Dermochelys* as directly in the line of turtle descent. Newman suggested that the extra laminae recall a stage in which the evolving turtle was adorned with a series of scale rows similar to the seven principal and seven alternating rows of the "tail-trunk" of the clearly primitive *Chelydridae;* and Hay's views regarding the homology of the twelve rows of shell laminae in alligator snappers and the twelve longitudinal keels of *Dermochelys* were mentioned earlier.

It seems reasonable to suppose that the roof of the shell of the primitive turtle was four-layered (see Figure 5), with the ribs broadened and overlaid by independent membrane bones (thecals), which in turn were covered by scales (the laminae) with bony cores (epithecals) like those on the backs of crocodilians. If these strata are numbered from 1 to 4, beginning with the uppermost, then the leatherback can be supposed to have retained 2 and 4, and the thecophoran 1, 3, and 4. The normal occurrence in thecophorans of stratum number 2 is limited to such isolated cases as the bony osteoderms of the fore legs of tortoises and the caudal scales of snapping turtles.

Hay (1922, 1928) and Noble (1923) carried on a spirited debate over the significance of certain splinters of bone discovered beneath the laminae of the shell of *Chelys.* Hay regarded these as vestiges of the epithecal bones of fossil turtles, and thus of the dermal mosaic of *Dermochelys,* but Noble was convinced that they were merely the result of local ossification at the sites of old injuries. There is little question but that such bones do sometimes form after injury, but Noble's argument that this was the case in Hay's specimens of *Chelys* were by no means conclusive. Moreover, there appears to be little room for doubt that an epithecal layer occurred in the shells of ancient turtles, as well as in the highly specialized armor of *Dermochelys.* Hay believed that in primitive forms the laminae corresponded with these epithecal ossicles, and that since modern thecophorans have lost the epithecals there is no reason for any correspondence in their scales and bones (and we find laminae covering parts of from two to as many as ten of the thecal bones).

It was pointed out by D'Arcy Thompson that the present arrangement of the laminae of the carapace, with their trinodal and equiangular seam junctions, is merely an expression of a widely applicable principle of conservation of border. Examples of the same type of hexagonal symmetry are seen in as apparently unallied phenomena as mud cracks, the cells of the comb of the honeybee, and the shells of extinct Eurypterids.

Hay suggested that the areola of the lamina corresponds with the original position of the epithecal bone beneath the scale, and Newman, in a detailed account of the development of the chelonian color pattern, gave evidence that there is intimate association between pigmentation and dermal ossification, summarizing his section on this subject thus:

"All of the carapace markings have thus been accounted for as the growth centers of existing or lost scutes [laminae]. This has been done in a species with a highly intricate color pattern (*Graptemys*), and could be applied successfully, I believe, to any other species."

Plate 5. Humpbacked specimen of *Pseudemys scripta scripta*. Mature when taken, this kyphotic individual lived a year at the Ross Allen Reptile Institute before dying of unknown causes.

That this belief was justifiable seems to be corroborated by pattern development in late embryos of *Pseudemys*.

A different type of shell abnormality, of consistent occurrence among turtles but most frequent in *Amyda,* is a freakishly humped condition known as "kyphosis." Nixon and Smith reviewed the literature on this subject and listed occurrences of kyphosis in fourteen species of turtles belonging to five different families. The condition is presumably caused by irregularity in the scheduling of processes that amalgamate costal plates and ribs during embryology, thus thrusting the backbone into an abnormally strong curve.

PART II

ACCOUNTS OF SPECIES

KEY TO THE FAMILIES

1   Limbs modified to form paddles, the wrist and ankle joints rigid; species wholly marine 6

1'   Limbs not modified to form paddles, the wrist and ankle joints movable except where modified in *Gopherus* for digging; no wholly marine species included 2

2(1') Shell covered with numerous horny plates; cutting edge of upper jaws not concealed by fleshy lips 3

2'   Shell covered with a leathery skin, its edges pliable; cutting edge of upper jaws concealed by fleshy lips ..... *Trionychidae* (p. 411)

3(2) Plastron small and cross-shaped, not nearly covering soft underparts; tail more than half the length of shell ..... *Chelydridae* (p. 48)

3'   Plastron not small and cross-shaped, forming a more complete covering for soft underparts; tail less than half the length of shell 4

4(3') Laminae of the plastron twelve, the pectorals in contact with the marginals 5

4'   Laminae of the plastron ten or eleven, the pectorals not in contact with the marginals .... *Kinosternidae* (p. 73)

5(4) Feet elephantine or shovellike; toes not webbed .. *Testudinidae* (p. 320)

5'   Feet not elephantine or shovellike; toes more or less webbed .... *Emydidae* (p. 111)

6(1) Shell covered with horny plates .... *Cheloniidae* (p. 341)

6'   Shell covered with a leathery skin .... *Dermochelidae* (p. 442)

Suborder THECOPHORA

THIS is the category that was provided to receive all the other living turtles when *Dermochelys* was placed in a separate suborder. All members of this group have a shell built in part by large bony plates of membrane bone that are fused with the ribs and vertebrae. The proneural bone is not free but firmly joined to the rest of the shell by an immovable suture. In some of the forms the bony structure of the shell has undergone reduction in one way or another as a secondary adaptation to aquatic or terrestrial life.

There are three superfamilies in this suborder, as follows:

Cryptodiroidea, in which the neck is vertically retractile and peripheral bones are present and connected with the ribs.

Pleurodiroidea, in which the neck is horizontally retractile and peripheral bones are present and connected with the ribs; there are no turtles of this group in the United States.

Trionychoidea, in which the neck is vertically retractile; marginal bones and shell laminae are lacking.

Family CHELYDRIDAE

The Snapping Turtles

THIS small family of large-to-immense, ugly, and aggressive turtles includes only two genera, both of which are American. In 1905, Douglas Ogilby described *Devisia mythodes* as a new genus and species of snapping turtle from the Fly River, New Guinea, and thereby provided an example of discontinuity in range which served zoogeographers as a classic model for years. The cherished anomaly was marveled at and cited by savants to prove or disprove all sorts of notions, but it apparently never occurred to anyone to go so far as to look at the type specimen again. Not long ago this image was thrown to the ground when Dr. Phil Darlington (an entomologist, of all things) turned his critical faculty on the situation. He grew suspicious that the locality from which Ogilby's specimen was allegedly collected might not have really been New Guinea at all, since during the subsequent forty years no additional specimens had come to

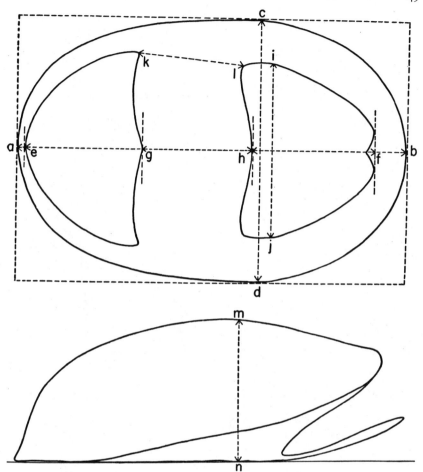

Figure 5. Method of measuring the shell of a turtle. Length of carapace, *a–b;* width of carapace, *c–d;* length of plastron, *e–f;* length of fore lobe of plastron, *e–g;* length of mid-part of plastron, *g–h;* length of hind lobe of plastron, *h–f;* width of hind lobe of plastron, *i–j;* width of bridge, *k–l;* depth of shell, *m–n.*

light. His idea encouraged Loveridge and Shreve (1947) to make a careful reappraisal of the characters of the type, with the result that these writers were able to show conclusively that the famous specimen is nothing but a mislabeled North American snapping turtle, which might well have come originally from Pennsylvania.

The snapping turtles have a reduced shell, the plastron being very small, cross-shaped, and only loosely joined to the carapace by a narrow bridge. The head is large and the jaws extremely strong, the upper having a hooked beak. There are paired barbels on the chin and several irregularly shaped

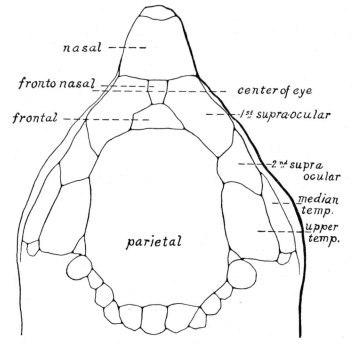

Figure 6. Dorsal head scalation of *Macrochelys temminckii*. Note position of eyes and compare Figure 8.

inframarginal laminae at the lateral underedges of the shell. The tail is long and is armed above with erect bony scales.

Fossils representing the family have been found in deposits as old as the Miocene.

KEY TO THE GENERA OF CHELYDRIDAE

1 Eyes lateral, the orbits not evident from dorsal aspect; a row of supramarginal laminae above the marginals along each side . . . . . .
. . . . *Macrochelys* (p. 52)
1′ Eyes dorsolateral, the orbits readily discernible from dorsal aspect; no supramarginals . . . .*Chelydra* (p. 60)

Genus *MACROCHELYS*

The Alligator Snappers

This genus has but one living species, which is confined to the southeastern United States. That the group was formerly more widely distributed is evident from the recent discovery of *Macrochelys schmidti*, an alligator snapper described by Zangerl (1945) from excellent fossil material from the middle Miocene of Nebraska and the early Pliocene of South Dakota.

These big turtles are similar in general appearance to the common logger-head snapper, but the two are easily distinguished. In the present genus the carapace is strongly elevated in three continuous and nearly even longi-tudinal ridges. There are from three to five supramarginal laminae between the first three laterals and the marginal series. The orbits open on the sides of the head, and the eyes are thus not evident to inspection from directly above. The snout anterior to the eyes is very long and is strongly hooked terminally. The shell is deeply and evenly dentate behind. An intergular lamina is sometimes present.

Hay (1898) attached the greatest significance to the presence of the supernumerary scale rows in the alligator snapper. He saw in the supra-marginals, submarginals and intergular vestiges of the three laminal rows needed to furnish complete representation of the 12 longitudinal keels

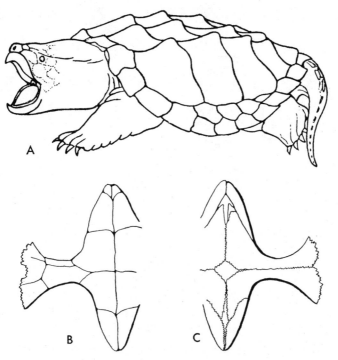

Figure 7. A: Macrochelys temminckii, the alligator snapper. *B:* Horny and *C:* Bony plastra of *Macrochelys*, showing relation of parts.

(7 dorsal, 5 ventral) of the leatherback turtle, and was thereby persuaded that *Macrochelys* occupied an important ancestral position.

Alligator Snapping Turtle *Macrochelys temminckii* (Troost)

(Plates 6, 7. Map 1.)

Range: Central Illinois, southern Indiana, the lower half of Missouri, and southeastern Kansas, southward to the Gulf. The western limit of its

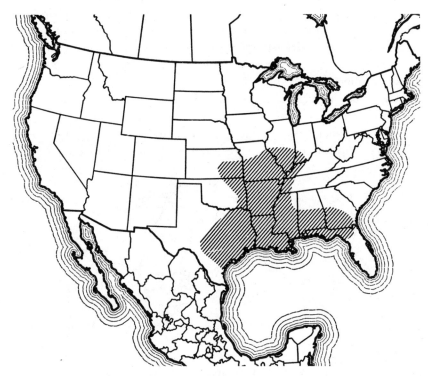

Map 1. Distribution of *Macrochelys temminckii.*

distribution traverses eastern Oklahoma and central Texas, where it reaches the Gulf in southern Texas, north of the Rio Grande. The eastern edge of its range passes from southern Indiana across the extreme western parts of Kentucky and Tennessee and into northern Mississippi, whence it runs eastward across central Alabama to southern Georgia, where, however, it evidently does not enter the Atlantic drainage. Below this line it is lacking only in peninsular Florida, where the Suwannee River appears to be the southern limit of its range.

Distinguishing features: Since there is only one living species in the genus *Macrochelys* the generic characters alone will serve to distinguish it, and a cursory examination will reveal a unique combination of features in the deeply ridged shell, the long tail, and the position of the eye sockets on the sides of the head where they are not evident from above.

Plate 6. Macrochelys temminckii. A and B: Male from Mandeville, Louisiana. *C and D:* Female; Okefinokee Swamp, Georgia. *E:* Specimen at Ross Allen Reptile Institute showing the "lure" in the floor of the mouth.

Description: The shell is broad and oblong, relatively straight-sided, moderately deep, and about ⅔ as wide as long. This is the largest of North American fresh-water turtles and it ranks among the very largest of modern species. Of five specimens measured by Cahn (1937) the largest had the following dimensions and weight: carapace length, 625 mm. (24.6 inches); carapace width, 533 mm. (21.0 inches); plastron length, 445 mm. (17.5 inches); weight, 148 pounds. Two hundred pounds is most frequently mentioned as the upper weight limit for this huge turtle, but Hall and Smith (1947) stated that in 1937 a specimen with the unheard-of weight of 403 pounds was caught in the Neosho River in Cherokee County, Kansas. The basis for this record should be carefully reappraised.

The three keels of the carapace are high and persistent even in old adults; the fore margin of the carapace is smooth, the rear margin deeply serrate. There are from 3 to 8 supramarginal laminae located between the costals and upper marginals. The color of the carapace is dark brown to mahogany, without markings. The plastron, like that of *Chelydra,* is small and leaves most of the soft parts exposed. The head is very large and pointed; and the upper jaw is strongly hooked at the tip. The head is covered by plates above, and the eyes are laterally placed and thus not visible from directly above the head. There are dermal projections of varying length on the chin and neck. The tail is very long, with three rows of tubercles above and numerous small scales below. The soft parts are dark brown above, lighter below, and there may be vague darker spots on the head, but these are usually lacking in large specimens.

Sex differences are not marked, and what few exist have been inadequately studied. The anus of the male is farther back on the tail than that of the female. The female is said to be the larger but I am inclined to doubt this.

Juveniles generally are much rougher than the adults, with the snout relatively longer and the dermal projections, barbels, and warts much more exaggerated. The tail may be longer than the carapace. Agassiz's specimen (1857, pl. 5), still in good shape 92 years later, in the Museum of Comparative Zoology, is of hatchling size though lacking the egg tooth. It has the following measurements: carapace length, 43.5 mm. (1.71 inches); carapace width, 41.0 mm. (1.61 inches); shell depth, 21 mm. (0.825 inches); plastron length, 28 mm. (1.10 inches). Young specimens show no more conspicuous markings than adults.

Allen and Neill (MS) gave the following measurements of a specimen newly hatched from an egg laid by a captive individual: carapace length, 44 mm. (1.73 inches); plastron length, 32 mm. (1.26 inches); distance from anterior edge of carapace to tip of tail, 104 mm. (4.10 inches).

Habitat: Rivers, sloughs, "dead lakes," oxbows, and canals. Although it appears to prefer the larger streams, the species is known from swamps and even marshes, although not usually far from running water. Individuals are sometimes seen in clear, shallow water on rock bottom in West Florida creeks tributary to larger rivers, but these have usually appeared to be migrating from one deep hole to another. The alligator snapper seems to be absent in the small rivers that drain directly into the Gulf in the

Florida panhandle, inhabiting only those that penetrate considerable distance inland.

Habits: Despite the huge size and relative abundance of the alligator snapper in some regions, there is very little published information concerning its habits. For many years my friend Ross Allen of Silver Springs, Florida, has kept these turtles in captivity and has observed them in the wild. During recent months he and his associate, Wilfred Neill, have extended these observations and have prepared an exceptionally interesting paper which constitutes the only important source of information on this extraordinary animal. These gentlemen have kindly consented to let me consult their manuscript in advance of publication and I have quoted from it extensively in the following account:

Alligator snappers are much less aggressive than common snappers and are also far less agile in striking. When disturbed, they are apt to sit passively with open mouth, lunging out only at objects that approach the head closely. A specimen that is lifted from the ground or is otherwise molested not only gapes widely but from time to time everts the mouth of the glottal opening for a few seconds, disclosing the bright white lining of the respiratory tract in sharp contrast with the dark mouth parts. The effect is surprising and probably serves as a warning mechanism.

The savage bite inflicted by *Macrochelys* has often been mentioned, and in fact overemphasized, in the literature. Particularly irksome is the frequently repeated story that a large individual can bite sections out of a broom handle. We find that specimens of 35 to 40 pounds' weight are scarcely capable of biting an ordinary pencil in two, although they can deeply score the wood. An enraged adult, biting savagely into the edge of a piece of soft wood, may pry off a splinter or even a small chunk. On seizing some yielding substance such as the human hand they retain their grip for some time, pulling and chewing to inflict a painful wound, but if a stick is proffered it is bitten viciously and then released.

Examples of *Macrochelys,* when molested, give off an odor which is neither as strong nor as unpleasant as that of *Chelydra.* In addition, alligator snappers, when lifted from the ground, may squirt a copious stream of clear liquid from the cloacal opening.

It is difficult to carry a large alligator snapper by the tail for the heavy creature must be held well away from one's legs. Furthermore, large specimens carried in this way may later die from injuries to the tail vertebrae. Fortunately, however, large *Macrochelys* may be carried conveniently in another fashion. One hand grasps the anterior margin of the carapace just above the head, while the other holds the posterior margin just above the tail. The turtle is thus lifted by the shell, and an individual carried in this way is unable to twist the neck enough to reach one's hand with the jaws, while the hands are also out of reach of the claws. It should be mentioned that a *Chelydra* cannot be carried safely in this fashion, since the hand grasping the anterior margin of the carapace is apt to be bitten.

Data on growth rates in this turtle are scarce. Two specimens kept at Silver Springs until they were five years old were respectively 90 and 84

Plate 7. Method of handling an alligator snapper. Lifting captive individuals by the tail eventually kills them. Here Ross Allen demonstrates the approved technique by lifting a 93-pound mossback from the Suwannee River.

mm. long (3.54 and 3.30 inches) at the time of their death. Another, hatched between September 11 and 19, 1942, had a shell 44 mm. (1.73 inches) long on that date and had grown to 85 mm. (3.35 inches) by January 9, 1949. Conant and Hudson recorded the 57-year residence of an alligator snapper in the Philadelphia Zoo.

What may or may not be a general tendency is the migration of one individual which Wickham (1922) observed over a period of three years in an Oklahoma stream and found to advance upstream at a rate of six miles per year.

Breeding: Several captive pairs of alligator snappers mate each year in the pens at Silver Springs, and Allen and Neill made the following observations on mating procedure:

The courting male follows the female, persistently trying to crawl up on her carapace. After successfully mounting, the male moves his body slightly to the right, and pushes his tail downward beneath that of the female. The female cooperates by moving her tail upward and to one side. Contact lasts from 5 to 25 minutes.

The female digs a nest by using the hind feet alternately, lifting the sand up and out of the hole and depositing it to one side. After oviposition, the female scrapes dirt back into the hole, packing it down firmly and then crawling over the site to smooth the surface.

Apparently nest building regularly takes place during the daylight hours. In the pens the females always laid their eggs as far from the water as possible.

A pair of large adults copulated February 28, 1946. The female deposited 44 eggs on April 21, 1946. Three eggs were broken during the laying process, eight were packed with wet sand at the bottom of the nest, and one was laid to one side of the nest. Twenty-five were placed in the main cavity, and seven near the opening of the nest. The eggs were nearly or quite spherical in shape. Oddly enough, the eight eggs laid in the bottom of the nest were larger than the others, the diameter ranging from 41 to 44 mm. (1.61 to 1.73 inches), while the remainder ranged from 38.4 to 40.9 mm. (1.51 to 1.61 inches), when measured seven days after deposition. [The egg figured by Agassiz was only 36 mm. in diameter.] The eight larger eggs were also somewhat softer than the others.

Another captive female laid 29 eggs on June 3, 1942. From this nest 11 young appeared on September 11, 1942, and from then until September 19 hatchlings emerged at a rate of two or three a day, all making their way to a pool of water in the pen. Other records of nesting at Silver Springs are as follows:

Date	Number of eggs
June 15, 1943	16
May 2, 1944	17
June 11, 1944	22
April 28, 1945	19

Nests measured at Silver Springs ranged in depth from about 14 inches to 20 inches.

Feeding: One of the most striking features of this remarkable turtle is the occurrence on the tongue of an irregularly cylindrical, distensible, vermiform growth that is different in color from the surrounding tissues and is capable of considerable movement. The possession of this structure, the snapping propensities of the turtle and its fondness for fish, and the fact that captive specimens have several times been seen to lie on the bottom, mouth agape and with the wormlike appendage in suggestive motion and quite the most conspicuous object in the murky vicinity, all naturally have led to the assumption that the structure is actually a lure to be overtly dangled for the enticement of potential victims in the shape of worm-loving fishes. Final corroboration of this assumption, however, has awaited the following observations of Allen and Neill:

The oft-repeated statement that the "lure" is worm-like or larva-like has obscured the fact that the structure is actually bifurcate, with an anterior and slightly smaller posterior section. The two portions arise from a rounded, muscular base, behind which there are many smaller papillae. In young examples, each part of the lure may bear a small extra branch while in larger specimens the entire structure is smooth, without branches or papillae. When the turtle is at rest, i.e. not "fishing," the lure is whitish or pale gray in color; but when set in motion it becomes pinkish, presumably being distended with blood.

Several baby *Macrochelys,* 3 to 4 inches in carapace length, were kept in an aquarium and supplied with live fish. The young turtles would hide between rocks in a corner of the aquarium and open their mouths widely. The muscular base of the lure would then pull down, first on one side and then on the other, imparting a wiggling motion to the two portions of the appendage. Sometimes the turtle would "fish" for hours without success, but often a *Mollienisia* or a *Gambusia* would swim into the open jaws and bite at the "bait." The turtle's jaws would immediately snap shut on the fish, which was next manipulated into position and then swallowed whole. Larger fish, caught in this fashion, were held in the turtle's jaws and torn apart by the front feet. It was noted that the turtle, while fishing, moved its eyes from time to time, obviously watching the movements of its potential prey. Sometimes the fishes would gather beneath the turtle's head, as though beneath an overhanging rock; and at such times the reptile would carefully alter its position, turning the head sideways so that the bait would be visible to the fishes. Although small fishes often nibbled at the algae that cover the shells of all adult alligator snappers, no attempt to catch these was made by the turtle.

It was interesting to note that when a new batch of fish was placed in the tank many would be captured by the turtles during the first day, while thereafter the fish would be wary, staying at the other end of the aquarium as far as possible from the turtles. At such a time a turtle would often move toward the fish and try to hide near them.

A young Florida snapper, *Chelydra serpentina osceola,* was for a time kept with the baby *Macrochelys.* When live fish were placed in the tank, the *Chelydra* would pursue them, usually catching them all before the *Macrochelys* could entice any into its mouth. Dead fish were usually discovered first by the Florida snapper, although the alligator snapper would eventually find and eat them if not forced to compete with *Chelydra.*

An adult alligator snapper weighing 100 pounds, for years on exhibit at the Reptile Institute, has accepted fish of all kinds, beef, pork, frogs, snakes, snails, worms, mussels, and various aquatic grasses. Captive specimens have been seen to stalk, catch, and eat other species of turtles, including *Deirochelys, Kinosternon, Sternotherus,* and *Pseudemys* spp. Specimens were also noted apparently trying to lure musk and mud turtles into their jaws. On one occasion a large individual killed but did not eat a smaller example of its own kind.

Specimens from the Suwannee River, during the first few days in captivity, voided numerous fragments of snail and mussel shells.

Alligator snappers are sometimes caught on trotlines. The best catches are usually made at night, and the only time I have ever seen them moving about was at night. Ross Allen has secured a number of specimens by a method in wide use among fishermen. A hook baited with meat is fastened to a line from a limb hanging over the water and is allowed to rest on the bottom. If the bait is visited at night and the site illuminated with a spotlight, turtles are often seen biting at the bait, and may be gaffed or gigged and hauled into the boat.

This suggests that the animals may actively forage by night and may reserve the "worm" to provide an occasional fish during the more passive daytime period, when the light would be better and fishes more likely to notice the bait.

Economic importance: The alligator snapper is in most places regarded as inferior to the common snapper for gastronomic purposes. Nevertheless it is eaten in nearly all parts of its range and is even shipped to nearby markets. It is frequently seen in New Orleans and was regularly sent to

St. Louis in the past, although whether this market still exists I do not know. Along the Suwannee River it is eaten locally.

Ross Allen often catches alligator snappers by swimming downstream and exploring the bottom with a diving mask, and, when a turtle is located, noosing its tail with a line and then lifting it into an accompanying canoe. Although probably the most efficient known means of hunting these creatures, this will probably never be widely adopted by run-of-the-mill fishermen.

Genus *CHELYDRA*

The Loggerhead Snapping Turtles

The present range of the loggerhead snapping turtles includes a large part of North America and extends southward as far as Ecuador. The genus is known from the North American Pleistocene, and a similar and nominally identical group existed in Europe in some abundance during Miocene and possibly upper Oligocene time. The carapace in this genus is keeled,

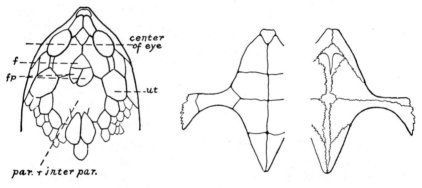

Figure 8. Left. Dorsal view of the head of *Chelydra*, showing scalation and position of eyes (Compare Figure 6).

Figure 9. Right. Plastron of *Chelydra s. serpentina*, showing laminae (*left*) and underlying bones (*right*).

but the keels are less regular than those of *Macrochelys* and the hind margin is less evenly dentate. The orbits are dorsolateral in position, and the eyes may thus be seen from directly above. The lower surface of the tail is covered with large scales. The snout is not markedly prolonged before the eyes.

KEY TO *CHELYDRA*

1 Width of the third central lamina less than ⅓ the length of the
   five centrals; knobs of dorsal keel located well behind centers of
   the laminae; lateral caudal tubercles much less conspicuous than
   median tubercles; eastern North America southward to northern
   Florida ....*serpentina serpentina* (p. 61)
1′ Width of third central ⅓ the length of the five centrals or more;
   knobs of dorsal shields located near centers of laminae; lateral caudal tubercles not much less conspicuous than median tubercles;
   peninsular Florida ....*serpentina osceola* (p. 69)

Common Snapping Turtle *Chelydra serpentina serpentina* (Linné)

(Plate 8. Figs. 8, 9. Map 2.)

Range: The common snapper is more widely distributed in North America than any other turtle. It ranges over the eastern part of the continent

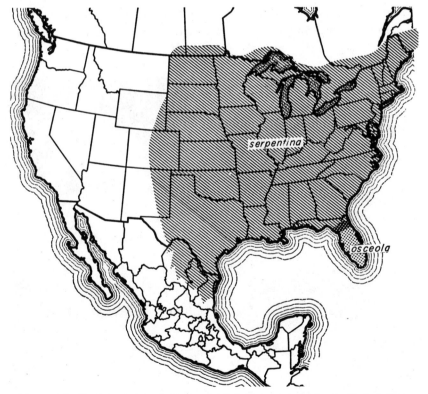

Map 2. Distribution of the subspecies of *Chelydra serpentina* in the United States.

from Nova Scotia, New Brunswick, and southern Quebec, westward across southern Ontario, where it extends northward to the Lake Nipissing area, and across extreme southern Manitoba to southeastern Saskatchewan. In the east it ranges southward to the upper part of the peninsula of Florida, where it meets a slightly different peninsular subspecies. The western limit of its distribution extends southward from Saskatchewan through western North Dakota, southeastern Wyoming, and the middle of Colorado and New Mexico into northeastern Mexico, where in an undetermined area it is replaced by one of a chain of races that extends southward to Ecuador.

Distinguishing features: There is little need for an extensive diagnosis of the characters of this big, aggressive, ubiquitous, and succulent turtle. In parts of its range it might be confused with the alligator snapping turtle, but it is easily distinguished by the position of the eyes, which may be seen from directly above in the common snapper. From the peninsular Florida race it differs in minor details of the form and disposition of the tubercles on the carapace and tail, and its comparatively narrow bridge is said to set it off from the tropical races, although this requires demonstration.

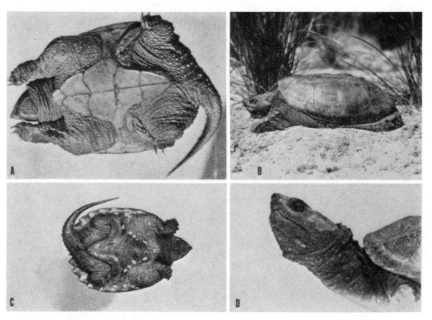

Plate 8. Chelydra serpentina serpentina. A: Female; Ithaca, New York. *B:* Female; Okefinokee Swamp, Georgia. *C and D:* Young; Ithaca, New York.

Description: The shell is broadly oval, usually widest posteriorly, moderately deep, and only incompletely protects the soft parts. The size is large, only a very few of our fresh-water turtles attaining comparable lengths and only one (*Macrochelys*) exceeding it in weight. Between 14 and 15 inches is generally stated to be the maximum shell length. The longest carapace measured by Conant (1938a) in Ohio was 322 mm. (13.0 inches); the largest specimen found in Maryland by McCauley (1945) was 300 mm. (11.8 inches) in shell length. Lagler (1943b) gave the average length of 24 Michigan specimens as 238 mm. (9.37 inches), with a maximum of 373 mm. (14.7 inches). Thirty pounds is most often given as the weight of large adults. There are, however, numerous records of 40 to 50-pound snappers, a few of 60-pounders, a couple of 70-pound specimens and one record of a hand-fed and abnormally obese individual of 85 pounds! Lagler and Applegate (1943) furnished equations for determining approximate weights of snapping turtles when the length is known.

The carapace is three-keeled except in old smooth-worn specimens, each keel usually having a knob or tubercle near the posterior edge of the lamina that it traverses. The shell is serrate behind and dark brown to black in color above. The plastron is very small and cross-shaped, not nearly covering the neck, legs, and tail, and is yellowish in color, without markings. The width of the bridge is usually one-tenth the length of the plastron or less. The head is large, dark in color, and pointed anteriorly, with a strong hook at the tip and two barbels on the chin. The eyes, located far forward on the head, are dorsolaterally placed and thus may be seen from a point directly above the head. There are numerous wartlike tubercles on the neck and on the powerful legs, which are dark above and light below. The tail is long and tapering, with three rows of tubercles above, the middle row much the largest.

When one casually looks over a group of snapping turtles from the dorsal aspect, one is struck by the consistent similarity between the sexes. If there are minor disparities in body proportions, they have yet to be described. The male is perhaps a trifle smaller and its vent opens farther behind the posterior edge of the carapace.

The young snapper is much rougher as to carapace than the adult, the keels being acute and with strong tubercles. The tail is relatively much longer, in the very young equaling, or even exceeding, the carapace in length. Radiating lines are often discernible on the shells of juveniles, and there may be dark mottling on the head and jaws and white spots on the lower marginals. A hatchling measured by Hoyt (1941) was 29 mm.

(1.14 inches) long and 27 mm. (1.06 inches) wide (carapace) and had, as do all newly hatched snappers, "a small whitish tubercle, hard and pointed, located just behind the two openings of the nares."

The average shell length of 5 hatchlings measured by Hamilton (1940) at Ithaca was 30 mm. (1.18 inches); average carapace width was 29 mm. (1.14 inches); and average tail length was 35 mm. (1.38 inches).

Habitat: This turtle shows almost no discernible habitat preference. All over its broad territory it is said by naturalists to be found in almost any kind of aquatic situation. One tendency that may be noted is its somewhat more frequent occurrence in bodies of water with soft, muddy banks or bottom in which it may bury itself in hibernating or possibly lie in ambush for the animals that form part of its diet. Engels recently (1942) noted its presence in brackish pools at the heads of tidal creeks on Ocracoke Island, North Carolina.

Habits: Although this extremely hardy turtle can tolerate a surprising amount of cold and has even been observed walking about both on and beneath the frozen surface of bodies of water, it appears to hibernate throughout most of the northern parts of its range from October to May. Hibernation usually occurs under water and preferably beneath a covering of bottom mud or vegetable debris. Whether through sociability or because of a scarcity of suitable sites is not known, but snappers often congregate in large numbers to hibernate in some restricted area of pond or stream bottom. Lagler (1943a) recorded a take of 500 pounds of these turtles from a little channel between two lakes, and muskrat houses are favorite winter quarters. After spring emergence, both sexes may wander widely or migrate from one body of water to another, at times covering distances of half a mile or more.

The disposition of the snapper is too notorious to require description here. It is sufficient to say that the common name is an apt and succinct characterization of the animal. In offense and defense it is truly a snapper, never reaching deliberately for an object to bite it but rather striking with the speed and power of a big rattlesnake. The innate urge to strike is present in the hatchling and may even be demonstrated by babies not altogether free of their eggshells, although, in general, the young are less irascible than the adults. An interesting vagary was mentioned by Pope (1939a), who pointed out that the most irritable old individuals, constituting a menace to the life and limb of the handler when on land, usually calm down and become quite pacific when placed in the water. I never had thought of this, but it appears to be altogether true, and for me it brought a disenchantment with Quonab and his "Fight with the Demon of the Deep," which had held me in more or less sustained awe ever since I first read Seton's *Rolf in the Woods,* at age nine. Except for this harrowing encounter I never heard of anyone's having any trouble whatever with snappers in the water, and this impresses me as doubly curious in view of the fact that the "strike" of the snapper must have developed mainly as a mechanism for seizing stalked, lured, or ambushed prey in the water. Besides defending itself with its jaws, the snapper secretes a powerful musk, not so vile as that of *Sternotherus,* but bad enough. This may be protective, although as Pope mentioned, its value in the water is problematical. I have also been unable to ascertain whether or not the glands that secrete this musk have any relation to the "four peculiar, serially-arranged organs" found by Zangerl (1941) under the edge of the carapace on either side of 20-mm. embryos. Zangerl said that these were not typically glandular but gave no opinion as to their possible fate or function.

Breeding: Mating may occur from late April till November. As in *Sternotherus,* the male is said to hold his position atop the female's shell by clinging to the underedge of her shell with the claws of all four feet; the tail is then twisted and curved upward and manipulated until contact between the vents has been established. No detailed courtship has been described except for Taylor's observation (1933) of two captive individuals which rested on the bottom of a tank with their heads together and for 10 minutes gulped and expelled water, producing a boiling at the surface. As Taylor remarked, this may or may not have been some kind of *liebespiel.* McCauley saw a fight between two males on April 29, 1938, in water about 30 inches deep. The cause of the fight was unknown, but McCauley supposed that it had come about after some falling-out over females, which seems reasonable. McCauley's field notes were as follows (1945):

I noticed two large turtles ahead of me making considerable commotion in the water. I went up to them carefully and was able to wade out very close to them. I observed that they were *Chelydra serpentina,* very large specimens, and that they were fighting. They made vicious lunges at each other, hissing and rolling over in the water. They would take hold of each other's necks and cling tenaciously. Their heads and necks were somewhat lacerated; . . . their throats were red and raw looking.

The nest of the snapping turtle is dug at varying distances from the water, the site sometimes obviously being determined by the character of the soil but occasionally depending on what appears to be a whim of the female, since she will cross large areas of likely-looking soil to dig finally in an identical piece of ground. This characteristic, however, is not peculiar to snappers but may be seen in most nesting turtles, and indeed may be homologous to the traditional urge in the human female to move the furniture about. Snapper nests vary widely in form and in manner of excavation, and the nesting process has actually never been adequately described. From the scattered information available on nesting it would appear that the cavity may at least sometimes be definitely formed and flask-shaped, that it slants at an angle, that it is dug by the alternately working hind feet, that the eggs are directed individually to the bottom by a hind foot, and that the nest is covered by raking movements of the hind feet. No mention is made of any excavation or covering by the fore feet. Cagle (1942a) gave the following data on six nests found by him in southern Illinois:

Distance from water	Depth of nest	Number of eggs
30 feet	6.7 inches	31
20 feet	6.7 inches	20
3 feet	3.9 inches	20
55 feet	——	30 (appr.)
55 feet	7.1 inches	8
75 feet	6.7 inches	18

The same writer found eight other nests all within 12 feet of the water on a ridge by a lake. The majority of females lay in June, but the season is long and there are records of nesting from May to October. The average number of eggs is probably close to 25, although there are numerous reliable reports of clutches of twice that number, and more infrequent mention may be found in the literature of complements of 70 eggs or more. Norris-Elye (1949), for example, watched a female lay 77 eggs on a creek bank near Winnipeg, Manitoba, and the same writer said that Mr. Eric Fisher found in the oviduct of a female that had emerged to lay, 80 perfectly formed eggs. Dr. James Oliver has suggested to me that what seems to be a more consistent occurrence of such large clutches as these in more northern latitudes may be due to the simultaneous ripening and deposition of the entire complement of a year, which must be laid all at once if all are to escape the early fall cold. This theory is interesting, although not presently demonstrable, since there is no adequate basis for the (probably correct) assumption that farther south two or more clutches are laid each year.

Finneran (1947) watched a 30-pound female lay 52 eggs at a rate of two per minute, near Branford, Connecticut (June 13, 1941). These eggs hatched after 81 days of incubation. Cagle found 90 days to be the incubation period in southern Illinois, and Hoyt (1941) hatched a clutch of eggs found June 11, 1939 (Ithaca, N.Y.) in a terrarium in about 85 days. Much longer periods have been reported, but usually in cases in which there is the possibility that the normal rate of development has been retarded by an artificial environment. Toner (1933) and others have found nests in which hatching has been delayed by the advent of cold weather until the following spring, but this appears to happen only in the extreme northern parts of the range. The eggs appear to vary in diameter from about 24 to 33 mm. (0.945 to 1.30 inches), and are sometimes perfectly spherical, sometimes slightly oblong. Brimley (1943) said that if the contents are emptied and the soft shell allowed to dry it always assumes an oblong shape, but I am at a loss to explain the significance of this.

Feeding: Because of its large size and forbidding appearance the snapper has gained a reputation for rapacity that is probably not altogether warranted. There can be no doubt that a big snapper will drag under a shore bird or a duck or strike a game fish on occasion, if the urge and the opportunity to do so coincide, but the sum total of such predation in terms of damage by any one individual snapper is probably not great. The most recent and important contribution toward an intelligent appraisal of the diet of the snapping turtle has been the work of Dr. Karl F. Lagler of the Institute for Fisheries Research, Michigan Department of Conservation. Table 1 is taken from one of his papers (1943b) and is based on an examination of 173 stomachs and 261 colons (the stomach contents give evidence of the most recent meal, while those of the colon may indicate the nature of a number of feedings).

Table 1. Contents of stomachs and colons of snapping turtles

Food Item	Stomachs		Colons	
	Composition by volume (%)	Frequency of occurrence (%)	Composition by volume (%)	Frequency of occurrence (%)
Game and pan fishes	34.2	32.4	2.1	51.7
Forage fishes	0.3	5.8	0.4	10.3
Unidentified fishes	0.9	12.7	2.1	35.2
Other vertebrates	1.1	2.3	1.1	7.2
Carrion	19.6	6.4	1.3	2.7
Invertebrates	7.8	49.7	23.9	82.4
Vegetable matter	36.2	74.6	69.2	93.5

In summarizing his important findings regarding the food habits of the snapper Dr. Lagler commented as follows:

If the remains of all the game and pan fishes found in the food of the snappers studied from wild waters were of fishes that were alive and healthy when taken, 302 turtles would have accounted for 275 live fishes, over the period of time represented by the remains in the digestive tract. Since this figure must be considerably reduced by the proportion of specimens taken as carrion, the minimal significance of this species to game and pan fish populations seems apparent. It seems a conservative estimate that on the average not more than one game or pan fish is eaten per day by the individual snapping turtle.

Most game and pan fishes eaten by the snappers were less than legal size, and such smaller fishes are much more abundant in nature than those of legal size. Populations of snappers were estimated by the mark and recapture method in Wolf and East Twin Lakes to approximate two snapping turtles per acre of surface area in these waters. These considerations further minimize the significance of the estimate that each snapper on the average consumes six tenths of a game fish per feeding. Thus there need be little concern as to the adverse relations of snapping turtles to game fish populations in wild waters.

Since forage fishes are often more abundant in waters than are game fishes, it is surprising to note their fewness in the food of the snappers studied. They averaged about one to every ten of the 462 sample meals. Consequently, this turtle does not appear to have a significant role in the current decline of forage fish populations, nor does it seem to compete significantly with game fishes for this type of food.

Pell (1940) found plants and animals to be about equally represented in stomachs of 18 specimens from Massachusetts and New York, and found considerable variation in diet from one habitat to another.

Actually, the feeding habits of the snapper are somewhat anomalous, since in spite of a most impressive equipment for active predatory feeding it is almost perfectly omnivorous. Pell found a preference for dead fish over live, but Lagler noted a distaste for fish so long dead as to be rotting. Babcock (1938) mentioned as a common custom in rural New England the placing of a snapper in the swill barrel to fatten, and referred to a specimen that reached a weight of 86 pounds through this regimen!

Economic importance: A detailed account of the economic value of this turtle was given by Clark and Southall (1920). More recent notes were gathered by Lagler, who commented as follows on the hunting of this species for the market in Michigan:

During the summers of 1937 to 1939 the market demand far exceeded the supply. Trappers found a ready sale for all turtles caught at wholesale prices ranging from five to eight cents per pound, live weight. These trappers have mobile units and obtain snappers mostly by means of traps made of seine twine on wire loops with a single funnel-shaped opening. Captured turtles are held and fed in a live-pen until several hundred pounds have accumulated. They are then shipped alive in a barrel or crate to market. The best markets for Michigan turtles are in Chicago and various cities in northern Indiana and Ohio. The greatest demand comes from the restaurant trade, but many turtles are sold to private individuals. Retail prices range from ten to twenty cents per pound live weight and twenty to thirty cents dressed.

A trapper and helper operating about five dozen traps in Michigan waters report an average profit of a thousand dollars for each of the last three years.

In summing up the evidence for and against the snapper in the attempt to decide whether it would be sounder conservation to control or to protect it Lagler made the following suggestions:

The food habits data here presented give some indication of harmful effects of snapping turtles in aquatic communities. Considering also the services rendered to man, it seems best to recommend their conservation, except in waters where special investigation proves them to be undesirable. Conservation measures would help to insure a sustained yield of these reptiles to professional trappers.

It should be stressed that the above remarks were based on a study of the situation as it is in Michigan waters, and that the relation between the snapper and man may be quite different elsewhere. In Florida, for example, the thought of protection for the snapper would horrify a number of people who have flocks of ducks on lake margins, and a snapper would probably make little worth-while contribution in a pond in which experimental or highly concentrated fish culture was being carried on. On the other hand, any threat to an uninterrupted supply of eating snappers would bring cries of protest from many a Philadelphian.

The shells of snapping turtles had wide ceremonial use among the Indians. Speck (1943) recently referred to the part they played, dried and mounted on handles with corn kernels inside, in beating out time in the Great Feather Dance and the Woman's Dance of the Iroquois.

Florida Snapping Turtle *Chelydra serpentina osceola* Stejneger
(Plate 9. Map 2.)

Range: This snapping turtle is confined to southern and central Florida, where it intergrades with the northern common snapping turtle in Alachua

Plate 9. Chelydra serpentina osceola. Upper: Young. *Center and lower:* Adult. All from Orlando, Florida.

and adjacent counties and occurs in some abundance down to the tip of the peninsula.

Distinguishing features: This is a weakly differentiated subspecies, and when someone makes a detailed study of the relative growth of the structures involved, and of the differences between the sexes, the differentiating characters may not stand up. So far, however, no one has demonstrated conclusively that the following differences do not hold in a majority of cases:

(1) The knobs on the dorsal keels appear to be located near the centers of the laminae in *osceola* and farther back toward the hind edges in typical *serpentina*.

(2) The lateral rows of tubercles on the surface of the tail are usually not much less conspicuous than the central row in *osceola* while in *serpentina* there is marked disparity in their size.

(3) The width of the third central is said to be often ⅓ the length of the five centrals in *osceola* and less than ⅓ their length in *serpentina*.

All these characters are in sad need of corroboration.

I believe the Florida snapper may be on the average slightly larger than the northern form. At least, I measured a male specimen from Orange County that had a carapace length of 382 mm. (15 inches), one millimeter more than the maximum among the literally dozens of "records" for *serpentina* that have been published. The shell width of this giant was 324 mm. (12.75 inches) and shell height 156 mm. (6.15 inches). Individuals approximating these dimensions are seen with some frequency in Florida.

Sexual differences are no more conspicuous in this race than in the common snapper.

The hatchling is much more brightly colored than the adult, with the ridges, keels, and tubercles of the carapace very high and narrow, those forming the three principal longitudinal keels being especially thin and bladelike. The ground color of the carapace is chestnut brown, with considerable black pigment at the seams, especially at those of the marginals. The under surface of the margin and the plastron are black, a large black spot occupying a variable area on each marginal lamina and similar spots being widely scattered about the plastron, mostly near its edges. The soft parts are grayish brown above and grayish black below, rather evenly mottled, with the mottling tending to form lines on the head and neck. The chin barbels are usually white. The egg tooth is very sharp and thornlike, its tip inclined slightly downward. Measurements of a newly hatched individual from Charlotte County are as follows: shell length, 31 mm. (1.22 inches); shell width, 26.5 mm. (1.04 inches); shell depth,

18 mm. (0.71 inches); head width, 11.5 mm. (0.45 inches); and length of tail beyond shell, 30 mm. (1.18 inches).

Habitat: This turtle shows broad tolerance in habitat selection. It is found in almost any body of water and is one of the few vertebrates found regularly in certain of the acid and very sterile muck-bottomed hammock streams.

Habits: Insofar as the habits of the Florida snapping turtle are known, which is not to a degree that should evoke pride, they appear to be no different from those of the northern race, except to the extent that meteorological differences modify the life history of all reptiles in the peninsula. I have seen them walking about on land during June and have found nests during the same month, but I never witnessed the laying process, and no one seems to have described it in print. Writers on the breeding habits of the northern form do not mention the excavation of a large basin prior to the scooping out of the actual nest cavity, but from the looks of the ground around several nests that I have found, both with and without eggs, and from the pile of sand usually to be seen on the backs of females that have just finished laying I would judge that such a preliminary cavity, similar to that dug by some sea turtles, is made, at least on occasion. Whether its function is to gain depth or to conceal the laying female is uncertain.

The Florida snapper is evidently as omnivorous as its northern relative. It has several times been seen eating crayfish and frequently takes fish, although whether, under natural conditions, it catches them alive with any frequency is a moot point and one deserving of a detailed investigation. The few stomachs that have been examined have all contained greater or lesser amounts of plant material.

This turtle appears to be largely nocturnal. When ponds dry up it may either migrate or bury itself in the mud, where it can remain for weeks (or perhaps much longer) without suffering damage.

Economic importance: For some reason the snapper is little favored as food by people in Florida. Perhaps it is because there is an abundance of turtles more easily prepared and more comely to look upon. At any rate I know of no commercial trapping of snappers in Florida. I have eaten them and to me they seem in no way inferior to the revered northern snapper, which I have also eaten, and under the most propitious possible conditions. However, I must admit that a soft shell impresses me as just as good and a Suwannee chicken far better. The role of the Florida snapper in the fish management program ought to receive early and energetic study.

Family KINOSTERNIDAE

The Musk and Mud Turtles

THIS strictly American family includes turtles of small to medium size of two very closely related genera. The principal character by which the family is distinguished from the allied Staurotypidae is the absence of the entoplastron, the unpaired plastral bone located between the two epiplastra. This feature is almost unique among turtles. Apparently it is not known whether the missing bone has been crowded out by the enlarged epiplastra, which form the entire anterior lobe, or has been absorbed by them. A vestige of the bone may sometimes be seen on the inner surface of the fore lobe of the plastron of *Sternotherus*. Other features of this family are as follows:

The frontal bone is small and does not enter the orbit. The masticatory surface of the upper jaw is without a longitudinal ridge; neither the bony edge nor the tomium of the jaws is serrate. The plastral laminae are ten or eleven in number. The limbs are not paddlelike; the toe bones are jointed, and there are four or five claws. There are barbels on the chin. The

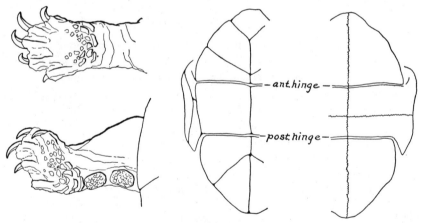

Figure 10. Left. Fore and hind legs of *Kinosternon*, showing the characteristic patches of sexual scales on the rear surface of the latter. These have several times been referred to (erroneously) as "stridulating organs."

Figure 11. Right. Plastron of *Kinosternon*, showing relation of the two plastral hinges to the laminae (*left*) and to the bones (*right*).

tail is less than half as long as the shell and lacks a median crest of scales. There are twenty-two marginals.

Fossil musk turtles are known from as far back as the Pliocene.

KEY TO THE GENERA OF KINOSTERNIDAE

1 Seam between the pectorals very short, the laminae nearly tri-angular in shape; plastron with the anterior and posterior portions nearly equal, movable on a transverse hinge .... *Kinosternon* (p. 89)

1' Seam between the pectorals not shortened, the laminae oblong, plastron with the anterior part shorter than the posterior part, not, **or** only slightly, movable .... *Sternotherus* (p. 74)

Genus *STERNOTHERUS*

The Musk Turtles

In this group the plastron is characteristically much smaller than in *Kinosternon* and the anterior hinge does not provide the mobility seen in the other genus. The pectoral laminae are quadrangular in shape.

KEY TO *STERNOTHERUS*

1   Side of head with light stripes; if no light stripes, then ground color sooty black .... *S. odoratus* (p. 82)

1'   Side of head without lines, or with dark lines; not light-striped 2

2(1')   Gular lamina lacking or vestigial; axillary and inguinal lami-nae very large, as broad as long or nearly so; second, third, and fourth centrals larger than broad; lower Mississippi Valley west of central Mississippi .... *S. c. carinatus* (p. 75)

2'   Gular lamina present; axillary and inguinal laminae small, much longer than broad; second, third, and fourth centrals broader than long 3

3(2')   Head with dark stripes; lateral keels absent; eastern Tennessee and south-central Mississippi into Georgia and extreme western South Carolina above the fall line .... *S. c. peltifer* (p. 81)

3'   Head dark-spotted, not striped; lateral keels present; Georgia to Alabama and northern Florida, in the coastal lowlands ....
.... *S. c. minor* (p. 77)

Mississippi Musk Turtle *Sternotherus carinatus carinatus* (Gray)

(Plate 10. Map 3.)

Range: Lower Mississippi Valley from central Arkansas and south-eastern Oklahoma southward to New Orleans and southwestward to Bexar County, Texas. In the area extending eastward from central Mississippi and eastern Tennessee, typical *carinatus* is replaced by the newly described race, *peltifer*. The group is also continuously represented toward the eastward along the Gulf coast, but where the range of *carinatus* ends and that of the intergrading subspecies, *minor*, takes up is not at present known.

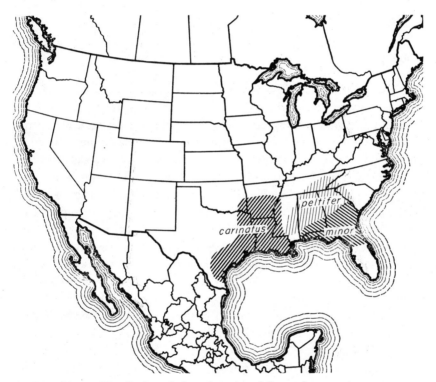

Map 3. Distribution of the subspecies of *Sternotherus carinatus*.

Distinguishing features: Typical specimens of this turtle are readily recognized by the high, sharply keeled back and steeply inclined, unkeeled sides, the lack of stripes on the head and neck, and the absence, in a majority of specimens, of a vestigial gular lamina.

Plate 10. Sternotherus carinatus carinatus. Left: Male. *Right:* Female. Both from southern Louisiana.

Description: The shell is very deep, with a sharp central keel and with the slope of the unkeeled sides very steep, although they may be either straight or curving in cross-sectional outline. This is the largest member of the family in the United States, reaching a measured maximum shell length of 150 mm. (5.91 inches). A male of good size from Lake Concordia, Louisiana, has the following measurements: length of carapace, 103 mm. (4.05 inches); width of carapace, 73 mm. (2.87 inches); height of shell, 45 mm. (1.77 inches). Each lamina of the carapace may overlap that behind it to a greater or lesser degree, although this is not always the case. The ground color of the carapace is light to medium brown with streaks and bars of darker brown, those on the centrals and on the upper half or third of the laterals running lengthwise, while on the lower parts of the laterals they radiate forward and downward from the upper hind corner of the lamina. The laterals are often conspicuously margined with dark brown posteriorly. The plastron is yellow, usually unmarked, and often with much skin exposed in the broad seams. The gular lamina is usually absent and never well developed. The head and neck are brownish, with small dark-brown dots and with the horny jaw surfaces often vertically streaked with brown. There are two barbels on the chin.

Differences between the sexes of this musk turtle have not been worked out in detail, but probably involve no unique features. The several mature males that I have seen had the usual thick, long tail and more distally located vent.

Young specimens are closely similar to the adults, differing in their greater tendency toward overlapping of the dorsal laminae, more accentuated markings of the head and upper shell, and a lighter ground color of the plastron.

Habitat: This is principally a river-swamp form. It probably occurs in nearly any kind of swamp-shored body of water in or near the flood plains of larger streams.

Habits: Apparently nothing has been written about the habits of this turtle in the natural state. Captive specimens have been found to be practically omnivorous, although decidedly inclined toward a carnivorous diet.

The longevity record for this species in the Philadelphia Zoo is five years and eight months.

I have noticed a curious erosive disease that attacks the margin of the shell of Louisiana specimens of this turtle and that I have never seen elsewhere. In ten of a series of twelve mature specimens from southern Louisiana, the posterior marginals and peripherals had been partly or entirely destroyed by this disease.

Loggerhead Musk Turtle *Sternotherus carinatus minor* (Agassiz)
(Plate 11. Map 3.)

Range: Georgia in the lower coastal plain southward to Lake County, Florida, and southwestward to coastal Alabama. A related form occurs in coastal Mississippi, but whether this is *minor, carinatus,* or the recently described *peltifer* (type locality, Jefferson Davis County, Mississippi) is not known.

Distinguishing features: This musk turtle is very similar to *S. c. carinatus,* the two agreeing closely in details of coloration of the shell and soft parts, in general body conformation, and in the relatively large size attained by both. In the loggerhead musk turtle, however, only an occasional individual lacks the gular lamina, and the characteristically steep, unkeeled sides of *carinatus* are rarely seen in *minor.* There occur specimens of *minor* in which the lateral keel is lacking and the shell is comparatively low and smooth, sometimes even lacking the middorsal keel and showing no imbrication of the dorsal laminae. In these individuals, usually of advanced age, the juvenile color pattern of dark seam borders and dark radiating lines is also usually absent. Measurements of a pair of mature loggerhead musk turtles from Ichtucknee Springs, Florida, are as follows:

Sex	Shell length	Shell width	Shell height	Greatest bony width of head
Male	123 mm. (4.85 in.)	81 mm. (3.18 in.)	50 mm. (1.97 in.)	38.5 mm. (1.52 in.)
Female	116 mm. (4.56 in.)	83 mm. (3.26 in.)	46 mm. (1.81 in.)	35 mm. (1.30 in.)

Several herpetologists have commented on Agassiz' choice of a specific name for this musk turtle, disturbed by the fact that it is the largest kinosternid that occurs within its range and not the smallest, as the name might seem to imply. However, Agassiz' term did not, as has been supposed, compare the form given the name *minor* with *odoratus*, which *minor* conspicuously exceeds in size, but with the Mississippi Valley form now known as *carinatus*, and this is unquestionably the larger of the two.

In his description of *minor*, Agassiz (1857, 1:424) commented as follows: "This species differs from the preceding [*carinatus*] by its smaller size, and more distinctly still by its arched sides and the low keel of the back." When he made this remark Agassiz had in his collection a specimen of *carinatus* with a shell no less than 150 mm. (5.91 inches) long.

This turtle shows considerable sexual dimorphism. The tail of the male is enlarged and muscular, horned at the tip and strongly prehensile, serving him as an effective grappling organ during courtship. In the female the tip of the tail barely reaches beyond the hind edge of the carapace, while in the male the vent is even with the rear rim of the carapace and the rest of the tail, when straightened, extends far beyond it, but is usually twisted and curved in two planes to be tucked for safekeeping under the overhanging shell margin. The middle laminae of the plastron of the male show a slight tendency toward concavity, but this is not marked. The most striking secondary sexual feature is to be seen in the enormous heads of the old males, to accommodate which the anterior third of the shell is often greatly swollen and the anterior lobe of the plastron permanently bent downward. Such individuals, domed and bulbous before and tapering sharply to small hindquarters, present a most astonishing and extraordinary appearance.

Habitat: The loggerhead musk turtle occurs in swamp-shore streams. and lakes with fluvial histories or connections; in cypress swamps in the flood plains of rivers; and in spring runs.

Habits: This turtle may be found actively foraging during the daytime, and sometimes at night, on the bottom of springs and spring runs and

Plate 11. Sternotherus carinatus minor. A and B: Old male from Ichtucknee Springs, Florida. *C:* Female from Alachua County, Florida. *D:* Male from Eureka, Florida. *E:* Female from Ichtucknee Springs, Florida.

wherever the water is clear enough to allow the bottom to be seen. The foraging turtles rarely swim, but more frequently crawl slowly about, often stopping to poke their noses into piles of trash or beds of submerged plants. Their swimming is labored, the short legs working furiously to attain only feeble speed. Populations of this turtle are sometimes so large that it is difficult to see how they manage to find sustenance. Marchand (1942) said of Ichtucknee Springs Run, Marion County, Florida: "The population of *Sternotherus minor* is the largest I have found anywhere. In a day's goggling, 500 or more of these animals may easily be seen." I have seen a dozen or more at once on the bottoms of several clear central Florida streams. As was mentioned earlier, where such concentrated populations of *minor* occur, it is usually very difficult to find a single specimen of *S. odoratus*.

A really noteworthy attribute of this stink-jim is its penchant for high basking perches. In taking a sunning station (and this species is very fond of basking) it appears to seek the very highest point to which it can possibly climb, and it has no mean ability as a climber. With the utmost patience and exertion, and great craning of its neck, an individual will sometimes climb the trunk of a partly submerged sapling and make its way far out over the water to the tips of the smallest twigs before stopping to take the sun. They are often seen on tall stumps, and I recall as one of the most ludicrous turtle postures that ever fell to my gaze that taken by a huge-headed male on the sharp tip of a smooth, conical cypress knee that rose six feet above the surface of the water. Only through the constant tension of all its muscles and the stretching of its neck to the fullest extent was this ambitious individual able to retain his place in the sun on the towering pedestal, but stay there he did, for an hour that I know of. I have seen this performance repeated several times on smaller cypress knees.

The large-headed males appear to be aware of the shell-crushing power of their freakishly enlarged jaws and often make every effort to bite an offending hand; and since the neck is very long and maneuverable, the only safe hold is at the rear third of the shell.

Breeding: The mating and nesting habits of this turtle are not known. At Blue Spring Creek, Jackson County, Florida, I found several females wandering in the woods beside the stream and scratched out numerous eggs laid singly or in twos or threes and shallowly buried at the bases of trees and beside logs. This occurred on April 14. During May and June I have frequently seen similarly placed eggs, almost certainly of this species, in the Suwannee drainage. Neill (1948b) wrote as follows concerning a pair of hatchlings found by him:

On August 12, 1947, I found two turtle eggs that had been unearthed by a plow on a hillside overlooking a lake near Louisville, Jefferson County, Georgia. When broken open, one of the eggs disclosed a young *Sternotherus minor,* well developed, but with considerable yolk material remaining. The little turtle was capable only of the feeblest movements. When handled it voided the contents of its scent glands, producing the vile odor that has earned the species its local name of "Stinking-Jim." The second egg also contained a little *S. minor* that reacted in similar fashion. It is interesting to note that the scent glands, and a behavior pattern involving their use, are well developed at a time in embryonic life when most muscular reactions are lacking. One wonders how effective the scent glands of these and other reptiles may be in discouraging the attacks of predators.

Feeding: No detailed study of the feeding habits of this musk turtle has been made. Random observations indicate that it is probably dependent on crayfish and snails for a large part of its diet, but that it is an inveterate scavenger and not averse to plant food on occasion, although the only plants that I have actually seen them eat were potato and tomato peelings and watermelon. They are easily caught on hooks with nearly any kind of bait of animal origin. They eat fish joyfully, but I never saw any evidence that they could catch a fish that was in a normal state of health.

Tennessee Musk Turtle *Sternotherus carinatus peltifer* Smith and Glass

Range: The original description of this musk turtle was based on one specimen from Jefferson Davis County, south-central Mississippi, the describers identifying as also *peltifer* the specimen on which an old record from Roane County, Tennessee, was based. More recently Neill (1948d) somewhat doubtfully referred specimens from near the fall line in eastern Georgia and adjacent South Carolina to this subspecies. If all these records are to be accepted, then it would appear likely that records of *"minor"* from the North Carolina–Tennessee line and of *"carinatus"* from central Alabama also represent *peltifer*. The range thus extended would be: eastern Tennessee southeastward to the fall line in west-central South Carolina and southwestward to south-central Mississippi.

The intermediate features and geographic position of this musk turtle and the overlapping of its distinguishing characters by variations in *minor* and *carinatus* indicate that these three forms are at most races of a single species and that trinomials should be applied to them, as has been suspected by a number of herpetologists for a long time.

Distinguishing features: This subspecies is as yet poorly known. It is described as having a high central keel, no lateral keels, and a pair of vestigial gulars. Although the free edges of the laminae of the carapace are dark-margined, there appears to be less tendency toward a radiating arrangement of the dark markings on the shell. The axillary and inguinal laminae are described as much reduced and much longer than broad. As in *minor*, the cross-sectional outline is not steep and is evenly curving. Unlike both *carinatus* and *minor*, this subspecies has the brown markings on the sides of the head and neck in the form of definite lines.

The type of this subspecies is a male with a shell 85.7 mm. (3.37 inches) long and 63.8 mm. (2.51 inches) wide.

Nothing has been published concerning age and sex difference in this turtle.

Habitat: This musk turtle has been recorded from rivers.

Habits: Nothing is known of the habits or economic importance of this race.

Common Musk Turtle *Sternotherus odoratus* (Latreille)

(Plate 12. Map 4.)

Range: Eastern North America, from the Parry Sound area of southern Ontario eastward through southern Maine and southward to Dade County, Florida. Toward the west it covers all of Michigan south of a line extending eastward and westward through Saginaw Bay, and reaches southeastern and north-central Missouri (apparently without entering Iowa), extreme southeastern Kansas, and the eastern halves of Oklahoma and Texas.

Distinguishing features: The presence of light stripes on the side of the head and neck of this small, dark stink-jim is usually sufficient basis for positive identification. In certain central Florida variant populations that lack light stripes the skin surfaces are deep glossy black.

Description: The carapace of this turtle is moderately elongate and narrow, and usually rather highly arched, although quite variable in this respect. Large individuals range in shell length from 114 mm. (4.5 inches) to 120 mm. (4.74 inches). The largest found by Conant (1938a) in Ohio and by McCauley (1945) in Maryland had shell lengths, respectively, of 127 mm. (5.00 inches) and 103 mm. (4.06 inches). Dimensions of a specimen from Stafford County, Virginia, were as follows: length of carapace, 86 mm. (3.38 inches); length of plastron, 64 mm. (2.52 inches); width of plastron, 26 mm. (1.02 inches). The carapace may be keeled or smooth in adults, and is sometimes flattened middorsally. It varies from dark brown to black in ground color and is usually unmarked, although sometimes with scattered spots or with some tendency toward a radiating pattern of streaks on the laterals of lighter-colored specimens. The dorsal laminae may or may not overlap slightly. The plastron is small and ranges in color from yellowish to brown. A gular is present and usually well developed. The sides of the head and neck typically bear a pair of conspicuous yellow

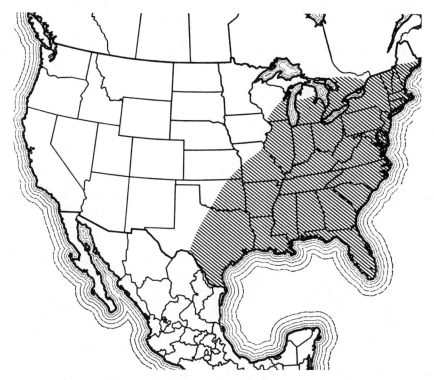

Map 4. Distribution of *Sternotherus odoratus.*

or whitish lines that begin on the snout and extend backward, one above the eye and the other below it. These lines may be faded or broken, and in some Florida specimens are lacking. The ground color of soft parts is olive to black, with or without light markings.

Differences between the sexes, as summarized by Risely (1933), are as follows:

(1) The tail of the male is longer and with a heavier and usually blunt terminal nail.

(2) There is more exposed skin area in the median seams between the plastrals in the male.

(3) There are two small patches of tilted scales on the inner surface of each hind leg of the male (the legendary "stridulating organs" of some other writers), interpreted reasonably by Risely as of use in grasping the female during copulation.

(4) The plastron of the male is somewhat shorter in proportion to its length.

In the young the shell is relatively higher and more sharply keeled mid-dorsally and usually with an auxiliary lateral keel. Both carapace and soft parts are darker than at maturity, and the markings are more definitely delineated and more intensely pigmented, the marginals being decorated with a more or less regular series of light spots. The shell measurements of three specimens, made immediately after hatching, were as follows:

Length	Width	Height
22.2 mm.	15.8 mm.	12.8 mm.
(0.875 inches)	(0.622 inches)	(0.505 inches)
21.3 mm.	15.5 mm.	11.5 mm.
(0.840 inches)	(0.610 inches)	(0.454 inches)
21.0 mm.	16.0 mm.	12.1 mm.
(0.827 inches)	(0.630 inches)	(0.476 inches)

It should be pointed out that the shell proportions of hatchling mud turtles (and probably of all other kinds of turtles) change rapidly and markedly owing to the lateral expanding and spreading of the periphery of the shell on being freed from the compressive force of the egg membranes and shell. Thus it is that Pope (1939) found a "very young" individual to have a shell 22.5 mm. (0.885 inches) long and 19 mm. (0.750 inches) wide. This nearly circular shell, contrasting strongly in shape with the three hatchlings that I measured, means merely that Pope's specimen had unfolded itself, and the process may have taken place within no more than a couple of weeks after hatching.

Habitat: The common musk turtle is very tolerant as regards its habitat. It is found in nearly any kind of water, although it is extremely aquatic and shuns temporary or fluctuating puddles and pools. In Florida it is only very rarely found in the same situations as *S. c. minor.*

Habits: In the northern parts of its range the common musk turtle hibernates in wet or damp debris, or preferably in mud under shallow

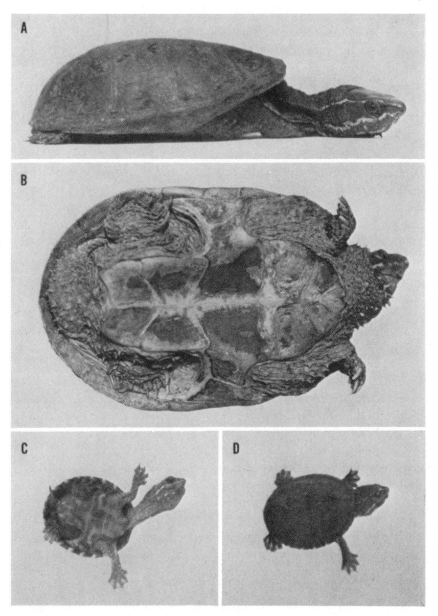

Plate 12. Sternotherus odoratus. A and B: Male. *C and D:* Young. Both from New York.

water, where, according to Cagle (1942), it begins to burrow into the bottom in a few inches of water when the water temperature falls below 10° Centigrade. Individuals observed by Cagle were near enough to the surface so that the heads could be thrust out for air from time to time, and the head hole thus left in the mud was the only evidence of the presence of an imbedded turtle. In Georgia, Richmond (1945) found this species hibernating in recesses in banks of streams and beneath logs and trash. Thomas and Trautman (1937) reported finding no fewer than 450 individuals on March 28, 1927, a foot deep in mud within an area of canal bottom of about forty-five feet by six feet, where the water had been partly drained from the canal. All but a half dozen of the turtles were dead, apparently from exposure to cold. The musk turtle emerges early and in Indiana has been seen active every month of the year except January and February. In Florida and southern Georgia it apparently only very rarely hibernates, and colonies inhabiting the numerous big springs of Florida probably detect little difference between summer and winter.

This is a highly aquatic species, being one of the few turtles that apparently never bask in the sunshine. Although active on occasion at all hours of the day, it shows in the South decidedly nocturnal tendencies, a characteristic noted also by Lagler (1943b) in Michigan individuals. It is most easily collected in clear streams by paddling or wading about and lighting the bottom with a headlight. It is rarely found swimming, more often being seen walking slowly about on the bottom, and sometimes at considerable depths. In the clear Florida springs I have often seen them in water thirty feet deep and slightly more.

By nature this turtle is pugnacious and ill-tempered, making strong and determined efforts to bite when caught and being capable of removing a sizable piece from an incautious hand. The unpleasant odor that it emits, and from which it takes its common and technical names, comes from four musk glands which open at the edges of the shell, one at each outer hind edge of the bridge and another on each side where the skin meets the carapace about halfway between the anterior ends of bridge and shell. What the primary function of this musk is has not been determined, but it is secreted when the animal is annoyed or angry and is presumably protective. Despite its initial misanthropy, this species becomes quite tame in captivity. In one case a captive specimen lived for twenty-three years.

Breeding: Mating appears to occur sporadically throughout the period of activity, musk turtles having been observed copulating in the spring from April on, and in the fall as late as October. Risely showed that the sperm are retained alive in the oviducts of the female throughout the winter. After a rather straightforward courtship in which the male merely follows and chases the female about until her resistance has been lowered, copulation occurs. Finneran (1948b) observed a pair copulating plastron-to-plastron near New Haven, Connecticut, on April 26, 1943, and in looking over other observations of mating in this form I can find no reference to the method by which the female is mounted except the statement that in one case the female was "lying prone as if dead." If the type of copulation seen by Finneran was the usual thing, it appears likely that the numerous observers who have witnessed the mating of this turtle would have noted it as something worth mentioning, and it thus seems probable that Finneran's pair were exceptional. Laying occurs in Florida from April to June and in the north from May to August. Hatchlings are found in Florida in August and September and in the north in September and October. The laying process is erratic, the nests being sometimes excavated in a highly desultory way and so shallow that an egg may lie exposed, although at other times the female may dig a well-formed nest three or four inches deep. The number of eggs deposited varies from one to five, although as many as nine shelled eggs have been found in the oviduct of a large specimen. In common with other members of the family, this species prefers to nest near or under some elevated mass such as a log, a pile of trash, or the base of a tree. In Florida I have found a majority of nests between the buttresses of cypress and black gum trees, where the accumulation of detritus is moved and the eggs are barely imbedded in the soil below, after which the leaf mold is either scraped back over the nest or not, according to the whim of the mother. In Tennessee, Cagle (1937) found individuals laying on top of stumps and also located sixteen nests under one small log. Risely measured seventy-five eggs of Michigan specimens and found the following range in dimensions: maximum length, 31 mm. (1.22 inches); minimum length, 24 mm. (0.95 inches); average length, 27.1 mm. (1.06 inches); diameter, 14.2 to 17.0 mm. (0.56 to 0.67 inches), average, 15.5 mm. (0.61 inches). He found the incubation periods for two hundred laboratory-hatched eggs to be from sixty to seventy-five days. Edgren (1949) found the lengths of forty-four eggs from Wisconsin specimens to be about half a millimeter shorter than those measured by Risely, and those of Florida specimens average about one millimeter shorter. The eggs are rather unusual in appearance, being hard and brittle-shelled, with a porcelainlike pebbled or undulating surface, and in strong light usually showing alternating areas of bluish, dead white, and pinkish white. On August 12, 1944, I happened to be looking at an egg

I had picked up at Bivins Arm, Alachua County, Florida, when hatching began, and I made the following note on the process:

At 6 P.M. noticed a small crescentic piece of shell, about 2.5 mm. long and 1 mm. wide that was loose but remained in place. By 11:00 P.M. this hole enlarged to 3.5 mm. in diameter and another hole, on the opposite side of the same end, has appeared, giving the egg two separate openings. From the larger (the first-made) of these the toenails of the hatchling protrude. August 13, 11:00 A.M.: no change except holes slightly larger, now irregularly so, and with jagged outlines; the inner membrane is not broken under the newer hole. Nine P.M.: the two holes are now continuous, the band of shell separating them having been fractured. August 14, 9:00 A.M.: several cracks now radiating from the single large hole in the end. 12:00 noon, hatchling out; it did not emerge by forcing its way through the initial hole but by breaking out the entire end of the egg in one large and two small pieces.

This seems a remarkably long period of time for the pipping and emerging process, and it is possible that some factor in the artificial environment provided hindered the hatchling, but it appeared perfectly normal in all respects; and as is usual with these turtles it was ready to bite the instant it saw daylight.

Feeding: The common musk turtle is mostly carnivorous in its diet, but not disinclined to scavenge or to consume small amounts of aquatic vegetation. It is able to feed on land and has been known to do so, but in keeping with its strongly aquatic life history it finds its food mainly in the water. The data in Table 2 showing results of an analysis of the stomachs of seventy-one Michigan specimens are taken from Lagler (1943b).

Table 2. Stomach contents of musk turtles.

Food item	Composition by volume (%)	Frequency of occurrence (%)
Game and pan fishes	6.2	0.8
Carrion	40.1	5.8
Crayfishes	6.2	3.3
Insects	16.9	34.2
Snails and clams	23.2	28.3
Plants	3.4	15.0
Vegetable debris	4.0	12.0

Economic importance: This species is of no value as a source of food for man. Regarding its relation to fish culture and management, Lagler (1943b) gave the following summary:

Although a few fishes are eaten, this so-called damage may be more than counterbalanced by feeding on other "enemies" of fishes. Such insects as the predacious belostomatids [water bugs] and diving beetles may be placed in this category. Seeds of the beggar-tick (*Bidens*) appear to be injurious to fishes [and these are sometimes eaten by musk turtles, as Lagler had mentioned earlier].

Excepting possible harm by predation at hatcheries, and on eggs and larvae of desirable species of fishes, it would seem that the principal adverse effect of the musk turtle on fish populations may be in the numbers of fish-food organisms that it eats. Benefits may accrue from its scavenging habits, its consumption of mosquito larvae, its destruction of mollusks that figure in the life cycles of parasites of fish and other animals, and possibly from its dissemination of undigested, viable seeds of aquatic plants.

To add an inconsequential point to the credit of the musk turtle with the fish culturist, it should be mentioned that specimens are sometimes found in the stomachs of largemouthed black bass.

Genus *KINOSTERNON*

The Mud Turtles

This genus is not strongly differentiated from the preceding. In the species within our limits the head is usually relatively smaller than in *Sternotherus* and the shell is more depressed, but these distinctions do not hold up in tropical species. The pectoral laminae are triangular in form or with a very short fourth side. The plastron is relatively larger than that of *Sternotherus* and its fore lobe is more movable. Some Central American species are able to close the shell as tightly as a box turtle.

In all the characters used to differentiate the two genera of kinosternids the two approach each other most closely in the Floridian *Kinosternon subrubrum steindachneri*.

KEY TO *KINOSTERNON*

1 Carapace with three longitudinal light stripes, these either distinct and definite or intercepted and more or less broken by other light bars and areas 2
1' Carapace plain, not striped 3
2(1) Laminae of carapace opaque, not showing sutures between

bony plates beneath; peninsular Florida except in the extreme

southern tip ....*baurii baurii* (p. 90)

2' Laminae of carapace thin and transparent, sutures of the underlying bony plates evident; Florida, in and south of the Everglades ....*baurii palmarum* (p. 94)

3(1') Ninth marginal lamina roughly triangular, its apex extending
much higher than upper border of eighth; Texas northward to
southern Nebraska and westward to southern Utah and Arizona; central Illinois .... *flavescens flavescens* (p. 94)

3' Ninth marginal oblong, its upper margin not much above that
of the eighth 4

4(3') Carapace smooth or with a single median keel 5

4' Carapace with three more or less complete longitudinal keels;
southern California westward to extreme western Texas .....
....*sonoriense* (p. 99)

5(4) Side of head with two distinct yellow lines; southern half of
the Mississippi Valley .... *subrubrum hippocrepis* (p. 106)

5' Side of head plain, mottled, or with spots arranged in linear
order; without definite light lines 6

6(5') Bridge wide, its breadth usually half the length of fore lobe of
plastron or more; Connecticut and southern New York southward to northern Florida and northward in the Mississippi
Valley from the Gulf to Indiana and Ohio ....
.... *subrubrum subrubrum* (p. 101)

6' Bridge narrower, its breadth usually contained 2.5 to 3 times
in length of fore lobe of plastron; peninsular Florida ....
.............................*subrubrum steindachneri* (p. 108)

Striped Mud Turtle *Kinosternon baurii baurii* (Garman)

(Plate 13. Map 5.)

Range: Peninsular Florida from Leon, Alachua, and Clay counties southward to Key West except for an area in the Everglades at the southern tip
of the peninsula, where it is replaced by *palmarum*. The following remarks
regarding the distribution of the striped mud turtles in extreme southern
Florida are quoted from my paper of 1940:

Garman says of this form: "Several collectors have secured specimens in
Key West. It was found to be tolerably abundant in the brackish pools, where
it seems the only tortoise, during our own collecting there." Garman's specimens seem to me to be much closer to typical *baurii* than to *palmarum,* which
is singular, though not so strange as the total absence of mud-turtles on Key
West today. I have spent many days searching for them without success, and

Plate 13. Kinosternon baurii baurii. A and C: Male. *B, D, and E:* Female. Both from South Jacksonville Alligator Farm, Florida.

none of the host of "Key Westers" whom I have questioned have ever seen them. Since the authenticity of Garman's observations can hardly be doubted it seems probable that somehow the turtles have been exterminated on the island. Garman's record of a specimen from Cuba was almost certainly based on an introduced individual.

Distinguishing features: This species is characterized by its low, broad shell, unkeeled but with three longitudinal light stripes, and its small head, on each side of which there are two light stripes.

Description: The shell is smoothly globose, its widest and highest points usually, but not always, posterior to the middle. It is a nearly perfect oval in lateral outline. Dimensions of a pair of specimens from Gainesville, Florida, are as follows:

Sex	Shell length	Shell width	Shell height	Head width
Male	84 mm.	57 mm.	34.5 mm.	18 mm.
	(3.3 inches)	(2.24 inches)	(1.36 inches)	(0.71 inches)
Female	97 mm.	68 mm.	40.6 mm.	20 mm.
	(3.82 inches)	(2.68 inches)	(1.6 inches)	(0.787 inches)

The laminae of the carapace are smooth; the central series may be slightly convex, flattened, or markedly depressed, rendering the carapace distinctly bilobed in appearance. The marginal series is shallow, its lateral and posterior portions almost or quite vertical. The ground color of the carapace varies from horn color to nearly black; it is marked by three more or less distinct light, longitudinal bands, two of which traverse the upper ends of the costal laminae while the third occupies the median line. These stripes are subject to considerable variation. The median line may be so broadened as to occupy the entire area of the central laminae; the dorsolateral bands may be wavy and vaguely outlined or, rarely, even lacking. The plastron is large, with freely movable anterior and posterior lobes; it is yellowish or olive in color, either unmarked or with an indefinite darkening of the seam borders. The head is small and somewhat conical in shape; in ground color both it and the other soft parts are usually slightly darker than the carapace. There are two light lines extending posteriorly from the margin of the orbit, one above and the other below the tympanum; these stripes may be the only markings of the head and neck or there may be extensive mottling of these regions with color like that of the plastron. The lower jaw is either wholly horn colored or marked with a light stripe that extends backward on either side of the terminal tooth.

Secondary sexual differences are not marked and are essentially those characteristic of the family as a whole. Disparity in head size is rather less than usual. The tail of the male is long and heavy, and is provided with a strong terminal spine; it is usually bent into a compound curve, the upper part bending downward, the terminal part forward. The tail of the female is extremely short and stubby. The male reaches sexual maturity at smaller size and never attains the dimensions of the larger female. Conspicuous patches of enlarged scales, presumably used in grasping the female during copulation, are present on the upper and lower posterior surfaces of each hind leg of the male.

The carapace of the very young individual is roughened by numerous irregular but more or less longitudinally directed ridges on all the laminae, and that part of the median area beneath the sagittal stripe is elevated to form a narrow keel. The marginals of the hatchling are each marked by a large light spot which lies chiefly on the lower surface but which may also extend up onto the upper surface. The median area and seam borders of the plastron are dark; the outer middle portion of each lamina is light yellow.

Habitat: This is probably the least confirmedly aquatic of the North American kinosternids. It is a characteristic inhabitant of small, shallow, usually quiet bodies of water, whence it frequently emerges to forage on land. It is especially fond of wet meadows. It likes to prowl on land during rains, and may frequently be seen in temporary rain pools after thunder-showers.

Habits: This is an easy-going and gentle species of mud turtle, and while it will bite if a special effort is made to provoke it, it is normally quite inoffensive. It has been found in a temporary sort of hibernation or, per-haps more correctly, a withdrawal from the cold, in piles of moist decaying hyacinths along the shores of Lake Newnan, Alachua County, Florida. Pope (1939a) mentioned a specimen in the possession of Dr. W. T. Davis of Jacksonville, that was taken September 7, 1913, and was still alive in captivity on June 16, 1938.

Breeding: The breeding habits of the striped mud turtle have been very inadequately studied, and I can add little to the sketchy account that I con-tributed in 1940:

The eggs are laid from April to June [one, two, or three at a time] in holes in the sand and in piles of dead water-hyacinths. On May 2, 1936, I saw two males fighting over a female on the bank of a little stream near Gainesville. After a struggle of five minutes or so [during which the main aim of each in-dividual seemed to be to overturn the other], one of the combatants lost his footing and fell off the three-foot bank into the water; the victor immediately turned to the female, who had been observing the conflict from a distance of two feet or so, thrust his snout beneath her plastron, and crawled under her, leaving her balanced upon his carapace. The two remained motionless in this singular position for half an hour, when I finally grew bored and put them in a bag. The eccentric *liebespiel,* if such it was, was not resumed in the laboratory.

Feeding: This is an omnivorous feeder, probably being not at all preda-cious and depending more on the fruits of scavenging than any other member of the family. It shares with *Terrapene* the name "cow-dung cooter" in some places, and I have several times seen individuals eating manure. I saw three individuals feeding in a garbage pile on the beach of

a lake in Lake County. They are easily taken on hook and line, and I have
seen them caught on various baits including liver, grasshoppers, worms,
and dough.

Paradise Key Mud Turtle *Kinosternon baurii palmarum* Stejneger
(Plate 14. Map 5.)

Range: The Everglades northward to West Palm Beach, Lake Okee-
chobee, and Collier County. Spurious examples occur occasionally throughout the range of the typical form.

Distinguishing features: This subspecies is very closely related to *K. b.
baurii*. The only moderately reliable difference between the two is evidently
the reduced opacity of the dorsal laminae of *palmarum*. This transparency
appears to be due to a reduction of both the pigmentation and the thickness
of the scales, which do not hide the sutures of the bones beneath them. Thus
the general effect is that of a shell crisscrossed with light bands, rather than
of a three-striped one. Although this difference in coloration has appeared
to some herpetologists as too trivial a basis to warrant giving the form a
name, it cannot be denied that specimens from the lower Everglades are
easily and pretty constantly recognizable as such, and until it can be proved
that the reduction of laminal pigmentation and thickness is a direct reaction of the individual to the environment it seems that nothing would be
gained by discarding the subspecies.

Sexual differences are those of the typical race.

I have not seen hatchlings, but the racial coloration is evident in very
young specimens.

Habitat: The habitat includes all the types of aquatic situations available
in the low, marshy, hammock-studded area occupied. These are chiefly
natural channels in grass marsh or drainage ditches in reclaimed muck-
land, canals cut into the oölitic limestone bedrock, and water-filled potholes in the same material.

Habits: Nothing is known regarding the habits of this mud turtle. My
friend Dr. Frank Young kept a specimen in captivity in Miami for 13
years, feeding it mostly on hamburger.

Yellow Mud Turtle *Kinosternon flavescens flavescens* (Agassiz)
(Plate 15. Map 5.)

Range: Southern Nebraska southward through western Kansas and all
but extreme eastern Oklahoma and Texas to the lower Rio Grande and to

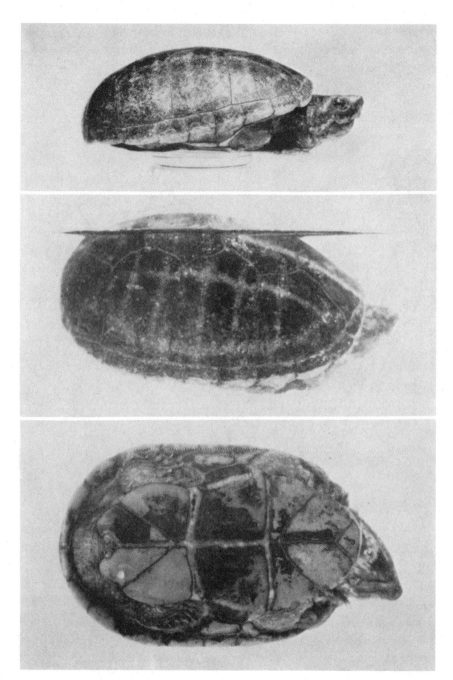

Plate 14. Kinosternon baurii palmarum; Collier County, Florida; sex undetermined. The center figure shows the transverse striping of the carapace that appears to be the only distinguishing character of this slightly differentiated race.

95

northern Coahuila, Mexico; southwestward through the southeastern corner of Colorado and New Mexico to southern Utah and southern Arizona. In Sonora and Durango, Mexico, *K. f. flavescens* is replaced by the subspecies *K. f. stejnegeri* Hartweg. Whether the apparently discontinuous distribution of this mud turtle in the northern Mississippi Valley is real or

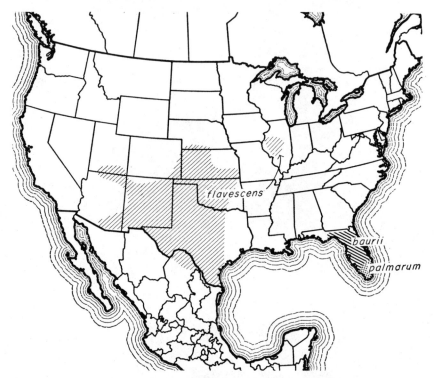

Map 5. Distribution of *Kinosternon f. flavescens* and the subspecies of *K. baurii.*

merely reflects the inadequacy of collecting and observations in the area is not known. P. Smith (1948) pointed out that records for northern Illinois are all from sandy areas noted for the occurrence of relict populations of western animals.

Distinguishing features: This large mud turtle is characterized by its smooth, unmarked shell with elevated ninth and tenth marginals, and by the presence of variably extensive areas of bright yellow on the throat, sides of the neck, or head.

Description: This is a comparatively large species of *Kinosternon* with a short, broad shell that is noticeably low and either flat above or depressed

Plate 15. Kinosternon flavescens flavescens. Males; Ridge Spring, near Marathon, Texas. *Lower right:* Male (*A*) and female (*B*): near San Antonio, Texas.

along the median line, often giving it a bilobed appearance. A male specimen of average size from Illinois had the following proportions: length of
carapace, 132 mm. (5.2 inches); width of carapace, 93 mm. (3.68 inches);
length of plastron, 114 mm. (4.5 inches); depth of shell, 50 mm. (1.97
inches); length of head, 29 mm. (1.14 inches); length of tail, 56 mm. (2.2
inches). The largest specimen of the yellow mud turtle on record is a male
from Illinois with a shell length of 146 mm. (5.75 inches). The carapace is
olive green in color, except for occasional black seam borders. The ninth
and tenth marginal laminae are markedly elevated and more or less triangular in shape. The plastron is generally light, either yellow or brownish
yellow, and some of its seams are usually broad, with much cartilage exposed. The head and neck are olive above, with a certain amount of bright
yellow laterally or ventrally. The jaws are horn color and may be mottled
with brown or olive. The soft parts are olive or grayish green, darker above
than below.

Sex differentiation follows the pattern for the family. The tail of the male
is longer and tipped with a heavy horn, and the two patches of tilted scales
are present on the inner surface of the hind leg. In addition, Cahn (1937)
found males to have a more exaggerated upper beak and to attain larger
average size than females.

The young are in general similar to the adults, but differ in one important feature, this being the relative size of the ninth and tenth marginals,
which are often lower than the seventh and eighth instead of being much
elevated. Cahn says the elevated marginals begin to show up in individuals
with a shell length of about 67 mm. (2.63 inches).

Habitat: Cahn, who knows this turtle as well as anybody does, described it as "preeminently a pond turtle" in Illinois. It is found in a
variety of smaller bodies of water but is seldom reported from streams. In
the West it is sometimes abundant in cattle tanks in dry prairie land.

Habits: Surprisingly little is known of the habits of the yellow mud
turtle. Burt (1935) found six individuals buried together in the bottom of
a dry pond, and Burt and Hoyle (1934) said that they often leave the water
to forage on land, especially during the height of the rainy season. Cope
(1892) said that they were frequently seen migrating overland during the
dry season as the pools dried up.

Breeding: Taylor (1933) observed in the laboratory that during courtship the male may tap the carapace of the female with his plastron, and
that the female is held during copulation by all four feet of the male, while
the tail is used as a grasping organ as in other species (October 20 and 30,

1932). The eggs may be 25–26 mm. (0.985–1.02 inches) long and 16 mm. (0.63 inch) wide. Very meager evidence indicates that the usual egg complement may include only two eggs, but no information is available concerning nesting dates, the laying process, or the incubation period.

Feeding: The species is apparently able to feed on land as well as in the water. The only notes on diet are those of Strecker (1927c), who found that in captivity Texas specimens accepted meat but preferred insects and small mollusks (*Planorbis* and *Limnaea*).

Sonoran Mud Turtle *Kinosternon sonoriense* Le Conte

(Plate 16. Map 6.)

Range: This mud turtle has a curiously circumscribed range, extending from the Imperial Valley in extreme southeastern California through northernmost Mexico, the southern half of Arizona, and southern New Mexico into western Texas.

Distinguishing features: The Sonoran mud turtle is differentiated by its relatively large size and long, narrowly elliptical, often three-keeled carapace; mottled, unstriped soft parts; and the relatively even height of the upper edges of the posterior marginals.

Description: The shell of this mud turtle is comparatively elongate, depressed, and usually elliptical in lateral outline. Van Denburgh (1922) appears to have measured the longest specimen on record; the dimensions of this were as follows: length of carapace, 144 mm. (5.67 inches); width of carapace, 101 mm. (3.98 inches); length of plastron, 134 mm. (5.28 inches); width of plastron, 79 mm. (3.11 inches). Two specimens from Arizona had the following dimensions:

Sex	Carapace length	Carapace width	Shell height
Male	107 mm. (4.22 inches)	75 mm. (2.95 inches)	36 mm. (1.42 inches)
Female	111 mm. (4.37 inches)	80 mm. (3.15 inches)	37 mm. (1.46 inches)

The carapace is either almost completely smooth or with a median keel that is often present only on the anterior and posterior centrals, or occasionally with three more or less distinct keels. Its ground color is dark, brownish or olive and unmarked. The plastron is flattened and brownish in color. The head, neck, and limbs are mottled dark and light.

Plate 16. Kinosternon sonoriense. A and D: Female. B: Young. C: Male. All from Arivaca, Arizona.

Except for sexual features common to all members of the group, nothing has been published concerning dimorphism in this species beyond a statement by Siebenrock (1907) regarding a difference in plastral proportions, and this should probably be discounted because of the meagerness of the material on which the supposed character was based.

The hatchling figured by Agassiz (1857, pl. 5) had the following dimensions: length of carapace, 28.5 mm. (1.11 inches); width of carapace, 24.0 mm. (0.945 inch); height of shell, 14 mm. (0.55 inch); length of plastron, 23.5 mm. (0.925 inch); width of plastron, 14.5 mm. (0.57 inch). The short, broad shell of the young individuals is feebly keeled above, the central keel being low and broad and the lateral keels often little more than a broken row of more or less elongate ridges. The hind edge of each upper marginal often has a black smudge, and a foliate dark plastral figure roughly follows the trends of the seams.

Habitat: The Sonoran mud turtle is found in creeks, ditches, ponds, and residual water holes in canyons and dry arroyos, whence occasional rains stimulate it to sporadic emigration. It occurs up to elevations of at least 5,300 feet in the Huachuca Mountains of Arizona, according to Stejneger (1902a).

Habits: The life history of this turtle is almost completely unknown. Van Denburgh said that it is very aquatic but that it sometimes leaves the water to bask in the sunshine. The same writer found that captive individuals ate meat voraciously and mentioned a specimen caught near Tucson on hook and line baited with meat.

Common Mud Turtle *Kinosternon subrubrum subrubrum* (Lacépède)

(Plate 17. Map 6.)

Range: Eastern United States from southwestern and east-central New York southward in the coastal plain and adjacent Piedmont through most of South Carolina and Georgia to the vicinity of Alachua County, Florida, and the Gulf coast of the Florida panhandle, Alabama, and Mississippi; from here northward in the eastern part of the Mississippi Valley through Tennessee, western Kentucky, and southwestern Illinois to northwestern Indiana. In extreme western Florida, western Mississippi, eastern Louisiana, and western Tennessee it intergrades with *K. s. hipprocrepis.*

Distinguishing features: The size is small and the carapace is smooth, unmarked, and unkeeled in adults. The line formed by the upper edges of the posterior marginals is relatively level, none of the marginals being markedly elevated. The bridge is relatively broad and the side of the head usually lacks well-defined light lines.

Description: The shell of the common mud turtle is depressed, either oval in lateral outline or nearly straight-sided and with precipitous posterior marginal region. The size is moderate for a *Kinosternon,* the shell

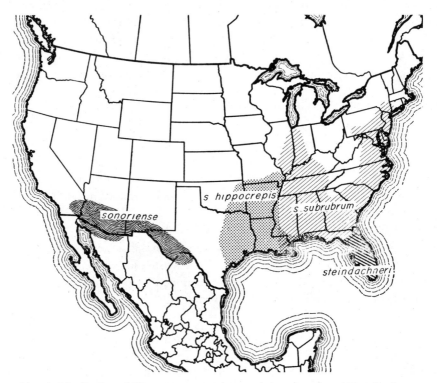

Map 6. Distribution of *Kinosternon sonoriense* and the subspecies of *K. subrubrum.*

length of the largest specimen found by Nichols (1947) having been 111 mm. (4.37 inches) and that of the largest seen by McCauley (1945), 123 mm. (4.85 inches). Dimensions of a medium-sized specimen measured by Lynn (1936) were as follows: length of carapace, 87 mm. (3.42 inches); width of carapace, 61 mm. (2.40 inches); length of plastron, 75 mm. (2.95 inches); width of plastron, 36 mm. (1.42 inches). The carapace is smooth in the adult, the middorsal region depressed along the centrals, giving to the shell a bilobed appearance. The carapace is undecorated in mature specimens, and varies in ground color from yellowish or olive to nearly

black. The plastron is usually yellowish or brownish in color, its anterior
lobe shorter than the posterior, with growth rings usually fairly well
marked, and with the bridge wide, being equal to ⅔ (or more) of the
length of the interabdominal seam. The head is of medium size and of
some dark shade that is nearly always broken by mottling which may
either be evenly dispersed or vaguely organized to form one or two indefinite light bands on the side of the head and neck. The other soft parts
are brownish or olive, with or without desultory markings.

Plate 17. Kinosternon subrubrum subrubrum. A: Male; Southern Pines, North
Carolina. *B* (female) and *D* (male): Specimens from Chesser's Island, Okefinokee
Swamp, Georgia; sex undetermined. *C:* Young; Kitrell, North Carolina.

Secondary sexual differences in this turtle are not strong and for the most
part involve features by which the sexes are differentiated in many other
species, such as the larger head and jaws of the male and its longer, heavier
tail, tipped with a blunt nail and the deeper posterior notch of the male.
Grant (1935a) gave the following comparative measurements of mature
specimens: length of males, 94–105 mm., average 102 mm. (3.70–4.14 inches,
average 4.02 inches); length of females, 88–102 mm., average 95.7 mm.
(3.46–4.02 inches, average 3.76 inches); width of males, average, 71.1 mm.
(2.80 inches); width of females, average, 70.9 mm. (2.79 inches).

The carapace of the very young individual is higher and more regularly
domed than that of the adult. It has a more or less well marked, narrow

central keel and usually a number of narrow broken, longitudinally directed ridges crossing the laterals and sometimes joining to form a nearly continuous lateral keel. Above, the shell is dark, often nearly black, with light spots along the margin. The plastron is irregularly mottled with orange or whitish and black. Both Nichols and Grant found hatchlings with 24 mm. (0.945 inch) shell length; the width of the latter's specimen was 10.5 mm. (0.414 inch).

Habitat: A great variety of situations has been named as the habitat of this turtle. It appears to show little strong habitat preference, its choice of a home being evidently influenced by a dislike of large, very deep, or moving bodies of water and by a tendency to puddle about in its foraging, half submerged or even on dry land. An optimum habitat is perhaps represented by a well-established, though fluctuating, shallow-water ditch with considerable aquatic vegetation. Ponds are often mentioned as collecting sites, and this turtle has been seen occasionally in any number of other kinds of places, even including brackish-water creeks and estuaries.

Habits: This species hibernates in mud or in decayed wood. The only definite emergence date available is that of Wetmore and Harper (1917), who on March 25, 1917, the first warm day of the spring, found near Alexandria, Virginia, a specimen that had just left its burrow in damp soil near a marsh. The hole that remained was 9½ inches (241 mm.) deep. The earliest date on which Nichols (1947) found active individuals on Long Island was April 11, 1925; the latest date, November 11, 1934.

Both the disposition and the potency of the musk of this turtle have been the subject of considerable disagreement. So evenly balanced, in fact, is the pertinent evidence that I am forced to consult my own limited experience in reporting on these points. I find the disposition mild and the musk rather nauseating, although far from being as utterly revolting as the stench that exudes from a hysterical stink-jim.

One of the more remarkable features of the biology of the common mud turtle is the frequence of its occurrence in brackish and even salt water all along the Atlantic coast within its range. It appears to be the most nearly euryhaline of any North American turtle except the Suwannee turtle, which is even more at home in the two environments.

Breeding: The common mud turtle has been observed mating in Indiana during the first half of May; copulation occurred under water. Nichols (1947) found a mated pair of which the female was buried in mud and the male had his head above the shallow water. Barkalow (1948) found a copulating pair in the edge of a bog in Cobb County, Georgia, on May 1, 1934.

June appears to be the principal month for laying, although it has been observed from the last of March through July in various parts of the range of the race, and Richmond (1945), who made a study of the nesting habits of this turtle in Virginia, found that there is a late summer nesting period that may last until late September. This writer's observations enabled him to give the following composite account of nesting:

The female may try several places before she finds a suitable site. Then she starts digging with her fore feet, thrusting the dirt out laterally until she is almost concealed. At this point she turns around and completes the nest with her hind feet. During this stage, and while laying, only the head of the turtle is visible. After the eggs are deposited, the turtle crawls out of the ground and may proceed directly to the water, or she may make a slight effort to conceal the nest cavity by levelling the site and scratching around it. Of the fourteen completed nests examined only three showed any indication that the turtles had tried to conceal them, and they were conspicuous as disturbed areas, not at all like the carefully concealed nests of *Pseudemys* observed in the same field.

The completed nest is usually a semicircular cavity from 3 to 5 inches deep, entering the ground at a 30° angle. This cavity extends above and slightly beyond the position of the eggs. The soil around and immediately above the eggs is firmly packed, indicating that the turtle carefully covers the eggs even though no effort is made to conceal the nest. In the loose sandy soil where these nests were observed the nest site is soon obliterated by rains.

The only hatchlings of *Kinosternon* observed in this area were unearthed by plowing during April. Therefore, some of them spend the winter in the nest. Whether or not any leave the nest in late summer has not as yet been determined.

The number of eggs laid is from two to five, three eggs being the most frequent complement; according to Richmond, five were next in frequency. McCauley (1945) found the size of the eggs to be very variable. He measured eighteen eggs selected at random from a total of thirty-two eggs obtained. The extremes in length of the longer diameter were 29 and 22 mm. (1.4–0.87 inches). The extremes in length of the shorter diameter were 16 and 13 mm. (0.63–0.51 inches). The average length of the above eggs was 25.3 mm. (0.99 inch); the average width was 14.7 mm. (0.58 inch).

McCauley observed that most Maryland hatchlings emerged under laboratory conditions during September, but Obrecht (1946) found a hatchling in South Carolina on June 15, 1945. It is conceivable that Obrecht's specimen might have been an individual that had passed the winter in the nest after a late summer emergence from the egg.

Feeding: Little or nothing concerning the food of the common mud turtle has been published. Insects are known to be eaten. The lack of a strict habitat preference and its ability to live in such divergent habitats as meadow brooks and salt marshes would appear to indicate a capacity for taking things as they come. In captivity specimens eat nearly anything edible offered them.

Mississippi Mud Turtle *Kinosternon subrubrum hippocrepis* Gray

(Plate 18. Map. 6.)

Range: Southeastern Missouri southward (mostly west of the Mississippi River) to the Gulf coast of Mississippi, and westward through Arkansas and eastern Oklahoma to central Texas. Intergradation with *K. s. subrubrum* occurs along the Gulf coast from Pensacola to New Orleans and probably to the northward in the eastern part of the Mississippi Basin.

Distinguishing features: This subspecies has never been adequately defined. The only differential character that appears to hold up to any extent is the presence of a pair of light stripes on the side of the neck, representing, in consolidated and more definitely delimited form, the light head markings of *subrubrum* and *steindachneri*. There appear, superficially, to be differences in shell shape which have as yet to be analyzed. In parts of its range identification of this turtle might be complicated by the presence of the yellow mud turtle, but the high, triangular ninth marginals found in the latter contrast strongly with the even upper edges of the marginals of *K. s. hippocrepis*.

Habitat: The Mississippi mud turtle occurs in ponds and ditches, and in shallow water generally.

Habits: There is little in print concerning the life history of this subspecies. Anderson (1942) unearthed three young individuals from beneath ten inches of soil, where they lay only a few inches apart; this occurred in Jackson County, Missouri. A concerted migration of these mud turtles was observed long ago near Waco, Texas, where forty-five individuals were seen leaving a shallow lake that was drying up; all were heading in the same direction.

The feeding habits in the natural state are not known. In captivity specimens accept meat, but much prefer insects and such small snails as *Planorbis* and *Limnaea*.

Plate 18. Kinosternon subrubrum hippocrepis. A, B, C, and F: Males. *D and E:* Female. All from southern Louisiana.

Florida Mud Turtle *Kinosternon subrubrum steindachneri* Siebenrock

(Plate 19. Map 6.)

Range: Peninsular Florida, from the neighborhood of Alachua County southward to Dade County and Cape Sable. Intergradation between this subspecies and northern *subrubrum* takes place in the area between Gainesville, Tallahassee, and the Okefinokee Swamp.

Distinguishing features: Although this subspecies is similar in general appearance to the common mud turtle, there is little difficulty in distinguishing between typical specimens of the two. The width of the bridge of the present race is contained two and one-half to three times in the length of the interabdominal seam; the nasal shield is forked behind; and the plastron is somewhat smaller than that of the common mud turtle, the fore lobe being as long as or longer than the hind lobe.

Besides the standard sexual differences, found to a greater or lesser degree throughout the genus, the sexes of *steindachneri* are strikingly differentiated by the very large heads of some of the males, which approach the

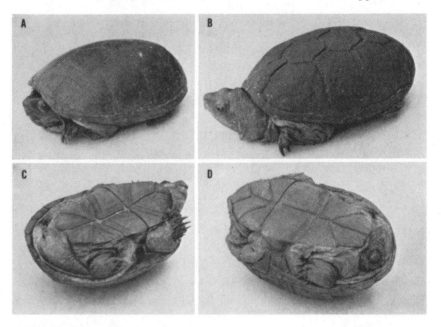

Plate 19. Kinosternon subrubrum steindachneri. A and C: Male. *B and D:* Female. Both from Collier County, Florida.

astonishing extremes of enlargement to be seen in *Sternotherus c. minor.*
The side of the head lacks definitely delimited stripes. Within its range the
Florida mud turtle would perhaps be more easily confused with *Sternotherus c. minor* than with any other turtle; indeed, it comes very close to
bridging the inconsequential gap between *Kinosternon* and *Sternotherus,*
but its triangular pectoral laminae nearly always serve to place it in the
right genus.

Habitat: In 1940 I described the habitat of the Florida mud turtle as
"small streams, sloughs, drainage ditches and marshes," and can add little
to this at the present time. It is found away from water much less frequently than the northern subspecies. Goin (1943) reported it from hyacinth marshes in the vicinity of Gainesville.

Habits: In view of the close relationship between this turtle and *K. s.
subrubrum,* it is curious to note in the former almost no inclination to
walk about on land, while *subrubrum* is found out of water with great
frequency. It should be mentioned, however, that Richmond thought all
this terrestrial wandering might be attributed to prospecting by females
ready to lay. Another point of divergence is seen in the belligerent bad
temper of male individuals of *steindachneri,* which contrasts with the relatively pacific nature of *subrubrum.*

Little is known of the habits of *steindachneri* beyond a few chance field
notes. I have found several specimens in horizontal holes dug in clay of
canal banks, and watched one large male dig such a hole just below water
level, using all four feet to remove clay. When it was finished the turtle
emerged, turned around, and backed into the cave.

Bogert and Cowles (1947) mentioned several specimens of *steindachneri*
that were ploughed up in mud around the margins of a small pond at the
Archbold Biological Station in Highlands County. In order to test the
tolerance of this species to severe conditions of temperature and humidity,
the above writers kept specimens for long periods in a desiccating chamber
in which the temperature was maintained at 38° Centigrade and the relative humidity at 37 per cent. During a period of fifty-one hours a specimen
weighing originally 117 grams lost 23 per cent of its body weight, while
after a period of thirty-seven hours a 90-gram specimen lost 22.2 per cent.
Neither showed any ill effects. Still another specimen was placed in a con-
tainer of wet sand in the chamber. After 48 hours it had lost no weight.
When this same turtle was confined in a wire mesh cage in the laboratory
and no water was provided, it kept its temperature at 27° Centigrade and
lost weight slowly.

Breeding: I have never observed the laying process but have found numerous eggs, buried in twos and threes in a desultory fashion, beside a stump, under the edge of a log, in leaf mold or in rotten wood, and sometimes almost wholly uncovered. The eggs have been found to range in length from 27.5–29.2 mm. (1.08–1.15 inches) and in width from 16.8–18.0 mm. (0.66–0.71 inches); they are pinkish or bluish-white in strong light (varying in color with age) and with a hard, lustrous, undulating surface. In Alachua and Lake counties the nesting season lasts from the middle of March to mid-June.

Feeding: I have only seen them eating dead fish left behind after a seine haul, but the masticatory equipment would appear to indicate a diet that includes much hard-shelled material, most probably mollusks or crayfish, or both.

FAMILY EMYDIDAE

Fresh-water and Marsh Turtles

THIS is the largest family of living turtles. It is cosmopolitan in its distribution, being unknown only in Australia. In all members of the group the head may be drawn into the shell, and when so retracted the neck is bent vertically. The temporal region of the skull is without, or with only a vestige of, a roof. The middle digit usually has three phalanges and the toes are webbed, at least to some extent. The species are aquatic or semiaquatic for the most part, although the box turtles have become secondarily adapted to a largely terrestrial existence. The differences between this family and the Testudinidae are slight but clear-cut, and the closest similarities between the two are attributable to parallelism. The family has been found with its characteristics already well developed in the Paleocene of Saskatchewan and is known from several Eocene localities.

KEY TO THE GENERA OF EMYDIDAE

1   Plastron in one rigid piece, not hinged, attached to the carapace by bony sutures 2

1'   Plastron hinged near middle, its halves movable, attached to the carapace by ligaments only 7

2(1) Neck strikingly long, the distance from tip of snout along extended neck to shoulder about equal to length of plastron . . . .
. . . . *Deirochelys reticularia* (p. 316)

2'   Neck not strikingly long, the distance from tip to snout along extended neck to shoulder approximately equal to one-half of the length of plastron 3

3(2') Alveolar surface of upper jaw with a ridge or row of tubercles extending parallel to its margin 4

3'   Alveolar surface of upper jaw smooth or undulating, not ridged 5

4(3) Carapace smooth, not wrinkled longitudinally and without a keel, its hind margin not serrate . . . . *Chrysemys* (p. 213)

4'   Carapace with longitudinal wrinkles, a more or less complete keel, or both; its hind margin serrate . . . . *Pseudemys* (p. 234)

5(3') Alveolar surface of upper jaw narrow, not forming most of the

anterior roof of the mouth, not narrowing anteriorly, its inner
edge parallel to cutting edge .... *Clemmys* (p. 112)
5′   Alveolar surface of upper jaw either very broad, forming most
       of the roof of the anterior part of the mouth, or broadened pos-
       teriorly and much narrower toward the symphysis 6
6(5′) Laminae of the carapace smooth, not concentrically striated or
       ridged; head and neck with longitudinal light stripes ....
       .... *Graptemys* (p. 186)
6′   Laminae of the carapace usually with concentric ridges or stria-
       tions; head and neck plain, spotted, or mottled, without longitu-
       dinal light stripes .... *Malaclemys* (p. 162)
7(1′) Upper jaw with a hooked beak, not notched at tip; carapace
       usually highly arched, not notched posteriorly .... *Terrapene* (p. 137)
7′   Upper jaw without a beak, notched at tip; shell low and nar-
       row, notched posteriorly .... *Emys blandingii* (p. 132)

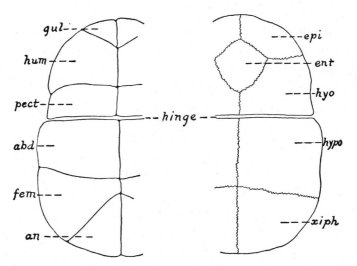

Figure 12. Plastron of *Terrapene,* showing relation of the plastral
hinge to laminae (*left*) and to bony plates (*right*). For explanation
of abbreviations see Figure 2, p. 38.

Genus *CLEMMYS*

This genus includes about a dozen forms occurring in Europe, Asia, and
Africa as well as in North America. It is the earliest representative of the
family to appear in the fossil record, being known from Paleocene and

later Cenozoic deposits. The turtles of this group are of medium size, and the shell is relatively low. The plastron is firmly united to the carapace by sutures. The alveolar surface of the upper jaw is without a median ridge, and the skull has a bony temporal arch.

The American forms show a curious distribution, one species being found on the Pacific coast, while three others are confined to the Atlantic drainage.

KEY TO *CLEMMYS*

1   Carapace marked by one or a few small but conspicuous circular light spots on each lamina ....*guttata* (p. 113)

1'   Carapace marked or unmarked but lacking small circular light spots 2

2(1') Temporal region of head with a conspicuous orange blotch ...
....*muhlenbergii* (p. 128)

2'   Temporal region of head without orange blotch 3

3(2') Carapace usually rough and strongly keeled in adult; much of skin reddish or reddish orange; eastern North America from Nova Scotia to Wisconsin, Iowa, and Virginia ......*insculpta* (p. 118)

3'   Carapace usually smooth, almost or wholly unkeeled in adult; skin not reddish; Pacific coast 4

4(3') Inguinal laminae usually present; throat lighter than sides of head; Pacific coastal area from southwestern British Columbia southward to San Francisco Bay ......*marmorata marmorata* (p. 123)

4'   Inguinal laminae usually reduced or lacking; throat and sides of head almost or quite the same color; southern California from near Monterey Bay southward into northern lower California ....*marmorata pallida* (p. 127)

Spotted Turtle *Clemmys guttata* (Schneider)

(Plate 20. Map 7.)

Range: Eastern United States, from Maine westward across southern Ontario (where it extends northward to the Georgian Bay area) and Central Michigan into southeastern Wisconsin and northeastern Illinois. In the southern part of its range it is restricted to the coastal plain and piedmont strips. Thus, though known throughout Pennsylvania, it has been collected only in extreme northeastern West Virginia and eastern Virginia, whence it ranges southward to eastern Georgia and, according to some old, unauthenticated records, to northern Florida. A good record for Riceboro, Georgia, in the lowlands, may lend weight to the Florida reports.

Map 7. Distribution of *Clemmys guttata.*

Distinguishing features: Except for occasional variant specimens, the spotted turtle is easily recognized. The broad, low, yellow-spotted carapace, the yellow spots on the head and neck and yellow bar behind the ear, and the extensive black area on the plastron serve to distinguish it at a glance from any species with which it might be found.

Description: The shell of this species is low, oval in lateral outline, and widest posteriorly where the marginals flare somewhat. It is a small turtle, rarely exceeding 125 mm. (4.95 inches) in length, although Conant (1938a) measured a female specimen with a shell 127 mm. (5.00 inches) long. A representative specimen showed the following dimensions: length of carapace, 98.4 mm. (3.87 inches); width of carapace, 73.0 mm. (2.87 inches); length of plastron, 88.9 mm. (3.50 inches); width of plastron, 54.0 mm. (2.12 inches). The carapace is usually smooth (reported as rough in brackish-water specimens and in a certain proportion of a population in northern Indiana) and keelless in adults and is nearly always decorated by a conspicuous scattering of small, rounded yellow spots. These are trans-

Plate 20. Clemmys guttata. A, B, and C: Young; Hammonton, New Jersey.
D, E, and G: Male; Richmond, Virginia. *F:* Female; Ithaca, New York.

parent windows in the laminae, overlying patches of yellow pigment, and they vary with locality (Richmond and Goin, 1938, for instance, found them smaller and less conspicuous in specimens from Virginia than in others from farther north), with age (young individuals have fewer), and possibly with sex, inasmuch as Yerkes (1905) found females to have 15 per cent more than males. Grant (1935b), who has given the best account of sex differences in the species, was unable to corroborate Yerkes' assertion. Incidentally, another curious claim by Yerkes would have more spots on the left side than the right and would even correlate this disparity with right-handedness and right-eyedness! Conant counted spots ranging in the aggregate from 14 to 114 in Ohio specimens. The plastron is large and immovable and is yellow and black in color, the yellow area usually lying within the black in adults but surrounding it in young. The dorsal surface of head and limbs is black, the latter spotted or not, the former with a few spots above. On the side of the head there are one or two more conspicuous spots or broken bands of yellow, one near the tympanum, the other sometimes extending backward from the orbit.

In addition to the sexual character noted above, Grant gave the following differences between males and females among 78 Indiana specimens: the yellow crown spots averaged 1.5 in males, 2.0 in females; the vent of the female barely reached the rear margin of the shell, and that of the male extended about 5 mm. beyond; the plastron of the male was slightly concave, that of the female flat or convex; the postmandibular yellow stripe was much more accentuated in females, which had, also, bright orange eyes, while those of the male were brown. Although Ohio specimens showed males to be slightly larger than females, this was not the case in individuals from Indiana.

The young spotted turtle is about 28 mm. (1.10 inches) long at hatching, and its width may be at least 95 per cent of its length. The hatchling usually has one spot per lamina, but gets more with advancing age. There may be a feeble keel in the very young.

Habitat: The spotted turtle is more aquatic than other members of the genus with the exception of the Pacific pond turtle. It is pre-eminently an habitué of small bodies of water, preferably meandering meadow brooks or little bog holes and ponds. Several writers have stated that it prefers a muddy bottom. It is nearly always rare in regions where ponds are few.

Habits: The hibernating habits of the spotted turtle are little known. Netting (1936a) observed what he interpreted as a spring exodus from a site of hibernation in Orange County, New York, where on April 3, 1936, he found four individuals leaving high ground and heading for a swamp, indicating an apparent reversal of the usual tendency in related forms to spend the winter buried in mud on the bottom of small bodies of water. Spring emergence is early. Conant found spotted turtles in Ohio every month from February to November, although most of his specimens were collected in the spring and he found them surprisingly scarce in the summer months. When in the water, spotted turtles apparently seek the companionship of their kind, or at any rate often bask and sun themselves in groups, but their sojourns on the land are spent alone except during mating season. Pope (1939a) recorded finding, during the winter, a dozen or so individuals in the water under the bank of a spring in southern New Jersey.

One observer's tests of the ability of the species to learn routes leading to attractive goals in a maze led him to conclude that its ability in this respect was about equal to that of a rat. This is a gentle species, and while it sometimes tries to bite, it is easily tamed and is usually quite even-tempered. Nevertheless, the males sometimes fight fiercely, and Babbitt (1932) described a harrowing struggle between two males in Connecticut which continued to fight after darkness had driven Babbitt away. I have collected this turtle only on two occasions and cannot evaluate the conflicting reports that hold on one hand that spotted turtles are "very shy and wary" and on the other that "they are not particularly wary and the seasoned collector seldom has difficulty in capturing nearly all he may see."

Breeding: The courtship of the spotted turtle, which takes place most frequently in April, apparently in all parts of the range of the species, appears to be rather unusually active and exciting. It sometimes involves long and frantic chases of the female by the male, which may cover considerable territory. Finneran (1948b) and others have suggested that the brighter stripe on the face of the female may serve as a beacon to the male during these involved pursuits. Copulation may occur on land or in the water, evidently most frequently the latter. Finneran found a pair copulating on March 21, 1943, near New Haven, Connecticut, in water with a temperature of 19° Centigrade and another pair May 2 of the same year in water of the same temperature.

The one to four (usually three) eggs are on the average about 30 mm. (1.18 inches) long and 17.5 mm. (0.69 inches) wide, and are laid in the ground during June. The nesting process has been little observed and the natural incubation period is unknown. Eggs taken by Babcock (1938) from a nest made June 16 and reburied, hatched on September 6; and Finneran said that hatching occurs in late September in Connecticut.

Feeding: This species is chiefly carnivorous, the major part of its diet being small invertebrate animals, most frequently insects. Vegetable matter is occasionally eaten, but rarely in any great quantity. Although vertebrate animals are rarely caught, both frogs and tadpoles have been found in stomachs, and fish also, but the latter had probably been found dead in most cases.

Economic importance: This species is too small to have any value as food for man, and its inoffensive habits make it no menace to any other economically important animal.

More frequently than most other eastern species, the spotted turtle is recognized specifically by people in general, and in many places it boasts its own private common name—not one conferred after deliberation by a body of herpetologists, but a real folk name, given to it by the people with whom it lives. Wright (1919) recorded the use in upper New York of the term "cranberry turtle," applied with reference to the predilection that this species shows for cranberry bogs as a habitat. Elsewhere it is called "scorpion," adding another and most extraordinary misuse to that much-abused term. I was much interested in Engels' mention (1942) of the name "highland hickety" as applied to the spotted turtle on Ocracoke Island, North Carolina. I have heard any number of versions of this term hickety, in both Spanish and Caribbean English (*jicotea, hicotea, icoté,* hickatee, higgedy), but have never run across it in the United States. Its occurrence at Ocracoke conjures up visions of early and colorful intercourse between the Carolina coast and the lands of the Spanish Main, where since time immemorial the Arawaks and Siboneyes had used an ancestral form of the word hickety for the Antillean turtles of the genus *Pseudemys.*

Wood Turtle *Clemmys insculpta* (Le Conte)

(Plate 21. Map 8.)

Range: From Nova Scotia and all New England southward to Virginia in the Piedmont and highlands; westward through southern Ontario to Wisconsin and northeastern Iowa. It is not known from Ohio, Indiana, or southern Michigan.

Distinguishing features: The keeled and extraordinarily rough shell, rigid plastron, notched upper jaw, and usually conspicuous orange color of the

Plate 21. Clemmys insculpta. A and E: Male; Bloomingburg, New York. *B:* Female; same locality. *C and D:* Young; Ithaca, New York.

throat and lower limb surfaces will readily distinguish the wood turtle
from any other chelonian in its range.

Description: The shell of the wood turtle is of striking appearance, being
quite low and broad, deeply sculptured, and with widely flaring posterior
marginals.

The carapace is keeled in most specimens and nearly always shows curiously strong relief, the growth areas in some of the dorsal laminae tending
to pile up in lopsided pyramidal blocks. It is brown in color, often with
black and yellow lines radiating from the upper rear corners of the laterals.
The plastron is yellow in ground color, most of its laminae and the lower
marginals as well having an oblong dark blotch on the rear outer portion.
The head is medium-sized and the upper jaw is notched at the tip. Head,
neck, and legs are dark, usually deep brown or blackish above and more
or less orange below.

This is a medium-sized turtle, the longest specimen on record being
about 228 mm. (9.00 inches) in shell length. The largest specimen found
by McCauley (1945) in Maryland was 190 mm. (7.5 inches) long. Wagner
(1922) gave the following measurements comparing dimensions of a specimen from Pennsylvania with one from Wisconsin (measurements in
millimeters):

	Wisconsin	*Pennsylvania*
Carapace		
Length	169	114
Width	123	87
Plastron		
Length	151	110
Width	104	72

Characters by which the sexes differ were given by Wright (1918b) as
follows: preanal part of tail of male twice as long as that of female; plastron
of male concave; claws of male longer and heavier; scales on front surface
of fore limb of male more prominent; hind edge of plastron more nearly
approaching rear margin of carapace in female than in male.

The carapace of the young wood turtle shortly after hatching (after losing
the globular shape produced by its cramped quarters in the egg) appears
to be usually about 32 mm. (1.26 inches) in length and is very low and
broad—sometimes as wide as long, or even wider, and smoother than that
of the adult; the disproportionately long tail may be equal in length to
the carapace. Pope (1939a) stated that the characteristic orange color of the

lower surfaces of the soft parts was not present in three hatchlings from New Jersey.

Habitat: The habitat requirements of the wood turtle are not strict. It wanders through nearly all the types of terrain within its range, spending most of the warmer part of the year on land, in woods and meadows alike, and turns up in or near streams and ponds when hibernating time approaches, during mating season, or when aestivation may be necessary. The adaptability of this attractive turtle when kept in the household affords some idea of its tolerance with respect to its environment.

Habits: After spending the winter hibernating in mud on the bottom of some small body of water, the wood turtle emerges during April, or occasionally as early as March. It is active during the day but not at night and from existing data appears to have a rather restricted home range, some individuals turning up year after year in the same place. Whitney (in Knowlton, 1943) in several cases found that hibernating individuals lost about two ounces of weight over the winter. One specimen that weighed two pounds when released by Whitney on November 6, 1928, weighed one pound, fourteen and three-quarters ounces when retaken on May 9, 1931. The same observer mentioned having found individual wood turtles in one circumscribed locality more than a dozen times.

Wood turtles are not gregarious, being usually found roaming alone except when mating or when induced by environmental conditions to band together in a good hibernating place, such as an old muskrat hole. It has several times been remarked that this turtle is unusually intelligent. From experimental tests in mazes Tinklepaugh (1932) decided that its ability to learn new routes was about equal to that of a rat, which, though it has a rather uncomplimentary ring, is really pretty good for a turtle; and anyway, in a personality contest a wood turtle would win from a rat in a walk. Mrs. Knowlton, who probably knows more about the subjective attributes of a number of species of turtles than does anyone alive, says it is the most intelligent species known to her. All who have kept them are agreed that the wood turtle makes a better pet than any other species. It rarely bites, and is pleasantly extroverted by nature; it is almost as completely terrestrial as its rival in human esteem, the box turtle, and unlike the latter has no plastral valves with which to shut itself up for hours when offended. A sprightly female that we kept in the house for eighteen months soon learned as much as interested her of our daily routine. She slept in a closet and showed up at the breakfast table every morning where she made for my wife's chair, craned her neck expectantly, and often stood on three legs with one front foot lifted and poised like a pointer dog that smells birds. This appealing gesture, incidentally, has been noted by others, as has the unusual extent to which this turtle uses its front feet in handling its food. Our wood turtle appeared to appease all its aquatic urges by sitting occasionally in a pie pan of water.

A wood turtle has lived in the Philadelphia Zoological Garden for twelve years and five months.

Breeding: Knowlton gave a very detailed account of the courtship of the wood turtle, which she has observed on a number of occasions. The procedure includes a sort of dance, previously noted in fragmentary form by other observers, in which the male and female approach each other slowly with necks extended and heads held high, to be suddenly lowered when the turtles are within some eight inches of each other and then swung from side to side for as long as two hours without stopping. After this strange procedure a pair observed by Mrs. Knowlton dived into a garden pool to attempt copulation, in which they were successful only after three days' intermittent effort. This presumably took place in summer, for the writer relates that the following fall courtship was resumed and was continued until December! A courting "whistle" has been referred to by several observers; it is said to be emitted by both the male and the female and to be quite loud and insistent.

Copulation takes place in the water. According to Knowlton the male holds the shell of the female with all four feet, the hands close together around the fore edge of her shell and the nails of his hind feet clamped under the hind edge. In attempting copulation the male is on the whole the more active and aggressive, but the female sometimes takes the initiative and during maneuvering observed by Fisher (1945) a female twice turned, tipped up the male from below, and crawled under him. This occurred on March 26, 1945, during a warm spell.

Nearly all the data on nesting and laying that are available come from captive specimens. In 1871 Gammons wrote that he saw a female dig a nest with her fore feet, pivoting on her hind feet until a circle thirteen inches in diameter had been completed and then reverse her position and with fore feet in the center throw out dirt with the hind feet, crawl out of the hole, and drop into it eighteen eggs. I regard this whole thing as some sort of hallucination, but it is the only existing account of the nesting of the wood turtle in nature. The eggs are usually laid in June—sometimes as early as May—and, according to Babcock (1938), in the afternoon. They have been found to number from four to twelve (if we reject Gammon's claim to have counted eighteen) and to be, on the average, 35 mm. (1.38 inches) long. The incubation period appears to be unknown. Hatchlings have been found in September and October, and Wright (1918b) found one April 20, 1913, which may have been a fall hatchling that overwintered in the nest.

Feeding: The wood turtle is by no means fastidious in its diet and has been found eating a wide variety of plant and animal foods. Of the specimens examined by Surface (1908), 76 per cent of all that had recently fed contained plant remains and 80 per cent animal matter. Lagler (1943b) gave the following list of items found by him in the stomachs of nine adult specimens from Michigan:

(1) 2.8 cc. of filamentous algae and willow leaves

(2) 11.0 cc. of plant material (mostly algae) and insect remains including three blackfly larvae

(3) 60.0 cc. of filamentous algae including several caddis larvae and many small mollusks

(4) 10.6 cc. of remains of a beetle, a tadpole, and a snail

(5) 37.5 cc. of a bluegill and 46.5 cc. of larvae of caddis flies and their houses

(6) 24 cc. of remains including fourteen *Brachycentrus* larvae, two limnephilid larvae, one neuropteran larva, and a leaf fragment

(7) 2.1 cc. of limnephilid larvae and their houses and a trace of leaf

(8) 4.8 cc. composed of five earthworms, one adult hymenopteron, moss, willow and grass leaves, and vegetable debris

(9) 1.8 cc. comprised of remains of one trout, insects, and vegetable debris, including some algae

It may be a fluctuation in food supply that stimulates the wood turtle to abandon the predominantly terrestrial existence that it leads in spring and early summer, when berries are ripe, and invade the streams later on to feed to a large extent on aquatic animals.

Economic importance: The wood turtle is edible, as are most turtles, but of no importance to anyone not lost in the woods. It is protected by law in some places. From its diet Lagler concluded that "this turtle can be of no concern to fish management."

Pacific Pond Turtle *Clemmys marmorata marmorata* (Baird and Girard)

Range: The Pacific coastal area from extreme southwestern British Columbia southward to San Francisco Bay. Specimens intermediate be-

tween this and the southern subspecies, *C. m. pallida,* have been found in
a crescentic area trending southeastward from the vicinity of San Francisco
Bay.

Distinguishing features: This relatively undistinguished-looking member of the small west coast turtle fauna has a low, wide, nearly smooth
shell, usually dark and often spotted, and a yellow plastron, with no vivid

Map 8. Distribution of *Clemmys insculpta* and the subspecies of *C. marmorata.*

markings on the head and legs, although these may be spotted or marbled
with darker color. The present turtle differs from its close relative to the
south, *C. m. pallida,* in its large inguinal laminae and its contrasting ground
colors on throat (light) and sides of head (darker).

Description: The shell is comparatively short and broad, with its widest
point usually behind the middle. The average length of a random selection
of mature specimens of both sexes from California was 149 mm. (5.86
inches) and maximum length appears to be about 180 mm. (7.10 inches).
Van Denburgh (1922) gave the following measurements of a good-sized

individual: length of carapace, 125 mm. (4.93 inches); width of carapace, 100 mm. (3.94 inches); length of plastron, 117 mm. (4.61 inches); width of plastron, 85 mm. (3.34 inches); length of tail, 48 mm. (1.89 inches). The carapace is usually smooth and keelless in grown specimens; its ground color varies from olive or horn color to nearly black, with or without darker mottling. The plastron is yellow, with irregular dark blotches or lines along the hind edges of the laminae. The head may be either uniformly dark above or reticulated, and the sides of the neck are usually marked with dark spots on a brownish or grayish background. The chin and throat are yellow, with or without dark dots, and the legs and tail may be either yellow with dark markings or brownish with yellow markings.

Seeliger (1945) was able to show no differences in the sizes of the two sexes, nor in the relative total tail lengths, but found concavity of the plastron characteristic of a majority of adult males (flat in most females). Despite comparable total tail lengths in the sexes, the position of the anal opening relative to the hind edge of the shell varies significantly, being located behind the rear margin in a majority of males and at or in front of it in most females. Over-all shell height is somewhat less in males than in females, and Storer (1930) found shells of males to show slightly more marbling than those of females.

The carapace of young Pacific pond turtles bears a keel down the middle. It is nearly circular in lateral outline, and the laminae often have brown centers margined by a region of lighter color. The markings on soft parts may tend to form longitudinal bands, and the tail is relatively longer than at maturity.

Habitat: This turtle is thoroughly aquatic and appears to move about on land only during laying season. It is of frequent occurrence in both ponds and lakes and in streams, but is typically an inhabitant of quiet, often muddy water. It is thus interesting to note that both Storer and Evenden (1948) report it from clear, swift trout streams on higher slopes, the former in the Coast Range of California, the latter in the Cascades. It approaches and occasionally enters the sea, and there are several records for brackish-water situations.

Habits: Evenden gave April 28 (1948) as the earliest date on which he found this turtle active in the Willamette Valley of Oregon, but thought it might move about the year around in the southern part of its range. The latest date of occurrence that he recorded was February 28 (1948), although elsewhere the hibernation period has been outlined as late September to late March. Most writers agree that the Pacific pond turtle is a wary animal, very difficult to approach when sunning on logs or rocks and easily caught only by means of hooks or traps or when wandering on land at nesting time.

Information on the breeding habits of this species has been summarized by Storer and by Pope (1939a), but since these accounts appeared before the separation of the two subspecies they are composite in nature. However, there seems little likelihood that any great differences in breeding behavior exist, beyond those imposed by disparities in the climates of the territories occupied. The females emerge from late April to late August to prospect for suitable nesting sites in sandy banks of the larger rivers or on hillsides rising from the higher streams. Dissections and erratic piecemeal laying by captive specimens indicate that the size of the egg complement may be quite variable, perhaps normally ranging from 3 to 11. The eggs vary in length from 32.8 mm. to 42.6 mm. (1.29–1.67 inches) and in width from 19.7 mm. to 22.6 mm. (0.775 inches–0.89 inches). The incubation period is unknown and the hatchling has not been described.

The diet is composed largely of animal food, a wide variety of small animals and of kinds of meat serving as good bait for hooks. However, Evenden, besides encountering an individual eating a dead mallard, found that the pods of the yellow pond lily (*Nuphar polysepalum*) were regularly eaten.

Economic importance: In past times (and, in some sections, down to the present) this turtle has been unrelentingly hunted for the market, and the consequent drain on the population must have materially reduced its numbers. As early as 1879 Lockington was writing that although it had once been abundant in the San Francisco area it had now become scarce because of the commercial demand. The same writer mentioned that dealers recognized two "forms" of the pond turtle, one from the Visalia–San Joaquin region and another from the vicinity of Sacramento, the two being distinguished on a basis of color and relative vigor and edibility. It would appear, thus, that these early turtlemongers had anticipated Seeliger's separation of the Pacific pond turtle into races by more than 60 years! According to True (1884), during the eighties this turtle was "still almost constantly for sale in the markets of San Francisco," and both Storer and Carl (1944) cited prices of from $3 to $6 a dozen as prevailing during the nineteen-twenties. Although these prices cannot be compared with the $60 and more per dozen being paid during the same prosperous decade for diamond-back terrapins in the East, they were evidently high enough to stimulate a continuous fishery.

Southern California Pond Turtle *Clemmys marmorata pallida* Seeliger

(Plate 22. Map 8.)

Range: Southern California in the coastal drainage from the vicinity of Monterey Bay southward into northern Baja California. According to Charles Bogert (letter), "it ranges up into the foothills, and is found on

Plate 22. Clemmys marmorata pallida. A: Male; San Diego, California. *B:* Female; same locality. *C:* Sex undetermined; same locality. *D and E:* Young; Perris, California.

the desert slope only in the Mojave River, where this emerges from the San Bernardino Mountains." Intergradation with the northern subspecies, *C. m. marmorata,* occurs between Monterey Bay and San Francisco Bay and, further inland, in Mariposa, Madera, Fresno, Tulare, and northern Kern counties. Apparently the identity of specimens from Baja California has not been established. Although Seeliger (1945) stated that they represent neither *pallida* nor typical *marmorata,* he did not offer an opinion as to what they are, and for want of a better disposition I group them here with the neighboring stock, *pallida.*

Distinguishing features: This subspecies is distinguished in a majority of cases from *C. m. marmorata* by the absence or reduced size of the inguinal laminae and by uniformity of ground color of the throat and neck.

Habitat: The southern California pond turtle occurs in "streams, ponds and reservoirs" (Klauber, 1934).

Habits: Life history data based definitely on observations of *C. m. pallida* are meager. July has been mentioned as a nesting date. Rüthling (1915) referred to winter hibernation in the form and March 1 has been given as a date of emergence. Klauber wrote that it "lives on animal food." In a letter, Bogert comments as follows:

When *pallida* occurs in shallow streams, as it does in the headwaters of the Sespe in Ventura County, it's easy to walk into the stream and pick them up. I've caught as many as a dozen by such methods, on a weekend, finding them browsing on some sort of aquatic plant that grows in shallow water where the stream is not too swift. Doubtless it also eats animal food, but isn't restricted thereto as Klauber would imply. I've fed captive specimens on a variety of things, and in particular recall watching one pick up a large sow-bug and carry it back to the pool before attempting to swallow it. It raises the question (but I'd not give any dogmatic answer) whether such turtles can swallow out of water.

Muhlenberg's Turtle *Clemmys muhlenbergii* (Schoepff)

(Plate 23. Map 9.)

Range: Rhode Island and central and southern New York to western Pennsylvania and southward through Virginia and Maryland to western North Carolina. Distribution over this area is evidently not presently continuous, there being a broad gap between the North Carolina colony and that in northern Virginia, another between the populations of eastern and western Pennsylvania, and possibly a third between the colonies in central and southern New York.

Plate 23. Clemmys muhlenbergii. A, B, and E: Females. *C and F:* Male. *D:* Young. All from New York.

129

Distinguishing features: The combination of a long, relatively straight-sided, somewhat rough and mottled (but not spotted) carapace with a lateral orange blotch on the side of the head (usually present and frequently very vivid and striking) will make identification of this turtle easy in a majority of cases.

Map 9. Distribution of *Clemmys muhlenbergii.* The range of this species is probably discontinuous (see text), but it seems pointless to plot the discontinuities until more systematic collecting in the territory may have defined the fragmented sections more clearly.

Description: The shell of Muhlenberg's turtle is moderately domed and quite long, with the sides frequently straight and either nearly parallel or slightly divergent behind. The size is rather small, maximum length of the shell usually being given as about four inches. Ditmars (1907) measured a male the length of which he said was exactly four inches (10.2 mm.). Ranges in dimensions of New York specimens were summarized by Wright (1918a) as follows (measurements are in millimeters):

	Length of carapace	Width of carapace	Height of carapace
Males	82–98.5	58.5–69	32–39
Females	82–93	61–68.5	37–39

The carapace usually is feebly keeled and the concentric rings of the laminae are often fairly deeply incised, each lamina likewise sometimes bearing more or less discernible light central areas on the general ground color of some shade of brown. The big plastron is dark brown to black in color, most frequently with light markings either irregularly dispersed or with some tendency toward symmetrical arrangement. The head is of medium size, dark above and with a variable but usually broad reddish, yellowish, or orange blotch above and behind the tympanum. The soft parts are mostly brownish above, sometimes mottled with reddish, and are lighter (often orange or reddish) below.

Besides having a much longer, thicker tail than the female, the male Muhlenberg's turtle has a deeper, wider head, longer snout, heavier fore claws, and more concave plastron; the female has a wider notch at the hind edge of the plastron, and the scales on the upper side of the tail are smaller.

The young have been only rarely observed. Wright said they are very broad, the width being equal to 83.8 per cent of the length. The shell of a hatchling measured by him was 34 mm. (1.34 inches) by 29 mm. (1.14 inches), evidently before postnatal expansion.

Habitat: This spottily distributed turtle, common in some places and rare in others, is partial to sphagnum bogs, wet meadows, and small streams. It is found frequently on land as well as in the water. Fragmentation of its range appears to have taken place both as a result of long-time geologic factors and of recent reclamation by man of much bogland terrain. It has been collected at altitudes ranging from sea level up to well over 4,000 feet in the Appalachians.

Habits: The main points in the life history of this turtle are still to be learned, there being available little published information beyond that on its habitat preference and on behavior in captivity. Various observers have mentioned a tendency to burrow, and McCauley (1945) recorded the finding of an aestivating individual imbedded in hard clay under a board on August 8, 1941. This specimen, when removed, was found to be in a torpid state and its shell and head were entirely caked with clay; it became active only after several hours had elapsed.

A captive specimen kept by Wright laid two eggs, one on June 5 and another one month later; a third was found in the oviduct. Two of these eggs measured: 30 by 16 mm. (1.18 by 0.63 inches) and 32 by 16 mm. (1.26 by 0.63 inches).

This turtle appears to be omnivorous, although complete reliance cannot be placed on the behavior of captive turtles in arriving at conclusions concerning feeding habits, because under artificial conditions they are often induced to eat things they probably would scorn in nature. They appear to feed both on land and in the water. Pope (1939a) found that a captive specimen swallowed earthworms more readily in water than out; Brimley (1943) stated that they eat on land and not in the water. In what appears to have been the only examination of the stomach of a specimen that had fed under natural conditions, berries and insect fragments were found.

Genus *EMYS*

This group of turtles includes two species, one of which is found in Europe, western Asia, and northern Africa and the other in eastern North America. The genus first appears as a fossil in the Miocene of Switzerland. It was evidently much more widely distributed during the Pleistocene than at present, and is thought to have become extinct in parts of its range, notably in Switzerland, almost during historic times.

Emys is characterized by a relatively elongate shell; by a flexible union between plastron and carapace, the former being more or less freely movable; and by the absence of a ridge on the alveolar surface of the upper jaw. There is a bony temporal arch and the free ends of the ribs are very long, as in *Deirochelys*.

Blanding's Turtle *Emys blandingii* (Holbrook)

(Plate 24. Map 10.)

Range: Northeastern North America, from northern Michigan and the Lake Nipissing area in southern Ontario eastward into eastern Massachusetts and central New Hampshire; southward to central Nebraska, and across the northern parts of Illinois, Indiana, Ohio, Pennsylvania, and New Jersey into coastal New York.

Distinguishing features: There will be found little difficulty in identifying mature specimens of Blanding's turtle. No other form in its range will

possess the hinged plastron, elongate shell, light-speckled carapace, terminally notched upper jaw, and conspicuously yellow throat and lower
jaw that characterize this species.

Description: The shell is long, relatively narrow, and moderately elevated,
suggesting in its proportions that of *Deirochelys*. This is a medium-sized

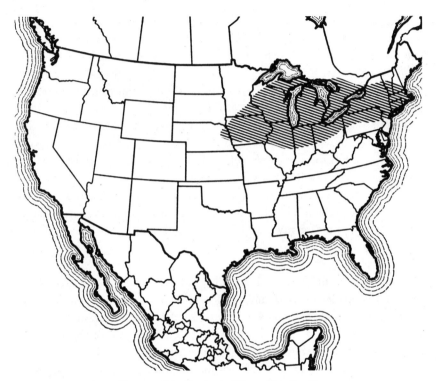

Map 10. Distribution of *Emys blandingii*.

turtle, the maximum shell length apparently being in the neighborhood of
250 mm. (9.85 inches). The largest individual measured by Lagler (1943b)
was 236 mm. (9.30 inches) in length, while Cahn (1937) had an "unusually
large specimen" that measured 240 mm. (9.45 inches) and Conant (1938a)
found the maximum length among Ohio examples to be 205 mm. (8.07
inches). The carapace is flattened middorsally and is oblong in lateral
outline. It is nearly always sprinkled with numerous small spots, flecks or
short streaks or bars of light color on a dark background. The plastron
is extensive and nearly as effective in enclosing the soft parts as that of a
box turtle, since it has a well-developed flexible hinge between the pectoral

and abdominal laminae. In color the plastron is similar to that of the wood turtle, being yellow with large dark blotches symmetrically arranged, although in some cases the dark elements are so extensive as to give the impression of a dark background for yellow markings. The head and legs are dark above, often speckled or mottled with yellow, and the throat is bright yellow.

Sexual dimorphism is apparently not marked. There appears to be no significant sexual disparity in size. Pope (1944), one of the very few writers who have mentioned the subject, found that "when the tail is extended, the anus of the female lies under or in front of the edge of the carapace, that of the male beyond it; the male's plastron is slightly concave, in contrast to the flat or barely convex plastron of the female."

Young Blanding's turtles are astonishingly rare and, accordingly, little has been written about them. Agassiz (1857), to whose superb engravings one still must turn with incredible frequency for information on young turtles, shows a hatchling with a shell 33 mm. (1.30 inches) long, 31.2 mm. (1.23 inches) wide, and 14.5 mm. (0.571 inches) high. The carapace is said to be dark, often nearly black and without the light markings of the adult. The plastron is uniformly blackish with a yellow margin. The disproportionately long tail of the juvenile may be seen in Conant's figure (1947a, opp. p. 25).

Habitat: Blanding's turtle is usually regarded as essentially a terrestrial species, but Lagler (1943b) suggested that this notion might require modification and stated that 99 specimens were taken in turtle traps set in two Michigan lakes during the summers of 1937 and 1938. Although individuals are frequently seen wandering on land in wet places, the real habitat of the animal appears to be a variety of pondlike situations, along smaller streams, preferably slow-moving or quiet, shallow, and with some growth of vegetation. While not commonly seen in the deeper or more exposed portions of large lakes, they turn up frequently in protected coves and bights.

Habits: Blanding's turtle hibernates in mud or under trash, usually under or near water, and is quite resistant to cold, having been seen swimming beneath ice, and (in Ohio) emerging from hibernation as early as January. This is a timid turtle; if surprised while sunning beside water it plunges in, goes to the bottom, and often remains there for hours; if startled away from water it closes itself up within its shell by means of the movable plastron. It appears to be quite well adapted to aquatic life, despite its tendency to wander about on land, and is a very good and rapid swimmer—better, for instance, than the wood turtle and far more skillful than the box

Plate 24. Emys blandingii. A, C, and D: Female; North Judson, Indiana. B: Male: same locality.

turtles. This turtle has a pleasant disposition and makes no attempt to bite when handled. A Blanding's turtle lived in the Philadelphia Zoological Garden for ten years and seven months.

Breeding: During copulation, which has been recorded for nearly all months from March to October, the male holds the edges of the shell of the female with all four feet and by pressure with his chin on her snout forces her to keep her neck retracted and her head within her shell. This apparently usually takes place in the water and may be preceded by considerable coyness on the part of the female; she is often chased for some time by the male before union is effected.

Nesting has been observed only in Canadian specimens. Logier (1925) saw a female lay eleven eggs in a flask-shaped hole about seven inches (178 mm.) deep; this occurred at Point Pelee, Ontario, on June 22, 1925. Other observers have extended the laying time to late July and have found that the number of eggs varies from six to eleven, that their dimensions are about 38 by 22 mm. (1.5 by 0.866 inches), and that the digging of the nest may take from forty-five minutes to almost an hour and the laying of the eggs a little less than half an hour. The incubation period is not known. Conant (1938a) found an evidently quite newly hatched individual on September 30, 1930.

Feeding: This turtle is both an aquatic and a terrestrial feeder and is practically omnivorous. Its efficiency in swimming permits it to capture small fish in large aquariums, and perhaps the same thing happens occasionally in nature. Lagler (1943b) gave the results (shown in Table 3) of

Table 3. Stomach contents of Blanding's turtle.

Food item	Composition by volume (%)	Frequency of occurrence (%)
Game fishes	1.6	5.9
Forage fishes	2.7	5.9
Fish remains	0.7	7.8
Bird remains	5.6	2.0
Carrion	4.7	5.9
Leeches	0.1	3.9
Crustaceans	56.6	74.5
Insects	21.4	54.9
Mollusks	2.6	17.6
Lower plants	1.2	21.6
Higher plants	0.5	31.4
Vegetable debris	2.2	39.2

the examination of stomach contents of fifty-one Blanding's turtles in Michigan, all taken in or near water. From these data it would appear that Blanding's turtle is decidedly more carnivorous than herbivorous in the aquatic phase of its feeding, and that the mainstay of its aquatic diet is insects and crustaceans. It seems almost certain that plant food would constitute a higher percentage in a similar analysis made from specimens taken on land.

Economic importance: Cahn (1937) said that this turtle is very good to eat, and Conant remarked that large numbers of Ohio specimens are caught and sent to market. Although Lagler mentioned the possibility that Blanding's turtle might do some damage in fish hatcheries, he saw as the only other way in which this species is of interest to the fish culturist the competition that it offers game fish for their food supply.

Genus *TERRAPENE*

The Box Turtles

Although the box turtles have become adapted to a more terrestrial habit than most of the other emydids, they nevertheless have retained features that show them to be more closely related to *Clemmys* and *Emys* than to the true upland tortoises, which they superficially resemble. The genus is exclusively North American and is known from a fossil record that extends back as far as the Pliocene. During the Pleistocene the group had its heyday in Florida, where one species attained a size far greater than that of any living form.

The most conspicuous feature of the box turtles is the bilobed, bridgeless plastron, which is so formed that complete closure of the shell is possible. There is no temporal arch and the alveolar surface of the upper jaw is without a ridge. The toes are only very slightly webbed.

KEY TO *TERRAPENE*

1   Carapace low, broad, usually flattened on top, little or no longer (or shorter) than plastron; median dorsal keel usually lacking; pattern on laterals usually composed of long yellow lines radiating from posterodorsal corners; Missouri and Mississippi rivers to Rocky Mountains ....*ornata* (p. 156)

1   Carapace usually longer than plastron, not flattened dorsally,

and with a more or less evident median keel; if pattern on laterals as described above, then carapace elongate, narrow, and very highly arched 2

2(1') Carapace with its highest point usually noticeably posterior to middle; pattern on laterals usually comprising long yellow lines radiating from posterodorsal corners; side of head usually with two stripes; peninsular Florida to the Florida Keys ....
.... *carolina bauri* (p. 147)

2'   Carapace with its highest point at middle; pattern on laterals usually not composed of long lines radiating eccentrically; side of head usually spotted or unmarked 3

3(2') Hind foot usually with three claws; anterior surface of fore limb usually with numerous small orange or yellow spots; western Georgia to northeastern Texas, Oklahoma, Kansas, and Missouri ....*carolina triunguis* (p. 153)

3'   Hind foot usually with four claws; anterior surface of fore limb usually without numerous small light spots 4

4(3') Shell elongate; markings frequently obscured by black or horn color in adults; hind margin or carapace usually distinctly flaring; northern Florida westward along the Gulf to Texas ....
....*carolina major* (p. 150)

4'   Shell short, broad; markings on shell usually not obscured by black or horn color; posterior marginals usually nearly vertical; Maine westward to Michigan, Wisconsin, Tennessee, southern Georgia, and extreme northern Florida ......*carolina carolina* (p. 138)

Common Box Turtle *Terrapene carolina carolina* (Linné)

(Plates 25, 29. Map 11.)

Range: Eastern United States, from central Maine southward to southern Georgia and Alabama, and westward below the international boundary, across central Michigan and southern Wisconsin; from here southward to include southern (but not northern) Illinois and all of Kentucky, Tennessee, and an undetermined part of Mississippi. In the area extending from southwestern Georgia and the Florida panhandle westward, and near the Gulf coast, there occurs a stock that is varyingly intermediate between *carolina, triunguis,* and the strictly coastal *major*. From southeastern Georgia into the northern counties of the peninsula of Florida *carolina* and *bauri* intergrade.

Distinguishing features: Since the subspecies of *Terrapene carolina* occupy ranges that are almost mutually exclusive and since only one kind of box turtle occurs in most parts of the United States, identification is usually not likely to involve great difficulties. In the case of specimens from the bewilderingly varied populations to be found where the ranges of two races approach each other, identification will often be impossible and, indeed, pointless, since all these subspecies are so closely related that no one need feel chagrined over admitting that his specimen shows characters of

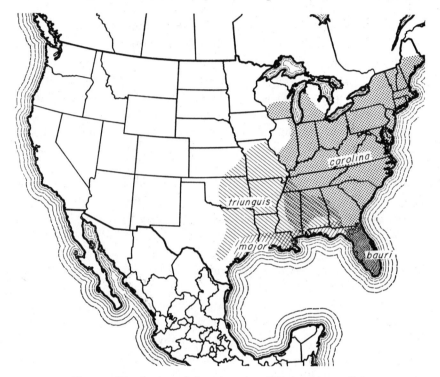

Map 11. Distribution of the subspecies of *Terrapene carolina.*

two, or even more, of them. Typical specimens of the common box turtle may be distinguished by the following combination of characters: the "box turtle" shell with broad, moderately high carapace, slightly keeled, and with its highest point near the middle and a rather steep rear margin; the four toes on the hind foot; the small size; the presence of light markings on the carapace but the infrequency of radiating narrow continuous lines on the lateral laminae.

Description: The shell of this turtle is rather short, moderately deep and wide, and broadly oval or nearly oblong in lateral outline. The size is small, even for a box turtle, maximum shell length being in the neighborhood of

152 mm. (5.98 inches). Conant (1938a) found a slightly larger one, a female with a 165-mm. (6.50-inch) carapace, but this must be considered an exceptional specimen. For example, the largest Maryland specimen found by McCauley (1945) was 134 mm. (5.28 inches) long; and the largest measured by Grant (1936) in Indiana was 160 mm. (6.30) in length. Nichols (1939c) measured the plastral lengths of 843 adult Long Island specimens. He found the great majority of individuals to range between 127 mm. and 142 mm. (5.00 to 5.60 inches) in length. Pope (1939) gave the following measurements of an average-sized female from northern New Jersey: length of carapace, 136 mm. (5.35 inches); width of shell, 105 mm. (4.14 inches); length of plastron, 133 mm. (5.25 inches); width of fore lobe of plastron, 71 mm. (2.80 inches); width of hind lobe of plastron, 83 mm. (3.26 inches); height of shell, 68 mm. (2.68 inches). The carapace is relatively high-domed, the highest point lying at or near the middle and with a more or less evident central keel, the margin not, or but slightly, flaring posteriorly. The color of the carapace is brown and yellowish, the disposition of the markings being exceedingly variable but often involving irregularly radiating spots and bars of light color on a dark background. Concentric growth furrows are usually evident but may be lacking in very old individuals.

The extensive plastron varies in coloration from an even, unmarked horn color to dark brown; there may be, and usually are, dark smudges or rayed blotches, or a central dark area that sends out branches along the borders of the seams. The plastral surface may be either concentrically ridged or quite smooth. The head is moderate in size. The upper jaw is hooked terminally. The head, neck, and legs are usually more or less extensively spotted, streaked, or mottled with yellow or orange. The tail is short. There are four toes on the hind foot.

Although there are no striking sexual differences in box turtles, there are certain minor external features by which sex may be ascertained. The most noticeable of these is a varyingly deep, smooth concavity in the posterior lobe of the plastron of the male. As usual, the vent of the male opens farther back, nearer the tip of the tail, and his hind legs are heavier and with shorter but stronger nails. The male is slightly larger on the average; in a group of mature specimens, most of those with the plastral length exceeding 140 mm. (5.50 inches) will be males and the smaller individuals females (Nichols, 1939c). There is a fairly reliable difference in eye color, the eye of the male being nearly always some shade of red, often bright red, and that of the female gray brown or very dark reddish.

Plate 25. Terrapene carolina carolina. Upper: Female. *Center and Lower:* Male. Both from Richmond, Virginia.

141

The hatchling box turtles are different in appearance from the adult. Allard gave the dimensions of seventeen newly hatched box turtles as follows: average length of carapace, 31.0 mm. (1.22 inches); extremes, 30 mm. and 33 mm. (1.18–1.30 inches); average width of carapace, 29.5 mm. (1.16 inches); extremes, 28 mm. and 33 mm. (1.10–1.30 inches). Frothingham (1936) measured two specimens in which the egg tooth was still present and found one to be 28 by 22 mm. (1.10 by 0.869 inches) and the other 26 by 19 mm. (1.02 by 0.75 inches). The carapace, thus, is more nearly circular in lateral outline than that of the adult. The shell is relatively lower, but the central keel is high and complete and usually conspicuously light in color. In general, light markings are fewer than in older specimens, those of the carapace comprising only one light spot per lamina. The outer rim of the shell is yellow-bordered and the plastron is dark centrally, with definite yellow margining. The plastral hinge is not functional and during the earlier stages prior to its development young box turtles are able to emit a strong odor when molested (Neill, 1948c), an adaptation that is lost later on.

Habitat: The common box turtle is found in open woodland, preferably in the vicinity of brooks or ponds, whence it roams widely into all kinds of places during rainy weather. The box turtle is by nature and specific adaptation terrestrial, but it is fond of water, as well as of the lush forage of stream banks and low, wet places. While it is capable of controlled locomotion in the water, the consensus of opinion seems to be that it is an ineffectual and bungling swimmer, although two observers, at least, say they have seen individuals swimming rapidly. The only common box turtles that I have seen in the water were placed there by me, perhaps against their wills, and they all floundered and flapped the water pitifully and got almost nowhere. I believe, however, that their efficiency and stability in the water may be markedly affected by hydrostatic adjustments which may take place only gradually, as is the case in other (and even in much more aquatic) species of turtles.

Habits: The common box turtle usually hibernates in the northern parts of its range from late October or November until sometime in April. It is useless to try to define limits for its cessation of activity because of the fact that any given turtle may spend two weeks or more in an intermediate state during which it may each day gradually burrow more deeply in the soil, or even emerge for a temporary resumption of its summer activities if warm weather intervenes. The most usual site for hibernation is loose soil, but individuals may burrow in vegetable debris in clear sand, or even in the mud of stream bottoms. According to Allard the burrows are dug by the fore feet, unlike the egg cavities, which are excavated by the hind feet. The depth of the winter burrows varies markedly, some individuals passing the period with part of their shells exposed and others allegedly digging to a depth of two feet. Allard found that his captive individuals emerged during warm spells at any time of year at Washington, D.C. Neill said that "vast numbers" freeze to death in Richmond County, Georgia, when they defer hibernating until too late or emerge prematurely during warm spells. Knowlton (1943) noticed that among her captive box turtles, which were kept under fairly natural conditions, males might often be seen walking about weeks after the last female had dug in for the winter, and Mrs. Knowlton interpreted this as evidence that the males were determined that all females be properly fertilized before calling this chore well done. In midsummer, box turtles often burrow into moist soil or leaf mold during the hottest parts of the day or even seek water, where they may merely sit in shallow places and pass the hot spells or may swim in a desultory fashion or crawl about on the bottom; some may even bury themselves in the bottom mud during droughts. Brimley (1943) mentioned a case of what appeared to be true aestivation under water, but Allard stated that there is no real aestivation comparable to the hibernation torpor of winter.

In disposition box turtles also vary widely, although the variety of opinion that has been published on this subject is perhaps due in part to the ease with which these animals are observed and the disproportionate abundance of data on their behavior that have accumulated. Box turtles rarely show any pugnacity toward man when handled, but some show little or no fear, either, while others are so affected by capture that they may retire within their shell and stay for hours and even days. Various differences in personality have been commented on by observers who have kept them in numbers.

Mrs. Knowlton pointed out that the familiar habit of elevating the off side of the shell and lowering the on side, when the shell or soft parts of one side are touched, is a reaction whereby a box turtle resists being upset when the snout of an aggressor is thrust under the edge of its shell. She found this overturning to be a common form of attack among her captives.

One of Mrs. Knowlton's pet box turtles learned to beg for food by lifting both fore feet from the ground, resting only on the hind feet and rear shell border while craning its neck upward. Since this feat appeared to require no protracted training process and is really only a step in advance of the common practice of lifting one front foot and reaching upwards for some object, one wonders if wild box turtles may not sometimes stand on their hind legs in the attempt to secure a high-hanging blackberry.

The work of Nichols (1939a) and the experiments of Ruth Breder (1927) have shown box turtles to have a surprisingly restricted home range, often comprising an area perhaps 250 yards in diameter or less. A large percentage taken from their home territory to spots as far as three-quarters of a mile away showed marked homing instinct, coupled with a good sense of direction and ability to choose the shortest route from point to point.

Until Brimley mentioned it in his paper of 1943, I had never seen printed reference to the fact that bird dogs frequently incur the displeasure of their masters by pointing box turtles. I have heard numbers of hunters complain of this and have observed it myself on dozens of occasions, involving all races of *carolina* and a Texas population of *ornata* as well. Whether the reaction in the dogs is due to deception by a similarity between the scents of box turtles and quail or whether the former possesses its own special point-evoking odor it is impossible to say. The only other creature that in my experience consistently causes a trained dog to haul up spasmodically in a rigid point is a small ground-dwelling sparrow which so regularly deludes bird dogs that it is known by quail hunters as "stink-bird." On finding himself fooled by a tortoise the well-indoctrinated pointer or setter usually backs off sheepishly and retires from the scene in humiliation. Hounds, on the other hand, often carry box turtles for long distances, perhaps stopping once in a while to gnaw ineffectually at the high, smooth shells and usually at last burying them in the ground. Individual hounds may become addicted to carrying box turtles, to their detriment as deer hunters, and one hound of my acquaintance habitually brought them in and buried them around the camp. While in the case of hounds one might suspect that the attraction stemmed from the chronic hunger that seems to torment them, the pointing reaction of bird dogs cannot be so explained. This seems to me another of the intriguing mysteries of the world of scent in which dogs live and of which humans know next to nothing through either personal experience or scientific investigation.

During the first four or five years of life, box turtles may grow at a rate of from half an inch to about three-quarters of an inch a year. Knowlton found that her specimens grew to a length of two and one-half to three inches in five or six years, but occasionally, at least, individuals may apparently attain a length of as much as five inches in this time. In specimens between three and four inches in length, Nichols found the maximum growth rate to be three-eighths of an inch per year. Sexual maturity is reached in about four or five years, and Nichols believed that full growth is attained in about 20 years and estimated the age of one of his specimens to be 60 years. The same writer suggested that the life expectancy of a box tortoise is between 40 and 50 years. The work of Nichols and of Flower (1925, 1937), who have carefully examined the abundant reports of great ages attained by these turtles, has led to the conclusion that there is no reliable evidence that they live to ages greater than 80 to 123 years.

Since box turtles do not shed the laminae of the shell, growth rings may be used to estimate age, although Nichols found that the accuracy of this method diminishes after the age of five years or so and that it is quite unreliable after fifteen years.

Breeding: Mating begins shortly after emergence from hibernation. Courting procedures involve the chasing of the female by the male, followed by much biting of her shell edges, head, and neck. According to Knowlton the male may tap the shell of the female with his plastron and may also repeatedly rotate her shell by pushing on it. Once mounted, the male hooks his hind feet under the back edges of the female's shell, and if acquiescent she holds them in place with her own hind legs. Apparently copulation is effected only when the male assumes a vertical position that allows the vents on the short tails to be brought into contact. There is no use in suppressing the comment that a pair of copulating box turtles is ludicrous to behold. Allard (1939) believed that the curious position assumed by the males leads to the death of large numbers of them, which on terminating copulation fall over backwards and on a soft substrate are unable to right themselves. It seems unlikely, however, that this could be the whole explanation for the occurrence of large numbers of shells in the woods, since in a few cases, at least, these have been the shells of females.

Both Ewing (1943) and Finneran (1948b) found evidence that females may retain fertility for two or three (and in one case even four) years after a single copulation. The former author had a box turtle that laid five eggs, all fertile, four years after contact with a male.

Nesting occurs mostly in June and July in the northeastern states and in May and June in the Mississippi Valley and in the southeast. Nesting nearly always takes place in late afternoon, often after sunset, and may extend well into the night. The female usually selects an open place on elevated terrain, preferably in loose or sandy soil but not invariably so. Cultivated lands or grassy lawns are often selected. The main points of the nesting procedure as described in Allard's unique and excellent account are as follows:

The fore feet are planted at the beginning of excavation and are not subsequently moved. The nest is flask-shaped, with a narrow neck and enlarged egg cavity. It is on the average about three inches deep, although its size and depth apparently are determined by the limits to which the leg and foot can reach. The hind feet work alternately, scraping, tearing at obstructions, and palming soil and even small pebbles to pile them at the rear edge of the hole, where they are pushed back out of the way from time to time. Allard found that the process of digging the nest often required four hours or more for completion. The eggs are dropped at intervals observed to vary from one to six minutes, and frequently each is arranged in the nest by a hind foot, although soil is not placed on top of each individually as is reported for other species. The nest is filled by broad lateral sweeping of each hind leg individually or by scooping soil forward by both hind feet drawn in at once. Filling involves much tramping and treading during which the toes, feet, knees, and plastron are all used. The digging and filling processes together may require as much as two hours or even more, and the whole nesting process may consume up to five hours of hard work and has been observed to last until nearly midnight. Some authors have referred to a wetting of the soil by the turtle to soften it before excavation begins, but Allard did not observe this and, indeed, was convinced that it was never a feature of the nesting operation as carried on by his large series of specimens.

The eggs are from two to seven in number, most often four or five, and are elliptical in shape, with average dimensions about 32.0 by 19.5 mm. (1.26 by 0.767 inches). The largest egg seen by Allard measured 40.0 by 21.0 mm. (1.57 by 0.827 inches). Most hatchlings turn up in September, although some appear to emerge as early as August or as late as October. The incubation period may be as short as 70 days, but is extremely variable, a period of 114 days having once been noted and other cases being known in which the eggs have passed the winter before hatching. Allard believed the average period to be between 87 and 89 days. Driver (1946) kept eggs laid July 17, 1939, in moist soil at temperatures above freezing, and after 230 days' incubation found live embryos in them. When the young hatch they either remain in the nest throughout the winter, emerge and go directly into hibernation, or emerge, move about for a few days or weeks, and then burrow for the winter. They apparently do not eat anything during the first summer and fall of their lives, at least in the northern part of the range of the subspecies.

Feeding: The common box turtle is an omnivorous feeder, showing marked changes in food preference from youth to maturity and from one season to another. The young are largely carnivorous and the adults mostly herbivorous. The young in captivity show preference for insect larvae and slugs and in nature are known to eat these and earthworms, snails, and a small percentage of vegetable material. The larger individuals eat fungi in quantities and are fond of berries and fruits of many kinds. They too take animal food with frequency, but the proportion of their total diet that it constitutes is far less than in the young. It has been suggested that the more aquatic tendencies noted in young box turtles may account for their predilection for animal food, since the types of small creatures they like to eat are more easily procurable in the water than on land. Babcock (1939b) kept a young box turtle for six years after hatching before it took any vegetable food at all. Creaser (1940) found a mature individual so stuffed from eating mushrooms that it could not close its shell. In this case the mushrooms were harmless, but it is said that poisonous ones are eaten with impunity and that this may account for the fact that people are sometimes poisoned by eating box turtles. The most extensive study of the food of this turtle was made by Surface (1908), who examined the stomach contents of forty individuals of various sizes, finding that animal material was present in 80 per cent of those that contained food and vegetable food in 62 per cent. It should be pointed out that young box turtles are so rarely encountered that any comment on their habits must be regarded as quite tentative.

Economic importance: Except for the poisoning of a small percentage of the few people who eat them, this turtle probably has as little material significance for mankind as any other. It is, however, an altogether appealing animal, a "quaint element in our fauna," as Pope puts it, and certainly one to preserve. There exists a curious lot of witless or psychopathic characters who love to run over box turtles on the roads to hear them pop, and there is probably nothing much that can be done about these people except to hope they skid.

Florida Box Turtle *Terrapene carolina bauri* Taylor

(Plate 26. Map 11.)

Range: Peninsular Florida, from Duval and Alachua counties southward to the Keys.

Distinguishing features: Several slight and intergrading differences may be seen in the box turtles of the Florida peninsula. The shell is comparatively narrow, elongate, and with its higher vaulting distributed posteriorly, mostly behind the middle, and with the rear margin flaring. The pattern of the carapace typically includes narrow yellow lines, mostly unbroken, radiating from the upper rear edge of each lateral. The size is intermediate between that of *major* and that of *carolina*, but nearer the former, the usual size for a breeding adult being about 110 mm. (4.34 inches); a specimen of this length from the Everglades had a shell width of 80 mm. (3.15 inches). Two more or less complete yellow stripes, nearly always present on the side of the head, are characteristic. There are usually three toes on the hind foot.

Plate 26. Terrapene carolina bauri. A: Female; Marion County, Florida. *B and D:* Male; Homestead, Florida. *C:* Young; Marion County, Florida. (Courtesy Ross Allen.)

Sexual differences are similar to those in *major*. The shell of the female is often broader than that of the male.

The hatchling is, astonishingly, unknown. Young individuals are broader, flatter, and more highly keeled than the adults and have much less conspicuous lateral markings.

Habitat: This turtle has been reported as widely distributed in flatwoods and upland and mesophytic hammock, being more rarely found in high pinelands (Carr, 1940). It appears to be most abundant in the limestone flatwoods of southern Dade County.

Habits: Perhaps the most striking feature of the natural history of this box turtle that has been noticed is the great disparity in abundance that it shows in various parts of its range. In some restricted areas it could be said to be really common, while in other places which would seem to offer equally suitable if not identical environments it may be so scarce that one wonders how the population maintains itself. The only region in which, in my experience, it attains the abundance of *carolina* in northern Georgia or of *ornata* in several parts of its range is the limestone flatwoods west of Homestead, Dade County, where after a rain a dozen or more individuals may sometimes be seen crossing the road within a distance of a mile or so.

I have never seen evidence that this turtle inters itself for a regular winter dormant period. Although individuals are inactive on cold days they may be found only partly covered by sand or soil or wedged under the edge of a log or a rock. Throughout the peninsula they have been observed active during all parts of the year on warm days.

Bogert and Cowles (1947) placed a 293-gram specimen of *T. c. bauri* in a thermal chamber in which the temperature remained constant at 38° Centigrade and the relative humidity at 37 per cent. After a period of 45 hours the turtle had lost 17.3 per cent of its original body weight.

The woods fires that are an annual occurrence in Florida are hard on box turtle populations and may perhaps be the most serious limiting factor in their environment, possibly even being in some way responsible for the remarkably spotty distribution of the turtles in the state. Thousands must be roasted in their shells each year, for they appear to have no notion of how to escape an advancing blaze. Nevertheless, they are remarkably fire-resistant, and it is not unusual to find an old veteran of repeated ground fires, and perhaps of a smashing or two on the highway as well, that is so scarred and deformed as to resemble more a clod or a strangely shaped oak gall than a live turtle.

Breeding: The eggs measure 35–38.5 mm. (1.38–1.51 inches) by 19.8–

21 mm. (0.78–0.826 inches) and are laid from mid-April through June. Complements of from one to nine eggs have been noted. The first extreme was the entire complement of a captive specimen and was probably abnormal; the last refers to the number of shelled eggs found in a dissected female, and possibly represented eggs to be deposited on two laying dates, although this seems unlikely. The average complement is probably five eggs. The only account of nesting is that of Ewing (1937), who saw a captive specimen excavate a nest "in the same manner as does *T. carolina*." The incubation period is not known, and since no one apparently ever sees the hatchling, it cannot even be given in approximate terms.

Feeding: I have seen individuals in the field eating blackberries and fungi on several occasions and have come across them scavenging around garbage dumps and once at the carcass of a dead cow. In captivity they are omnivorous, and one batch of captives that I kept showed much the same enthusiasm for lettuce as for chicken entrails.

Economic importance: This is an innocuous animal, aesthetically pleasing to view in the field but never, to my knowledge, prepared for the table.

Gulf Coast Box Turtle *Terrapene carolina major* (Agassiz)
(Plate 27. Map 11.)

Range: From the panhandle of Florida in the Cape San Blas area westward near the coast into eastern Texas.

Distinguishing features: Although the occurrence of geographic intergradation between this box turtle and the other subspecies illustrates the closeness of the relationships involved, a typical specimen of the Gulf coast box turtle is a well-differentiated, handsome, and even imposing animal. Its shell is elongate and fairly high, with the highest point in the middle and its sides either nearly straight or slightly constricted. In size it exceeds all box turtles of modern times, although its Floridian ancestors of some thousands of years ago were somewhat bigger. A species described from the Florida Pleistocene as *T. caniliculata* appears to have differed from modern *major* only in its great size (up to 266 mm. [10.5 inches] in carapace length). Dimensions of a mature male from Apalachicola are as follows: length of shell, 179 mm. (7.05 inches); width of shell, 118 mm. (4.65 inches); height of shell, 73 mm. (2.87 inches). Specimens with six-inch shells are not unusual. The carapace is keeled, with flaring rear margin (more so than in *carolina*) and is dark brown to nearly black, with or without a pattern of scattered spots or of lines that may approach the radiating disposal seen in

Plate 27. Terrapene carolina major. A: Female. *B, C, and D:* Male. Both from Jackson County, Florida.

bauri. The plastron may be nearly solid black in color. The head and soft parts are dark brown to deep black, either spotted or plain; the hind foot has four claws.

The male Gulf coast box turtle often has a very deeply concave plastron which may be accompanied by lateral constriction of the shell margins. It has, as well, bigger hind legs and a more distally placed vent than the female.

The hatchling has not been described. I have seen several young of 60 mm. (2.36 inches) shell length or slightly more, and in these the juvenile characters were as described above for *T. c. carolina.*

The relatively generalized nature of certain of its skeletal characters— the temporal arch of the skull, the form of the scapula, and the elongate cervical vertebrae, besides a more strongly webbed foot and the larger size attained—have led to the conclusion that *T. c. major* may be the most primitive of the box turtles, and the close similarity between *major* and the fossil form *T. caniliculata* may be further indication of this.

Habitat: In Florida, *major* roams widely but appears to be most numerous in pine flatwoods and upland hammocks. Allen (1932) thought it preferred wooded swamps in southern Mississippi, while Viosca (1931) described the Louisiana habitat as "wooded alluvial lands, cane brakes and palmetto thickets." This subspecies enters water with what appears to be about the same frequency as *carolina*.

Habits: Although the territory inhabited by this subspecies has a climate that sometimes permits box turtles to remain more or less active all winter, Penn and Pottharst (1940) found that specimens went into hibernation at temperatures lower than 65° Fahrenheit. These observers also learned that a six-day warm spell was sufficient to bring about temporary emergence of some individuals and that the females were less inclined to make such interim appearances, once buried, than the males. Cases were observed in which male turtles hibernated in the water, although Penn and Pottharst suggested that this may have been due to their having been accidentally overtaken there by paralyzingly low temperatures. Allen also mentioned several captive specimens that remained frozen solidly in the ice of a small pond for twenty-four hours, to emerge after thawing none the worse for the experience. This subspecies appears to be more fond of water than any of the other box turtles. The contrast between it and *bauri* in this respect is particularly striking.

Breeding: Penn and Pottharst found mating to be coincident with temperatures above 85° Fahrenheit in Louisiana. They stated that scars usually present on the precentral and anterior marginals of males are produced by the frequent fights in which they engage, and that similar scars on females are due to "post mating caresses" by males. They saw many cases in which the male "nose-rolled" the female—that is, thrust his snout under her shell and turned her over—this usually occurring after copulation but sometimes before. Two pairs were seen mating in water.

The eggs measure 37–38.4 mm. (1.46–1.51 inches) by 21.5–22.4 mm. (0.85–0.88 inches) and, in Florida, are laid in loose soil and decaying wood in April and May.

Feeding: Captives are omnivorous. A small individual that my mother kept for three years ate hamburger with more relish than anything else offered it. This individual was allowed the run of the house. It ate very sparingly during winter, even though the temperature of the house varied only slightly. During the warmer months it fed two or three times a week and soon learned to show its hunger by taking a station in front of the refrigerator where the hamburger was kept.

Economic importance: According to Löding (1922) this box turtle is often eaten in Alabama and "makes an excellent dish." Other things being equal, its relatively larger size would make it more suitable for such use than any of the other box turtles.

Three-toed Box Turtle *Terrapene carolina triunguis* (Agassiz)
(Plates 28, 29. Map 11.)

Range: Mississippi Valley from northern Missouri southward across southeastern Kansas and eastern Oklahoma into south-central Texas; and southeastward across western Tennessee and Georgia to the coastal lowlands (where it intergrades with *major*). Lockley (1948) recently found a three-toed box turtle, certainly introduced, on the California coast near La Jolla. Shannon and Smith (1949) described what they felt certain was a hybrid between *T. c. triunguis* and *T. ornata* from Wayne County, Missouri.

Distinguishing features: This midland race is characterized by its narrow shell, keeled above and flaring at the posterior margin; the usually conspicuous spotting (orange or yellow) of the head and fore limbs; and the three-toed hind foot. The plastron is frequently solid yellow, unmarked.

Sexual differences are presumably those of *carolina,* although for some reason neither they nor the hatchling have been described. This subspecies intergrades with the common box turtle where the ranges meet in the eastern part of the Mississippi Valley.

Habitat: It is found in forest, meadow, and second-growth scrub alike, often near streams and not infrequently in them. Most observers are agreed that it is more of a woodland form than *T. ornata*. Woodlots and fringe forests in the vicinity of water might be regarded as an optimum habitat.

Habits: Published observations on the biology of this box turtle are extremely meager. The breeding habits are unknown, although eggs have been described as measuring 35 by 23 mm. (1.38 by 0.906 inches) and 38.5 by 20 mm. (1.51 by 0.787 inches). In captivity the feeding habits appear to be similar to those of *carolina,* and a great variety of animal and plant foods are accepted. Rodeck found it especially fond of earthworms and grasshoppers.

Regarding the three-toed box turtle's capacity for water, Rodeck commented as follows:

The inclination of arid-lands turtles to imbibe large quantities of water after deprivation was illustrated by a female three-toed box turtle in captivity [a Colorado specimen]. After having had no water for about two weeks it was given a dish containing approximately half a pint. It drank steadily for nearly

Plate 28. Terrapene carolina triunguis. Females from Louisiana.

Plate 29. Intergrade specimen between *Terrapene carolina carolina* and *T. c. triunguis*. Male; Auburn, Alabama.

an hour, exhausting the water in the dish, and during the latter half of the period, a constant stream of water issued from the anus.

A specimen lived in the Philadelphia Zoo for six years and two months.

Economic importance: This turtle has been known to poison people who have eaten it, although I have heard of no very serious illnesses resulting. My father recalls that several of these box turtles that had been accidentally roasted in burning brush piles were eaten by a group of boys in Mississippi, with the result that all later became ill. Also in Mississippi, this turtle is

regarded by quail hunters as an undesirable animal because of its alleged
fondness for quail eggs and because of its susceptibility to being pointed
by bird dogs. I cannot imagine that any of these counts is of much significance, however. The importance to game management of the alleged
egg-eating habit has not been the subject of systematic study, and I daresay
it is negligible in comparison with the inroads made by the warm-blooded
enemies of ground-nesting birds.

Ornate Box Turtle *Terrapene ornata* (Agassiz)

(Plate 30. Map 12.)

Range: Southern Wisconsin, northern Iowa, and southern South Dakota
southward to the Gulf coast of Texas. To the westward it reaches southeastern Wyoming and eastern Colorado, and from Texas it extends across
northern Mexico and southern New Mexico to southern Arizona. To the
eastward *ornata* enters a restricted area in northwestern Indiana and crosses
Illinois, most of Missouri, and the northwestern half of Arkansas to enter
eastern Texas. Schmidt (1938) saw in the restricted occurrence of the
ornate box turtle in Illinois and Indiana, where it is known mostly in
sandy areas, additional evidence for an important general postglacial extension of the western plains fauna.

Distinguishing features: This species may be distinguished by its flat-
topped, unkeeled, ray-lined carapace and by the striking pattern of light
lines on the dark plastron. Osteological features appear to indicate that this
is the most specialized of modern box turtles.

Description: The shell is oval and broad, the width about three-quarters
of the length, and rather high but flattened on top. Maximum shell length
is around 145 mm. (5.70 inches); perhaps somewhat less in the eastern part
of its range. Table 4, using data given by Cahn (1937), compares features
of nine specimens from Illinois.

The low, short carapace is usually without a keel and is chocolate to
reddish brown in ground color, with light lines radiating from the upper
hind corners of the laterals and with a broken central light line. The
plastron may be as long as, or even longer than, the carapace and has a
characteristic and quite a constant color pattern of light lines extending in
all directions on a dark background. The head and legs are spotted with
yellow, especially above.

The sexes differ in the more distally located vent of the male and in the
anteriorly directed first toe on the hind foot of the male. Cahn said there

Table 4. Measurements (in millimeters) and weights (in grams) of
nine ornate box turtles.

Sex	Shell			Plastron		Weight
	Length	Width	Depth	Length	Width	
Female	112	91	52	111	73	308
Female	103	85	49	103	66	219
Female	96	77	52	99	78	203
Male	96	83	48	98	71	202
Male	99	85	48	102	. . .	210
Female	105	89	51	106	68	320
Female	83	71	41	85	54	120
Female	109	96	58	110	69	295
Female	99	89	50	101	65	231

are no sexual differences in eye color nor in the conformation of the plastron.

Marr (1944) gave the following localities and dimensions for young
T. ornata measured by him (measurements in millimeters):

Locality	Length	Width	Date (1942)
Kansas			
Finney Co.	37	34	May 26
Finney Co.	31	33	May 30
Ford Co.	33	34	May 31
Colorado			
Prowers Co.	35	34	June 6
Nebraska			
Red Willow Co.	30	28	July 2

The same writer contributed the following notes on the characters of
these young specimens:

The young are almost round and not elongated as are the adults. They appear
to be relatively flatter than the adults. A rounded dorsal ridge, not a crest, is
present in all five specimens. They are darker than the adults; the dorsal ridge
is yellow and there are faint, light spots on the carapace, but the adult pattern
is not yet distinct. The plastron is light margined with no pattern of light spots
present in the dark interior.

Habitat: Although regarded through much of its range as a prairie
species, tolerant of extremes of heat and aridity, the ornate box turtle turns
up in many other types of terrains. Smith (1947) found it in savannas and

woodlands and it is known from sand hills and even from swamps. During dry weather it may take refuge underground in the middle of the day. It has been recorded from the mountains of New Mexico at altitudes of from five to six thousand feet. It evidently almost never enters water voluntarily.

Map 12. Distribution of *Terrapene ornata.*

Habits: The ornate box turtle is an inhabitant of sandy and, in many cases, semiarid regions and is rarely found in wooded areas. In this preference for open terrain it differs from the more eastern *carolina,* with the subspecies of which its own range overlaps. It is also said to show less interest in water than *carolina;* I can find no mention of an individual's voluntarily swimming in any body of water. From observations on a number of specimens of *carolina* and *ornata* that he kept in a pen, Cahn (1937) made additional comparisons between the habits of these two box turtles. He noticed that *ornata* did not go into hibernation by degrees as is characteristic with *carolina,* but burrowed with the first cold weather and did not emerge until the following spring, although it did go deeper as the temperature of the soil became lower, reaching finally depths as great as 22 inches. There was thus a period of two weeks or so during which the common box turtles burrowed and emerged with the diurnal temperature cycle but during which the ornate box turtles were in a continuously dor-

Plate 30. Terrapene ornata. A: Sex undetermined; El Paso, Texas. *B and D:* Female; Arizona. *C and E:* Female; Brownsville, Texas.

mant state. Since the latter also remained dormant longer in the spring, emerging as much as two weeks later than *carolina,* its period of hibernation lasted a full month longer. Anderson (1942) observed that in Jackson County, Missouri, hibernation usually begins in October and terminates in May. Individuals that the same writer dug up in March were buried beneath two feet of soil.

In disposition Cahn found the ornate turtle to be even milder than the common box turtle. He referred to the Illinois stock, and it is interesting to note that Grant (1936a) regarded this eastern form as much less vicious than representatives of the same species in the western part of its range.

This species attains an abundance in parts of its range that is apparently unequaled by any other box turtle. Numerous writers have mentioned exceptionally dense populations and the fact that they turn up sporadically or in restricted localities. Rodeck commented on a recent occurrence of this sort in Colorado as follows:

During the summer of 1947 there were unusual opportunities to observe box turtles. A heavy rain in northeastern Colorado during July brought them out by the thousands, and literally hundreds of medium-sized specimens were killed along the highways by passing vehicles.

The question as to where they go between these sudden outbursts of abundance was perhaps partially answered by the discovery of a good-sized adult resting just within the mouth of a burrow in the sand dune area north of Campo, Baca County, Colorado, on the 19th of August. Upon being disturbed, it turned quickly about and disappeared into the burrow. Among the sand dunes along the Cimarron River a number of box turtles were found abroad just at sunrise, suggesting that they may forage at night. Although specimens are sometimes seen during the day, they probably usually go into subterranean retirement with the coming of daylight. Diurnal activity may possibly be correlated with rainy weather.

Although these turtles live in arid regions and presumably derive much of the necessary moisture from their food, they do not disdain water when it is available. Captive specimens drink deeply after having been deprived of water for some time. The mud flats of the Cimarron River bore numerous trails of turtles that had emerged from the vegetation of the banks and proceeded a hundred yards or more to the stream to drink. Some of them turned around and returned the way they had come; others crossed the shallower rivulets; and still other trails indicated that the animals had willingly or otherwise entered the stream, since the trail ended at the water's edge. Experiments with captives indicate that they can float buoyantly and an involuntary voyage downstream probably would affect them little or not at all. The writer, following one trail, found the turtle at the water's edge. Its head and neck were submerged, the head being not only in the water but beneath the surface of the soft mud. Active throat movements showed that the turtle was drinking.

In the same paper Rodeck made the following interesting suggestions:

These observations might permit one to propose a conjectural account of the life history of the ornate box turtle. It is possible that the young are even more subterranean than the adults. Perhaps they spend their early years in rodent or other burrows where there is a fairly abundant insect fauna. Increasing size might force them to the surface for feeding, with a daily return to a burrow for resting and protection. Heavy rains may either force them to emerge from flooded hiding places, or increased air humidity may permit their emergence in comfort and make nomadic wandering possible.

Breeding: Courtship and mating have been observed from May to October. Brumwell's observation (1940) of the process as demonstrated by a pair which he placed in a fenced orchard appears to indicate that courtship may be an energetic and in some respects unusual procedure. On being released in the orchard (on April 4, 1937) the male immediately walked up to the female and pushed at the margins of her carapace with his snout. An attempt to mount her caused the female to move away, and the male pursued.

Every now and then he would overtake her sufficiently to raise himself high on his front legs and hurl the fore part of his plastron against the hind part of her carapace, at the same time emitting a stream of fluid from each nostril, which sprayed over the female's back. This ceremony was repeated five or six times in as many minutes.

Copulation occurred after about a half hour of this play. On another occasion (May 9, 1939) Brumwell saw four males pursuing a female and each time one of these overtook her he tapped her carapace as described above. Although the female at first snapped when overtaken, she showed a noticeable preference for one of the males and was later observed copulating with this individual. Brumwell suggested that the tapping and biting at the shell may afford cues for sex recognition. Laying is said to occur in June, and according to Anderson, in the evening. The process of nesting has apparently not been described. The eggs are about 36 mm. (1.42 inches) long and 23 mm. (0.905 inches) wide, and clutches of three, four, and five have been noted. Rodeck recorded the mounting of a female *T. c. triunguis* by a male *T. ornata,* but apparently copulation was not completed. A supposed hybrid between these two box turtles was found in Wayne County, Missouri, by Shannon and Smith (1949).

Feeding: The feeding habits of this species have not been studied in detail, but scattered observations and examination of a few stomachs indicate that it is omnivorous and perhaps that it may be slightly more inclined toward predaciousness than the races of *carolina.* At any rate, there is a relatively great number of reports of individuals catching and eating live insects and other small animals. Marr (1944) found them eating trapped rodents, among other things, and Eaton (1947) saw a captive specimen catch, hold with the fore feet, and partially consume a horned toad (*Phrynosoma cornutum*). I have kept large numbers of these box turtles in captivity and have offered them literally dozens of different kinds of foods, both animal and vegetable, and I never found any that they would altogether refuse.

Economic importance: Although there have been mild complaints of damage done by this species in vegetable gardens, there can be little doubt that its value as a destroyer of insects far outweighs its importance as a garden pest.

Genus *MALACLEMYS*

The Diamondback Terrapins

This genus includes a single species distributed in salt and brackish coastal waters from New England continuously to Texas and possibly well into Mexico. The diamondbacks and the following genus, *Graptemys,* are very closely allied, forming a small group with no especially close relatives among other recent emydid turtles and with almost no fossil record. For differences separating these two genera the reader is referred to the account of the genus *Graptemys* (p. 186).

In both genera the plastron is firmly joined to the carapace by sutures, and the axillary and inguinal laminae are small or lacking. The skull has a long, bony temporal arch. The alveolar surface of the upper jaw is broad and smooth. The hind legs are relatively very large and the toes are webbed beyond the bases of the nails. Disparity in size of the tiny males and the sometimes huge females is the greatest to be found in any North American turtle. Members of both the genera may have markedly elevated projections from the central laminae.

KEY TO *MALACLEMYS*

1	Keels of centrals not expanded distally	2
1′	Keels of centrals with terminal expansions	3
2(1)	Sides of shell divergent posteriorly; Cape Cod south to Cape Hatteras	............................*terrapin terrapin* (p. 163)
2′	Sides of shell nearly (or quite) parallel; Cape Hatteras south to southern Florida	....................*terrapin centrata* (p. 174)

3(1′) Laminae of the carapace with conspicuous light centers; west
coast of Florida from the panhandle southward . . . .
. . . .*terrapin macrospilota* (p. 178)

3′ Laminae of the carapace without (or with only vague or desul-
tory) light centers 4

4(3′) Seams of lower marginals and of plastrals frequently bordered
with black; shell strongly oblong; Florida Keys . . . .
. . . .*terrapin rhizophorarum* (p. 177)

4′ Seams not regularly black-bordered; shell more oval 5

5(4′) Carapace uniform in color, usually black or dark brown; top of
head and upper lip dark; Louisiana to the Florida panhandle
. . . .*terrapin pileata* (p. 182)

5′ Carapace uniform or with concentric markings, usually light
brown; upper lip and top of head usually white; coast of Texas
southward, possibly to the Yucatan peninsula . . . .
. . . .*terrapin littoralis* (p. 184)

Northern Diamondback Terrapin *Malaclemys terrapin terrapin* (Schoepff)

(Plate 31. Fig. 13. Map 13.)

Range: From the northern shore of Cape Cod southward to the vicinity of Cape Hatteras, where it intergrades with *M. t. centrata.* The exact orig-inal northern limit of the range of the terrapin is not known. Stejneger and Barbour (1943) gave it as Buzzards Bay, while Pope (1939) and Babcock (1926b), respectively named Eastham and Barnstable, both on Cape Cod, as the northernmost localities. I have a copy of an old letter (January 31, 1921) from Dr. L. C. Jones of Falmouth, Massachusetts, to Dr. Robert E. Coker of the U.S. Bureau of Fisheries that has bearing on this point and I quote it below:

I was interested today in a reprint of your article on the diamondback terra-pin. I think, however, there is an inaccuracy in the distribution, which gives the northern limit as Buzzards Bay.

I was born at Sandwich, Mass., and during my boyhood I knew of a man at West Barnstable, on the northern part of Cape Cod, who caught and sold many terrapin from the so-called Great Marshes, contiguous to Barnstable Har-bour.

I have also known of specimens taken at Sandwich and no doubt they have occurred throughout the Cape. It should be an easy matter to confirm my testi-mony from the reports of men now living at West Barnstable and the fact may be worthy of record.

Distinguishing features: The northern diamondback may be recognized by its relatively wedge-shaped carapace (as seen from above) with the marginals behind the bridge rarely and only slightly curled upward, by its lack of a tuberculate central keel, and by the less convergent edges of the hind lobe of the plastron. As in all *Malaclemys,* the head and legs are not striped, but are blotched or spotted.

Map 13. Distribution of the subspecies of *Malaclemys terrapin.*

Description: The shell is fairly broad, with the widest part posteriorly and the deepest point in the middle. The size is medium, the maximum shell length being very uncertain because the literally hundreds of measurements of diamondbacks that have been published have in most cases given the length in terms of the greatest length of the plastron along the middle to the center of the posterior notch. It is the length as expressed by this measurement by which terrapins are sold—not by weight—and nearly all writers on turtles, if they have occasion to mention lengths of diamondbacks, give this dimension; in a discouraging number of published accounts

Plate 31. Malaclemys terrapin terrapin. A, C, and E: Males; Crisfield, Maryland. *B and D:* Females; same locality.

the plastral length has evidently been confused with total length. Even in the cases in which it is clearly stated that the figures given refer to the median plastral length, the over-all shell length can only be very roughly estimated, since the ratio of the former to the latter varies markedly and erratically. A good-sized female terrapin has a mid-plastral length of 152 mm. (6 inches) and a carapace roughly 170 mm. (about 6.7 inches) in total length. Individuals of this size or larger are referred to in the trade as "counts," and it is at this six-inch plastral length that the fancy prices begin. Terrapins with 178-mm. plastral length (7 inches) are not rare, or at least were not rare in former days; a very large individual measures 200 mm. (about 8 inches) along the plastron and perhaps three-quarters of an inch more in length of carapace. Hay (1904) gave nine inches (228 mm.) in plastral length as the upper limit but did not indicate in which of the races he found this record size, and I strongly suspect it was in one of the Gulf subspecies, which average a little bigger than those of the Atlantic. The maximum shell length for the much smaller males appears to be in the neighborhood of 140 mm. (5.5 inches). Dimensions of a large female from Chesapeake Bay are as follows: length of carapace, 206 mm. (8.1 inches); width of carapace, 152 mm. (6.0 inches); height of shell, 78.5 mm. (1.99 inches); length of plastron along median line, 184 mm. (7.25 inches).

The carapace of the northern diamondback is noticeably widest posterior to the middle; it is usually conspicuously roughened by concentric growth furrows in the laminae and the keel is low and not tuberculate, or even nearly completely absent in some specimens. The color varies from uniform black to light brown and, if the latter, then the laminae are ringed concentrically with darker color. The posterior marginals are rarely slightly upturned. The plastron is oblong, its hind lobe not tapering posteriorly and its color very variable, ranging from orange or honey-colored to greenish gray; in very dark specimens the plastron is usually marked with dark blotches. The head is comparatively small (on the average, smaller than that of *centrata*), flat-topped, narrow, and with the snout more pointed than that of the southern subspecies. The ground color of the soft parts varies from black to gray-green; lighter specimens are usually erratically marked with dark dots and short crooked lines; the lips and top of head are either uniform white or dusky. Regarding variation in coloration Hay (1904) said the following:

It seems very probable that some of the types of coloration indicate local variation, but the species has been so thoroughly mixed by the shipping of large consignments from one place to another that it is doubtful whether anything of this kind could be proved. All the specimens from Connecticut and other northern localities, so far as I have seen, are very light in color, with conspicuous concentric markings but very smooth shells; Potomac River specimens are similar, but have rougher plates; those from the ocean and inclosed bays of the Atlantic coast of the Maryland-Virginia peninsula are, in more than 75 percent of the specimens, very dark and without markings of any kind. The terrapin from Delaware Bay are more like those of more northern localities, but usually present very little contrast in the color markings on the plates of the carapace.

Sexual dimorphism is marked in this turtle. Aside from the striking disparity in size mentioned above, the head of the female is heavier and blunter, the shell deeper, and the tail shorter, with the vent located nearer the body. The carapace of the male is somewhat more pointed behind and the posterior marginals show more tendency to curl upward.

Young individuals are similar in appearance to the female parent, although the shell is slightly more circular and the head relatively larger. The skin of the soft parts is often more densely strewn with dark markings than the adult, and the lips and top of the head are usually dark. The concentric markings of the laminae of the carapace may be very conspicuous and the plastral laminae may be bordered within by one or more black lines. McCauley (1945) measured two hatchlings from Maryland which had the following dimensions: (1) length of carapace, 30.5 mm. (1.2 inches); length of plastron, 26.5 mm. (1.04 inches); (2) length of carapace, 31 mm. (1.22 inches); length of plastron, 26 mm. (1.02 inches).

Habitat: Diamondback terrapins are pre-eminently inhabitants of tidal shore waters, where they range from the open sea into and up the coastal rivers for considerable distances. The preferred habitat is salt marsh, in which the terrapins frequent the deeper tidal creeks and estuaries, hide in the mud on occasion, and range widely over the flats at flood tide. They sometimes make surprisingly long journeys on land, but since in most cases the season is early summer and the wanderers are females, it seems probable that the terrestrial jaunts merely attest to an uncommonly fastidious sense of the fitness of laying sites. Although they are able to tolerate fresh water for indefinite periods, I know of no record of the natural occurrence of a diamondback in completely fresh water.

Habits: There is a tremendous amount of data available on the biology of captive diamondbacks. For nearly forty years the U.S. Bureau of Fisheries, chiefly through its station at Beaufort, North Carolina, has conducted investigations aimed toward the development of effective propagation techniques, and any individual interested in starting a terrapin farm will find the results of these researches invaluable in giving him some notion of his chances for success. The zoologist who searches the literature for data on the life history of these famous animals under natural conditions, however, finds astonishingly little. Moreover, the mixing in the Beaufort pens of the northern and southern stocks reduces the significance of observations made there from the standpoint of strictly natural natural history. As was pointed out by McCauley (1945), almost no details of the life cycle of the northern subspecies are known.

Hay gave a summary of habits of "the northern species" which might refer to either *terrapin* or *centrata,* but since the localities in which he made his own observations were all within the range of the former, it seems probable that most of his data may be taken as applicable to that form.

Hibernation takes place on the bottoms of streams or ponds, where with the first cold weather the terrapins bury themselves in the mud, and whence they may emerge temporarily during any warm spells that occur before winter sets in. Hatchlings evidently go into hibernation immediately or shortly after emerging from the egg and may remain buried until well into their second summer. With the first warm weather of spring the adults emerge, and mate at once.

Breeding: Copulation takes place at night or in the early morning, always in the water, where the small male may be carried about by the much larger female. According to Hay, the eggs are laid during May and June. They are usually five to twelve in number and are buried in nests excavated to a depth of from five to six inches by the hind feet of the female. McCauley measured twelve eggs that had average dimensions of 31.1 by 21.2 mm. (1.22 by 0.826 inches). These were of two complements, one of which was laid in a pound on July 22, 1938. The same writer recorded the hatching of pen-laid eggs on September 15, 1936.

Feeding: The natural food consists mainly of crustaceans and mollusks, although on high tides the terrapins are said to invade flooded grassy lowlands and eat insects. They have been seen to chew tender shoots and rootlets of marsh plants, but the real importance of plant food in the diet is not known. They probably engage in a certain amount of scavenging, to judge from the fondness of captives for cut fish.

Economic importance: This is the most famous of the terrapins, the gourmet's delight, and pound for pound the most expensive turtle in the world. Originally so abundant that eighteenth-century tidewater slaves once struck for relief from a diet too heavy in terrapin, the diamondback gradually found a place on the tables of the privileged, and during the roseate period that extended from the heyday of Diamond Jim Brady to the close of the First World War it came to be surrounded by an aura of superlative elegance as synthetic as a latest Paris fashion. I don't mean to say that diamondbacks aren't good eating. I merely suggest that the difference between a seven-dollar diamondback and a forty-cent soft-shell of the same weight is to a considerable degree a difference in the state of mind of the consumer. The terrapin has an edge on the soft-shell, I suppose, but not that much of an edge. When I moved to Savannah in 1921 I remember Mr. Barbee, the noted terrapin culturist, saying that he had the previous year received top prices of ninety dollars a dozen for terrapins from his huge pound at Isle of Hope. At the same time any number of Negro families on the Georgia islands from Ossabaw down to Cumberland were depending on terrapin as a staple in their diet. By these people the diamondback is regarded in its proper light—as decidedly high-grade victuals and one of the more succulent breeds of turtles, but not conspicuously superior to a number of other creatures that a man can catch in the water.

As difficult to account for as its canonization during the second half of the nineteenth century is the decline in the popularity of the diamondback during the late nineteen-twenties. This question has been much discussed and variously explained. The most frequently suggested factors are the advent of prohibition, liquors of various sorts having been a necessary adjunct both to the preparation of terrapin and to the eating thereof; and the depression, with its attendant curtailment of florid entertaining generally. I think the real explanation lies in the fact that the veneration of the terrapin was an irrational fad and, like any fad, it passed. The fact that it has not passed entirely is due to the innate and incontrovertible succulence of the diamondback.

Although the name of Baltimore has always been most often associated with terrapin cookery, Philadelphia has apparently been the best market. In the early days Delaware Bay terrapins were regarded as the best, and the ascendance of the fabulous "Chesapeakes" was probably due to the growing scarcity of the former. Besides the "Delaware Bays" and "Chesapeakes," dealers used to claim to recognize "Long Island Terrapin," "Connecticuts," and several other local varieties of the northern subspecies, but the consistency with which the varieties were actually correlated with locality is open to question. Certainly, however, market men have always been adept at detecting adulteration of terrapin shipments by addition of representatives of any of the southern forms which today bear technical names and which have always been rated as inferior.

Attempts have been made to introduce the terrapin in at least two places. As long ago as the eighteen-thirties the Prince of Canino tried to transplant them into Italy and twice they have been stocked in San Francisco Bay, the last time as recently as 1943 (Taft, 1944). Coker (1920) mentioned two attempts to establish them in a pond fed by the Mississippi River at Fairport, Iowa, but these failed, as have all other ventures in the culture of terrapins in fresh water.

Although the northern subspecies has unquestionably been reduced in numbers to a pitiful vestige of its primitive abundance, it has probably never faced actual extinction as was once feared, and within recent years it has recouped slightly. Since Chesapeake Bay has for a long time been a center of activity in the industry, it is of interest to see what McCauley has to say concerning population level and conservation measures in that area (1945).

In 1891 the catch had been estimated at 89,150 pounds with a value of $20,559, while in 1920 the total catch was 823 pounds, valued at $1000. At this point the Maryland terrapin appeared to be in serious danger of extinction. The laws were, therefore, revised in 1929 to prohibit the take, catch or possession of "any saltwater terrapin, diamondback terrapin, skillpot or sliders between April 1 and October 31." Terrapin may not be purchased or placed in pens during the closed season, though live terrapins may be placed in pens in the water for propagation during the closed season. The size limit continues to be five inches measured on the plastron. It remains unlawful to interfere with, destroy, or have in possession terrapin eggs from the wild state. No individual is allowed to catch terrapin outside the county of which he is a resident.

Since the passage of this law the amount of terrapin produced has shown a gradual though erratic increase up to the present with a corresponding decrease in price owing to the decreased demand. Today, for various reasons, the terrapin industry in Maryland, though never large, has greatly diminished and the terrapin themselves are probably on the increase.

Finneran (1948a), rather surprisingly, reports the diamondback as fairly common at present along the Connecticut coast, stating also that there is no law there protecting it and that none is needed.

It should be emphasized that any recovery the diamondback may seem to have made is only a relatively feeble trend. The fact still remains that, as one writer puts it, two or three terrapin may nowadays be found in a week in places where formerly hundreds might be seen in a single day. The most encouraging sign is not the slight gains in numbers made by the terrapin, but rather its decline in popularity, since, if the demand should ever again equal that of the Gay Nineties, there would be no hope at all for its survival.

Propagation. For many years it was the paternal aim of the old U.S. Fish Commission to carry on investigations that would place terrapin culture on such a firm basis that the demands of the market might be supplied, the natural populations might be saved from extinction, and numbers of people in the tidewater areas of the United States might be provided with an extra source of food and income. Since 1902, at first on Chesapeake Bay and later at Beaufort, North Carolina, experiments have been carried on by the Fish Commission and by its successor, the Bureau of Fisheries, and an enormous amount of information on the care and breeding of captive terrapins has been collected. Despite the fact that terrapin culture may now be said to rest on a scientific basis, there seems at present little likelihood that it will ever be feasible or popular as a smalltime undertaking, or that there will ever be a terrapin for every pot. The utopian state foreseen in the following manuscript notes, written by Hugh M. Smith in 1919 when he was U.S. Commissioner of Fisheries, today seems a long way off:

Professional gourmands and confirmed epicures may soon be confronted with a question that to them at least will be a momentous one. If the choicest item in their repertoire of things aquatic becomes so common that the way-faring man and other equally impecunious persons may easily purchase and consume it, will the diamond-back terrapin continue to enjoy its vogue? Will gastronomic fashion sanction the further use of the diamond-back as the scintillating gem in the dietary of the elect? Will the millionaire wish to have this creature served on his table if the cost does not exceed that of a baked potato of the crop of 1918?

Actually Commissioner Smith's question had been answered long before, when during an earlier phase of the government's planning on terrapin culture the New York *Morning Telegraph* (May 7, 1912) commented as follows:

The Bureau of Fisheries, in Washington, is so confident that the diamond-back terrapin can be cultivated in the United States for commercial purposes that it will seek an appropriation from Congress for the employment of a terrapin cultivator. It is the idea of the Bureau that diamondback terrapin can be placed within the reach of everybody at a cost less than that of beef. This possibility is a good idea, but to interfere with the present exalted position of the diamondback and lower him to mediocrity will be a distinct crime against epicureanism. Diamondback terrapin never was intended for vulgar palates.

Anyone interested in raising terrapins should consult the various articles by Hildebrand, Hildebrand and Hatsel, Barney, and others in which a huge amount of information gathered by the government stations is made available. As I mentioned earlier, the stock with which most of the government researches were carried out is a mixture of *terrapin* and *centrata,* which since the beginning of the experiments have been combined in the pens. A brief résumé of the results of these investigations follows.

The inclosure: The terrapin pen should be located on a gently sloping shore where high tide leaves a small amount of dry land and low tide leaves a small amount of water, although to allow occasional complete drainage and to insure ease of cleaning the seaward wall should be located just above mean low water with a drain installed in its base. The walls should be of concrete or masonry, wooden walls being subject to damage by shipworms and often lasting only a short while. Partitions within the main walls, for segregating young from adult terrapins, are best made of wood. The outer walls should be solid up to a point high enough to retain the desired low-water level in the pen, and from here on up gates should be made to permit tidal movements. The interior should be divided into a large area for mature terrapins and one or two sections for young. In the section for adults there should be on the shoreward side an egg-bed, in which well-retained and firmly packed sand stands about one foot above the level of extreme high tide, and with a gradual slope leading up to the surface of the sand; this should provide one square foot of surface for every adult female. Since the young are expert climbers, in their enclosure boards or strips of sheet metal should be laid along the tops of the walls, projecting inward three or four inches from the inner wall. Fresh water should be provided, although the necessity for it has not been definitely demonstrated. The principal enemies of the terrapins at Beaufort have been rats, which at every opportunity eat the eggs and young; and crows are sometimes harmful. The possibility of muskrats burrowing into the pen and releasing terrapins exists.

Reproduction: The captive females at Beaufort lay between May 6 and July 31, a given individual laying from one to five times during a season (usually one to three times) and depositing an average of eight eggs each time. The number of eggs laid during one season by a single terrapin varies from five to about thirty, young females producing fewer than old ones. The incubation period varies with the season and weather but is usually about ninety days. The percentage of eggs hatching may be as high as ninety, although this varies considerably; best fertility may be expected when the sex ratio in the pens is about one male to five females. One mating may produce fertile eggs for four seasons, but the percentage of fertile eggs declines after the first year following copulation.

Feeding: Terrapins have been fed fish, blue crabs, oysters, corn meal, cabbage, and turnips. Salt fish may be given to the young in winter. The equivalent of a year's growth may be obtained by placing hatchlings in a brooder house in which a temperature of about 80° Fahrenheit is maintained and feeding them throughout the winter.

Growth: The period of development leading to sexual maturity is very variable, in nature possibly being as much as seven years, but it may be considerably accelerated in pen-raised terrapins by winter feeding. Under this regimen a female may lay at the end of her fourth year, and females regularly reach a salable size (five inches) within five years. The males tend to mature a year later, and since their maximum size is less than the marketable minimum of five inches they are of little or no commercial importance. The winter feeding not only accelerates development but increases the survival rate, and if combined with good care in other respects may allow as many as 60 per cent of a hatching to grow to maturity. In captive terrapins the maximum egg production is estimated to occur at an age of about twenty-five years, and the upper age limit is probably well over forty years.

Causes of loss: Terrapins are lost through attacks by such enemies as rats, crows, herons, and snakes; by the advent of floods; through the remarkable capacity of the young to climb vertical surfaces; and through two diseases, one a sore-producing malady thought to be of bacterial origin and the other a dietetic disorder called "soft-shell." There is no known treatment for the former, but the latter is nearly always associated with a disinclination to eat or with an inadequate food supply, and is readily corrected by providing proper diet.

Hybrid strains: In a sense, the entire Beaufort stock is hybrid, since, as noted above, the northern and southern races were combined there a long time ago. For lack of comparative data it is impossible to say whether this crossing has had any effect on the culturability of the animals involved. A more drastic cross was made in 1914 at Beaufort when the stock there was bred to the larger *M. t. littoralis* from Texas in an effort to gain size. Results of the attempt were not encouraging, however, since the hybrids had a much slower growth rate and matured later than the original strain, over which they showed no advantages.

Southern Diamondback Terrapin *Malaclemys terrapin centrata* (Latreille)

(Plate 32. Map 13.)

Range: From the vicinity of Cape Hatteras southward to southern peninsular Florida, where it meets and intergrades with the mangrove terrapin, *M. t. rhizophorarum.*

Distinguishing features: Although very closely related to the northern diamondback and morphologically interconnected with it by all sorts of variations and intergradations, the terrapin population of the southern Atlantic coast is on the whole a recognizable race. In it the sides of the shell are more nearly parallel, the carapace having from above a less wedge-shaped and more broadly ovoid appearance than that of the northern subspecies. The sides of the plastron posterior to the bridge tend to converge behind, while the plastron of *terrapin* is almost perfectly oblong. The head of the female, at least, averages heavier than that of the northern race and the snout is blunter and the central keel lower. In both these terrapins the dorsal keel is without tubercles. Most terrapin dealers (and several zoologists) agree that there is less tendency in *centrata* to retain the concentric furrows of the dorsal laminae, and that as a result the shell of this form is smoother above in a majority of cases. Another tendency that appears to characterize the southern subspecies, but which varies widely both sexually and individually, is the degree of upward curl of the posterior marginal rim, this being on the average more noticeable in *centrata* than in *terrapin.* The average size of the former is slightly less than that of the latter.

Sexual dimorphism is essentially the same as in the northern terrapin. The male shows marked difference from the female in the shape of the shell, the margins converging behind in almost straight lines instead of meeting to form a curved posterior edge as in the female. Differences in the tails, heads, and curvature of the upper marginal surface were mentioned in the discussion of sexual characters of *terrapin.* Males of this subspecies are said to mature at a plastral length of about 80 mm. (about 3.2 inches) and females begin to lay at plastral lengths of about 127 mm. (about 5 inches).

The young are similar to those of *terrapin* but perhaps somewhat smaller at hatching, the usual lengths of hatchlings being given as less than 30 mm. (1.18 inches).

Habitat: This is typically an estuarine turtle with apparently the same preferences and tolerances regarding habitat as were mentioned above for

Plate 32. Malaclemys terrapin centrata. A and B: Male; Isle of Hope, Savannah, Georgia. *C:* Hatchlings; Beaufort, North Carolina. *D and E:* Female; Isle of Hope, Savannah, Georgia.

the northern subspecies. I have set stop nets for this terrapin many times in the marshes of coastal Georgia, and best results were nearly always obtained when the net was placed across the deeper parts of good-sized tidal creeks (50 to 200 yards wide) over shell bottom and near oyster banks.

Habits: In Georgia and North Carolina the diamondbacks usually begin to bury themselves in the "pluff mud" of tidal flats, often in or near little trickles of water from the exposed marshes, in late October or early November, although any warm day is sufficient to bring them out again and they may be active off and on all through the winter. From the first of March on they are rarely in hibernation, although they may cover themselves shallowly with mud during low tides, since their feeding activities are curtailed at this time. This habit has given rise to one of the more popular methods of catching terrapins, by which a boat is poled up the smallest tidal creeks that it can navigate at low water while the operator probes likely-looking spots with a long pole or iron rod.

Breeding: Copulation takes place in the water in early spring. The female is not overly fastidious in her choice of a nesting ground and may lay anywhere from an open beach to a low tidal bank only a few inches above high water in a brackish estuary or river and where the soil has only a slight admixture of sand. At Wolf Island, Georgia, terrapins often come up from the heavy surf to lay from May through July; I used to find many nests there incidental to quests for loggerhead eggs. The female usually roams for a long distance among the dunes before finally selecting a place to dig. If she finds herself in marsh-bordered creeks at laying time she sometimes strikes out across country to look for proper nesting soil, and I have found such individuals at least a half mile from water. I have found 7 and 9 eggs in nests; in four nests Coker (1906a) found respectively 4, 5, and 6 eggs. The natural incubation period is apparently unknown.

Feeding: Southern diamondbacks are omnivorous, with a definite preference for animal food. They feed mostly at high tide, at which time they cruise the flooded marshes picking snails from grass stems, foraging along the bottom for small (or dead) crabs and thin-shelled mollusks, and nibbling at various kinds of aquatic plants. In captivity they eat any kind of sea food. I once had the job of feeding a thousand terrapins and more than once chopped for them the bodies of sharks and rays, which they ate readily; on one occasion when a protracted storm made fishing impossible we kept them in good spirits by opening for them dozens of big cans of war-surplus sardines in tomato sauce! I have elsewhere (1940) mentioned the development of an interesting conditioned response in this colony of terrapins, and it may bear repetition here:

The inclosure in which they were kept included a section of a shallow salt-water creek and a hundred feet of the steep bank of the stream. Each day when I came in with a boatload of fish for the terrapins it was my habit to chop the day's ration on a little dock three hundred yards from the pen. Although I rarely came in at the same time on two consecutive days, I always found the creatures scrambling over each other a foot or more deep around the gate when I opened it to feed them. I eventually discovered that the distant noise of the chopping was responsible for the enthusiastic reception, and that the terrapins could be made to crowd around the gate at any time by pounding on the dock.

I have subsequently noted that Coker recorded a similar reaction by terrapins in the Beaufort pens.

Economic importance: This race has always been regarded with less esteem than the northern form, but it is nevertheless sent to market wherever procurable and has been greatly reduced in numbers in many areas. Engels (1942) regarded it as "still numerous enough to form a staple in the natives' diet" at Ocracoke Island, North Carolina; while Jopson (1940) said that although it is fairly common in the marshes of Georgetown County, South Carolina, it is ardently hunted and only the size limit has saved it in that region.

Mangrove Terrapin *Malaclemys terrapin rhizophorarum* Fowler
(Plate 33. Map 13.)

Range: Florida Keys westward at least to the Marquesas Keys.

Distinguishing features: Although this subspecies is in a sense merely a population of intergrades between the diamondback terrapins of the Atlantic coast and those of the Gulf, it nevertheless appears to be a fairly homogeneous stock occupying a considerable territory in the tropical waters of the long archipelago of the Florida Keys. It has the freakishly bulbous central keels of the Gulf terrapins, but is very unlike the neighboring subspecies, *macrospilota,* on the west coast of the peninsula in lacking the yellowish translucent windows in the laminae of the carapace which so strikingly differentiate that form. The nodose central keels set it off from the Atlantic diamondbacks. Perhaps more or less fortuitously, it resembles most closely the terrapin of the coast of Texas, but from this it may usually be distinguished by the pattern of the under surface, which includes bands of black pigment along the seams of the lower marginals and frequently of the plastral laminae as well. The shell is strongly oblong, apparently more so than that of either *centrata* or *macrospilota*.

Sex differences have not been studied in detail, but they appear to be essentially as described for *M. t. terrapin*. The hatchling and young are unknown.

Plate 33. Malaclemys terrapin rhizophorarum. Female; Card Sound, Dade County, Florida. Preserved specimen.

Habitat: I have seen the mangrove terrapin only in mangrove swamps and mangrove-bordered creeks and around the shores of the hundreds of little keys in Florida Bay. Several people have called my attention to its occurrence in spring-tide and hurricane pools inland on dry keys and isolated from permanent water.

Habits: Nothing is known of the habits of this race. It is occasionally used for food locally.

Florida Diamondback Terrapin *Malaclemys terrapin macrospilota* W. P. Hay

(Plate 34. Map 13.)

Range: Gulf coast of peninsular Florida, intergrading with *pileata* in the panhandle and with *rhizophorarum* in Florida Bay.

Distinguishing features: This terrapin is similar in appearance to both *rhizophorarum* and *pileata,* differing, however, in one constant and very conspicuous character, this being a bright yellow or orange central area in each of the laminae of the carapace. These yellow spots are due to unpigmented patches in the laminae overlying areas of light color, and on the generally deep black carapace they produce a most striking and distinctive effect. Because of them the Florida diamondback is the most easily identified of all the subspecies of *Malaclemys*. In the size and shape of the head, the form and sculpture of the carapace, and general body proportions this race is quite similar to the Mississippi diamondback. The average size may be slightly less than that of *pileata*, since of the hundreds of specimens of *macrospilota* that I have seen none has attained shell lengths equal to those

Plate 34. Malaclemys terrapin macrospilota. A: Female. *B and C:* Male. Both from Englewood, Florida. *D:* Female; Cedar Key, Florida. (Courtesy Ross Allen.)

of some of a much smaller series of *pileata* examined by me. I can detect no difference in size range between *macrospilota* and southern *centrata*. A large adult female from Hillsborough County, Florida, has the following dimensions: carapace length, 195 mm. (7.66 inches); carapace width, 138 mm. (5.43 inches); carapace depth, 81 mm. (3.18 inches); carapace length (midline), 165 mm. (6.50 inches); head width, 26.6 mm. (1.05 inches).

Sexual divergence is apparently identical with that in the other races.

The knobs on the central keels are more pronounced in the males, which in all these turtles tend to retain juvenile characters to a greater extent than the females.

The hatchling is, as far as I can make out, indistinguishable from that of *M. t. pileata.* The carapace does not show the characteristic yellow-spotted pattern, since in the young each entire lamina is transparent (representing thus the "yellow spot" of the older lamina) and its whole shell is an even light horn color, somewhat lighter below than above and with each dorsal lamina narrowly bordered with black. The middorsal elevation is barely discernible on the first central; it is an obtuse keel on the second, a high and nearly complete hemisphere on the third, and a nearly perfect, slightly flattened spheroid on the fourth; on the fifth there is no elevation. The greater portion of each of these dorsal tubercles is pigmented with black. The soft parts are mostly grayish, sprinkled finely with black dots except on top of the head. The horny coverings of the jaws are yellowish. Dimensions of a hatchling from Charlotte County, Florida, are as follows: shell length, 32 mm. (1.26 inches); shell width, 27 mm. (1.06 inches); shell depth to top of keel, 16 mm. (0.63 inches); head width, 9.8 mm. (0.386 inches).

Regarding the discovery of this striking race of the diamondback, Hay (1904) commented as follows:

It is quite surprising that this beautiful species has escaped the naturalists for so many years, but perhaps the growing scarcity of diamond-backed terrapin in northern waters has only recently led to the appearance of this animal in our markets. My first specimens were selected from a barrelful which had been sent from Charlotte Harbor, Florida, to a dealer in Washington. In the summer of 1903 I noticed a considerable number of both males and females in one of the pounds at Crisfield, where they had been received from Sand Key, Florida.

Habitat: This terrapin is found in passes and mangrove-bordered lagoons and creeks, wandering offshore and well up the coastal rivers. It is interesting to note that the extent of shoreline along which the Florida diamondback intergrades with the Mississippi race begins where the northern range of the red mangrove terminates. Both this subspecies and *rhizophorarum* are essentially mangrove-swamp terrapins.

Habits: Apparently nothing at all has been published concerning the habits of this turtle, but I have recorded some random observations.

On the sea beaches of Collier and Charlotte counties I have seen dozens of nests of *macrospilota* in April and June, mostly among or behind medium-sized dunes and in soil ranging from fine white sand to coarse shell fragments.

The great majority of these had been dug out by raccoons, which had sucked the eggs and had left the shells beside the nest. The number of eggs varies from four to ten, and the nest is a rather crudely formed flask-shaped cavity of from four to six inches in depth. As in the case of *centrata* on the Atlantic beaches, the females of the present race often wander for astonishing distances behind the dunes before laying, and frequently dig one or more trial holes before making one to their satisfaction. In the fresh nests that I have dug out, the sand was more moist than that of the surrounding area, perhaps indicating that cloacal bladder water is released in the cavity. It is possible that the majority of females lay on the inner beaches and shores of the lagoons, but here, due to the presence of grass and hard soil, their trails are less evident and my own observations were very few.

I once had the opportunity of observing the last stages of an illegal stop-netting spree in Charlotte County waters in which two miles of mangrove-bordered tidal creeks were dragged with a seine while the far end was closed by a set net. I was astonished to find in the varied and abundant haul no more than 14 diamondbacks. I am quite sure that this is not a valid representation of the population of the water seined and imagine that several times this many terrapins escaped under the lead line of the net. The stomach contents of this series consisted almost wholly of rather finely crushed fragments of shells, about 90 per cent of which was made up of *Anomalocardia cuneimeris,* according to the identification of Dr. William Clench of the Agassiz Museum.

Economic importance: Although this terrapin has for years been sent north to market, it is regarded as inferior to the "Chesapeakes" and to the "Biloxi Terrapin" and does not find a ready sale. It is easily spotted as something different by its striking coloration. Those who have eaten it say that its flesh is "somewhat gelatinous and entirely lacking in the qualities which have made the northern species famous." All this appears to me to be unwarranted calumny. As to the flesh being gelatinous, I have found the meat to be no more so than that of the Atlantic form, after eating far more than my share of both, and the abundance of gelatinous substance in the shell and other parts, instead of being a ground for deprecation, is actually the predominant feature that distinguishes turtle from other, inferior, edible animals. As far as the other "qualities that have made the northern species famous" are concerned, as I have said earlier, I think they exist largely in the minds of aspirant gourmets who never tasted *macrospilota*.

Mississippi Diamondback Terrapin *Malaclemys terrapin pileata* (Wied)

(Plate 35. Map 13.)

Range: From eastern Louisiana, where it intergrades with *M. t. littoralis,* to the Florida panhandle, where it intergrades with *M. t. macrospilota.*

Distinguishing features: This subspecies is identified by the tuberculate central keel, the uniform brown or black carapace, and the regularly dusky upper lip and upper surface of the head.

Description: The shell is somewhat narrower and even more straight-sided than that of the southern diamondback. The size is a little greater than that of the Atlantic forms and probably a little smaller than that of the Texas subspecies, although the largest specimen of *Malaclemys* I ever saw was a huge female *pileata* from Grand Island, Louisiana, in the Louisiana State Museum; the measurements of this specimen were sent me through the kindness of Stanley C. Arthur, director of the Museum. These measurements are as follows: total length of shell, 238 mm. (9.37 inches); greatest width of shell, 180 mm. (7.08 inches); greatest depth of shell, 104 mm. (4.10 inches); greatest length of plastron, 180 mm. (7.09 inches); greatest width of head, 60.2 mm. (2.36 inches).

The above specimen shows a most extraordinary exaggeration of a tendency in *Malaclemys,* as well as in *Graptemys* and even rarely in *Pseudemys,* toward a monstrously enlarged and swollen head and facial region in the old female. The carapace of the Mississippi diamondback has a strongly revolute rear margin, and the central keels are well marked and, except in the oldest individuals, tuberculate, the knobs usually being expanded at their tips. The laminae are usually deeply grooved concentrically, and the notch in the precentral region is deeper than in the Atlantic races. The color of the carapace is uniform dark brown or black except for the upturned edges of the marginals, which are orange or yellow. The plastron is yellow, sometimes clouded with dusky. The ground color of the head and neck is gray or gray-green, brightest on the sides and lower part of the head and everywhere strewn with black or dark brown spots and specks. There is a black patch on the top of the head that covers the area from nose to nape and from orbit to orbit. The broad horny covering of the upper lip

Plate 35. Malaclemys terrapin pileata. A, B, D, and E: Males. C: Female. All from southern Louisiana.

is sometimes white but more often heavily marked with darker color. There is usually a dark streak extending from nose to orbit between the lip and the dark cranial patch. The skin of the legs and tail is nearly black, due to the heavy sprinkling of black spots on its surface.

Sexual differences are as described for *M. t. terrapin*. In addition, the knoblike expansions of the central keels are stronger in the male and his plastron is more often clouded with brownish black. The dark lip markings, compared by Prince Maximilian zu Wied (who originally described this terrapin in 1865) to a moustache, is a more constant feature of the male than of the female (appropriately so, one might remark).

Habitat: The Mississippi diamondback occurs in marshes and estuaries. Like the Atlantic terrapins, this race is partial to channeled tidal waters. It is surprisingly rare in the enormous areas of shallow grassy flats that lie between the mainland and the offshore islands.

Habits: Apparently no study of the habits of this terrapin has been made.

Economic importance: This subspecies is called by dealers "Biloxi Terrapin" or "Gulf Terrapin" and is said to rank next to the northernmost race in the esteem of expert victualers, among whom it has for many years commanded a good price. It seems strange that this form should be regularly preferred in the trade to the terrapins of the coast of Georgia and the Carolinas, but such is the case.

Texas Diamondback Terrapin *Malaclemys terrapin littoralis* W. P. Hay

(Map 13.)

Range: Western Louisiana and the coast of Texas, at least as far south as Brownsville and almost certainly extending well along the Mexican coast. Agassiz stated that it reached "South America." While this extension of the range only gives one to wonder what the basis of the inclusion of South America might have been, one hears frequent and apparently reliable verbal reports of the presence of diamondbacks in Mexican waters. Published records of their occurrence in Yucatan and Guatemala derive from misidentifications of a small specimen of *Geomyda* from El Petén, Guatemala. It seems quite possible, however, that the diamondback may someday be found all the way around to the Yucatan peninsula.

Distinguishing features: This subspecies is distinguished by a tuberculate, knobbed keel (present in young, absent in old); a higher shell with deeper bridge and with its highest point farther back than in the other races;

a deeper gular notch; more obtusely rounded median marginal edges and strongly revolute hind marginals; narrower plastron, with more convergent rear edges; a consistently white upper lip; and a uniformly colored carapace and plastron. This turtle shows a high incidence of a laminal abnormality, one or more of the centrals very frequently being divided into two parts by a longitudinal seam. In some batches a majority of the individuals show this condition. The inguinal lamina is also quite variable, in a series of 50 terrapins having been present in 26 and absent in 24.

There appear to be no sexual differences in *littoralis* that have not been described for *pileata*.

The young of the Texas diamondback are most unusual in appearance. In the words of Hay (1904), they

are very remarkable, and in the absence of any other distinctive character would serve to separate the species from *M. centrata*. They are much larger than the young of *M. centrata*, having probably twice the bulk, and seem much more vigorous and lively. The first vertebral plate is raised to form a broad, low carina; on the second plate the elevation is greater and stands out as a smooth boss on the otherwise smoothly wrinkled plate; the elevation of the third plate has the form of a hemispherical button with a well-marked constriction around the posterior half of the base so that it stands up prominently above the plate posteriorly but anteriorly slopes into it; on the fourth plate the elevation rises into a knob-like protuberance from a base which is constricted all around; the tubercle of the third plate is usually the broadest, but the one on the fourth plate is the highest of the three; all are smooth and polished while the plate upon which they rest is finely wrinkled. The fifth vertebral plate is flat, or with only a trace of an elevation. The color of these young specimens is brownish yellow or horn color, and the margins of all the plates of the carapace are thickened and darker than the remainder of the plate. The centers of the costal plates usually bear a small dark dot, around which there is sometimes a narrow dark ring.

Habitat: The habitat of this terrapin appears to be the same series of salt and brackish channels and marshes inhabitated by *pileata*.

Habits: The habits of this subspecies have apparently not been studied. Brood stock used for hybridization experiments at the Beaufort station (see above under *M. t. terrapin*), including 11 females about 30 years old, produced 270 eggs during the 1919 season after being bred to local males.

Economic importance: Although a larger animal than the Atlantic terrapins and although sold locally, *littoralis* is said to be decidedly inferior to *terrapin* in eating qualities and is not in demand in northern markets.

Genus *GRAPTEMYS*

The Map Turtles and Sawbacks

The relationship between this group of fresh-water turtles and the diamondbacks is so close that the two should perhaps be placed in a single genus. It is probably their mutually exclusive habitats that have led to what appears to me to be overemphasis on extremely slight group differences and the retention of separate genera for them.

The characters of *Graptemys* are those given above for *Malaclemys;* the two may be distinguished as follows:

In *Graptemys* the neck and sides of the head nearly always have light stripes on a darker background instead of dark markings on a light background as in *Malaclemys*. The hind edges of the shell are usually flaring in *Graptemys* and curled upward in *Malaclemys*. The laminae of the carapace frequently show more marked concentric sculpturing in *Malaclemys* than in *Graptemys*.

KEY TO *GRAPTEMYS*

1   Markings on top and sides of head involving large yellow areas, the median stripe from the snout very broad and tending to fill the space between the orbits, and joining at its upper rear edge a large postocular patch that extends downward and forward beneath the eye and which, from its upper rear border, sends a stripe back along the neck; centrals with a tuberculate keel greatly pronounced in males and young; plastron without an intricate many-lined dark pattern 2

1′   Head markings composed of stripes, or of stripes and spots, but not arranged as described above and not filling nearly all the space between the orbits, the median stripe on the snout thin and not confluent posteriorly with a postocular blotch; plastron with or without a complex dark figure 3

2(1) Each lateral lamina with a thin-lined, U-shaped mark occupying its central part, the open end of the figure against the rear seam of the lamina; masticating surfaces of upper jaws very broad, even in males and young, almost if not quite meeting in the median line; each upper marginal, if marked at all, with a crescentic, thin-lined light mark extending from the lower

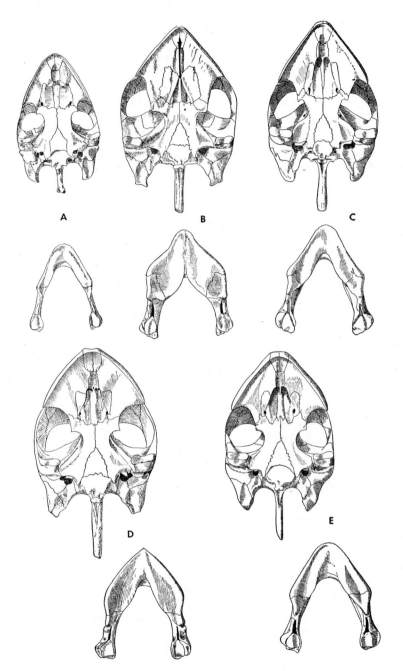

Figure 13. Ventral views of skulls and dorsal views of mandibles of some turtles of the genera *Graptemys* and *Malaclemys.* A: *Graptemys p. pseudogeographica.* B: *G. geographica.* C: *G. pulchra.* D: *Malaclemys t. terrapin.* E: *Graptemys p. kohnii.*

187

fore corner to the upper rear corner of the lamina; Apalachicola
drainage in northwest Florida and southern Georgia . . . .
. . . . *barbouri* (p. 189)

2′   Pattern of carapace not of isolated U-shaped figures but reticulate, two or three right-angled intersections of its light lines
occurring toward the center of each lateral lamina; masticating
surfaces of upper jaws not markedly broadened and not nearly
meeting in the median line; each marginal with a many-lined
ocellate figure; southern Alabama westward . . . . *pulchra* (p. 212)

3(1′) Chin with a median stripe or ocellated spot 4

3′   Chin with a broad transverse stripe; each lateral with one complete light ring; southern Louisiana . . . . *oculifera* (p. 200)

4(3) Side of head behind eye with or without a crescentic yellow
mark extending downward from one of the dorsal neck stripes;
if there is such a mark, it does not continue so far downward
and forward that all other stripes are excluded from the orbit;
young with or without a complex dark plastral figure; second-
order serrations of the rear margin of the shell not so deep as to
impart an extraordinary, spiny appearance 5

4′   Side of head with a crescentic yellow mark behind the eye; this
is the forward and downward extension of a dorsal neck stripe
and it extends far forward beneath the eye, to exclude from the
orbital rim all other head stripes that approach it from the rear;
young with a complex plastral figure and with the emargina-
tions between the rear marginals and those in the centers of
each marginal so deep and so nearly equal as to give the appearance of a row of marginal spines; lower Mississippi Valley
north to northern Arkansas and northwestern Tennessee, westward into eastern Texas . . . . . . *pseudogeographica kohnii* (p. 207)

5(4) Head of female large, its width usually contained in length of
carapace fewer than 5½ times; alveolar surfaces of jaws broad,
those of the upper jaws approaching each other in the median
line and those of the mandibles greatly expanded, scoop-
shaped; plastral pattern of young not a complex dark figure;
keels of second and third centrals present or absent, but if present, then convex forward; postocular spot usually oriented
longitudinally; precentral elongate and narrow; Mississippi
Valley north to Lake Superior and Quebec, east to eastern Tennessee, western Virginia, Pennsylvania, and New York . . . . . .
. . . . *geographica* (p. 195)

5′   Head of female small, its width usually contained in length of
carapace more than 5½ times; alveolar ridges relatively nar-

row, those of the upper jaws not nearly meeting in the median line and those below not greatly expanded in the middle; plastral pattern of young usually a complex dark figure like that of *Chrysemys p. bellii;* keels of second and third centrals usually strong and concave anteriorly; postocular mark, if present, not a spot that is elongated longitudinally; precentral usually as broad as long; central Texas and upper Mississippi, mostly in its western part, north to Wisconsin, east to Indiana 6

6(5') With a transverse expanded blotch or line behind the eye that is the downward extension of a dorsolateral head and neck stripe; upper Mississippi Valley, from Kentucky northward, westward into Oklahoma, Nebraska, and South Dakota ....
.... *pseudogeographica pseudogeographica* (p. 202)

6' With an expanded spot behind the eye that is not continued backward as a dorsolateral head stripe; central Texas ....
.... *pseudogeographica versa* (p. 210)

Barbour's Sawback Turtle *Graptemys barbouri* Carr and Marchand

(Plate 36. Fig. 14. Map 14.)

Range: Southwestern Georgia and the Florida panhandle, in the Apalachicola River drainage. Until recently known only from the Apalachicola and Chipola rivers in Gadsden, Jackson, and Calhoun counties, Florida, this turtle was taken in 1948 in a tributary of the Flint River, Baker County, Georgia, by Crenshaw and Rabb (1949), thus confirming an old record of Brimley's (1910) for the same county.

Distinguishing features: This sawback is distinguished by its very broad alveolar surfaces, extensive yellow areas on the head, strongly tuberculate central keel, simple or nonexistent plastral pattern, reduced markings of the marginals, and by the pattern of the carapace.

Description: The shell is rather high and tapered behind, with the highest point anterior to the middle and with slightly serrated hind margin. The largest specimen measured, a female, had a shell 267 mm. (10.5 inches) long; measurements of the type specimen, a mature female from the Chipola River, Florida, are as follows: length of shell, 231 mm. (9.10 inches); width of shell, 184 mm. (7.25 inches); height of shell, 94 mm. (3.7 inches); width of head, 59 mm. (2.32 inches). The carapace is smooth and with a prominent, tuberculate keel in most specimens, the part on the second and third centrals slightly concave outward. Ground color of the carapace is dull olive, with pale yellowish white markings on the lateral

Plate 36. Graptemys barbouri. Upper and center: Old female; Baker County, Georgia. (Courtesy John W. Crenshaw, Jr.) *Lower:* Young; Chipola River, Florida. (Courtesy Ross Allen.)

and upper marginals. A roughly U-shaped figure occupies the central portion of each lateral, the open side of the U against the posterior seam. On each upper marginal a crescentic, thin-lined mark extends from the lower fore corner to the upper hind corner, these markings tending to fade behind. The centrals are unmarked except in very young specimens, in which a suggestion of the lateral figure is evident. The strong tubercles of the dorsal ridge are dark brown in color. The plastron is very flat and truncate anteriorly, yellow to whitish in ground color. The juvenile plastral pattern consists of a narrow black band along the posterior margin of each lamina except the anals, there being no pigmentation along the longitudinal seams. These markings tend to disappear in old females.

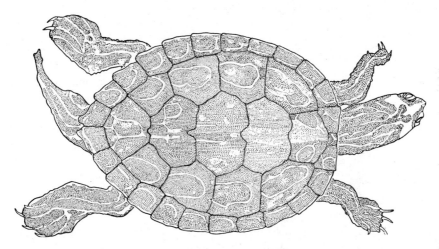

Figure 14. Graptemys barbouri. Male from Chipola River, Florida, showing markings of head and carapace. (Drawing by Norma Marchand.)

The head of the male is small and narrow, that of the female greatly enlarged and with the snout short, and the rami of the upper jaws meeting at an obtuse angle. The alveolar surfaces of both jaws are extremely broad in the female, the symphysis of the upper extending posteriorly almost to a point between the angles of the mouth. In males and juveniles these surfaces are broad but are only slightly in contact, if at all. In old females the cutting edge of the lower jaw may wear away; the edge may be flush with the masticating pavement.

The ground color of the soft parts is very dark brown or black; the markings are cream-colored or nearly white. The markings of the head comprise three interconnected, usually broad but variable areas of light color, one of these on the snout between the eyes and one behind each eye. The median blotch is continued anteriorly as the sagittal stripe, while posteriorly it is in most specimens broadly or narrowly connected with the postocular color areas, each of which sends anteriorly beneath the eye a rapidly narrowing stripe. The 12 to 14 stripes on the nape are narrowly connected with the postocular blotches. The ventral horny surface of the lower jaw is marked transversely with a broad band. There are six stripes on the anterior surface of the fore limb; that going to the second toe is the broadest.

The sexes are very different, the male being small and with more of the juvenile features of coloration and shell shape, in addition to having a much bigger tail. Some body proportions are compared in the following table of ratios from Carr and Marchand (1942):

Sex	$\dfrac{length}{height}$	$\dfrac{length}{width}$	$\dfrac{length}{head\ width}$
Male	2.0–2.5	1.1–1.3	5.5–6.2
	average 2.2	average 1.2	average 5.7
Female	2.0–2.4	1.1–1.3	4.0–5.4
	average 2.2	average 1.2	average 4.5

Markings of young females are similar to those of males, but these become more or less obsolescent with age. Males appear to mature at a length of 95 mm. (3.74 inches) or somewhat less, and the shell of the largest male on record was 118 mm. (4.65 inches) long. The largest female known is the 10.5-inch specimen mentioned above.

The young have the tubercles of the second and third centrals markedly elongated. The shell is relatively shorter and higher than at maturity and the markings are more intense. The hatchling is not known.

Habitat: This sawback prefers creeks and rivers, ideally with clear water and rocky bottom in the vicinity of snags and fallen trees. This turtle is very aquatic and shy, and ill-at-ease when sunning.

Habits: Although abundant in a few restricted localities, this turtle was unknown until recent years and the only data on its life history are random field observations.

I recall as one of my most pleasant zoological experiences our discovery of a populous colony of this then-undescribed sawback. It was during the period when Lewis Marchand was developing his remarkable talent for water goggling. The introduction of this, to us wholly new, technique had opened up unsuspected and thrilling vistas of zoological exploration, and for a time it seemed that the greater part of the Department of Biology at the University of Florida might succumb to the "big shakes" in the effort to keep up with Marchand, who is as little affected by low temperatures as a sulphur-bottomed whale. But before the war came along and interfered with our goggling (and probably saved several of us from chronic rheumatism and sinusitis) we had a wonderful time with water goggles.

The day we found *barbouri,* three or four of us were standing shivering and enviously watching Marchand as he wallowed and snorted about, searching the bottom of the Chipola River through his goggles. Suddenly he broached with a whoop and held aloft a small turtle with the high keel of an unmistakable *Graptemys.* This was not only the first *Graptemys* I had seen in Florida but it held high promise of being something strikingly new, and we all plunged forthwith into the river in a great state of frenzy. Downstream and on the far shore I saw a turtle slide into the water from the limb of a half-submerged tree. I thrashed across the intervening space, went under, and there, some six feet down, saw the turtle perched on the forked end of a twig like a bird, its neck craned and its eyes staring. Curbing the impulse to seize the prize, I hung suspended in the water for as long as my breath would allow, gloating, and returning the rather stupid gaze of the animal with the utmost affection. When I finally reached my hand toward it the turtle made no effort to move. From this spree we took home 16 sawbacks.

Several weeks later Marchand went with William McLane, Alfonse Chable, and Bert Schultz on a four-day canoe trip to learn something of the distribution of the new turtle along the river. A part of his field notes follow:

November 27–30, 1941. Put in at Bellamy's bridge, about 10 miles north of Mariana, Jackson County. During the first two days caught two saw-backs. The morning of the third day we took 18 within a distance of 50 yards. Good collecting water must be at least four or five feet deep and with a particular type of limerock bottom in which there are numerous pits and depressions; under these conditions the turtles are very numerous. Where the bottom is covered with large boulders or is broken into big fragments turtles are scarce and we saw almost none on sandy bottom. This tends to concentrate large numbers of turtles in small stretches of river, and within a hundred yards on the right type of bottom we saw about 30 individuals.

The turtles were resting on the bottom, usually in the depressions, with their heads and feet pulled in but with eyes open. In many cases silt covered their backs and gave the impression that they had not moved for days. The first turtles were all goggled up, but the temperature of the water soon made this unpleasant [an unprecedented admission for Marchand] and subsequently we brought them up by a light poke with a sharp gig. About 60 were taken and perhaps that many more seen. Of the turtles seen on the bottom, as most of them were, all but one were motionless. These turtles seem lethargic and when under water make almost no effort to escape when approached. The Mobilians seen were quite active, and were present in the ratio of about 1 to 2 or 3 *Graptemys* [the Mobilian is *Pseudemys floridana mobiliensis*].

Turtles were caught from 10 miles north of Mariana downriver to the Clarksville-Blountstown road.

It seems probable that the large female specimen of this species found dead on a road in Newton County, Georgia, by Crenshaw and Rabb (1949) may have been looking for a nesting site, but the authors mention no eggs. In Florida we have been after *barbouri* only in the wintertime and know nothing of its feeding or breeding habits. Although the form of the jaws of the mature females would appear to indicate a diet of hard-shelled animals—probably mollusks—the only definite information on this point is contained in a letter from Mr. John Crenshaw, dated February 1, 1950:

Over the past weekend I was fortunate enough to obtain two adult female *G. barbouri* from the Flint River adjacent to Newton, Georgia. The intestinal tract was removed from both of these within 12 hours. Mollusc shells and shell fragments constituted by far the greatest part of the material found. These were present in such abundance as to cause distension of the alimentary canal at intervals. Besides these shell particles and molluscan soft parts in various stages of digestion, three or four minute plant fragments and a single quartz fragment were found.

The shell fragments seem to indicate that at least two species of gastropod and one pelecypod were represented. From the size of the shell pieces I would judge that most of the shell-fish consumed were small. Two pelecypod valves measured 3 cm. in greatest length and another 1.9 cm. These are probably a little over half as large as the average bivalve shell seen along the Flint. Gastropod shell particles made up a greater proportion of the contents than did the pelecypod fragments. One uncrushed snail shell measured 7.5 mm. from apex to operculum; another was 7 mm. and a third 5 mm. From the general conformation of the other shell remains I would estimate that the largest was slightly over 1 cm. It was interesting to me that the smaller terrapin had not crushed its food as completely as had the larger.

Hundreds of what appear to be a single species of nematode were found in the intestine of both guts.

The jaws of the male and juvenile are much more lightly constructed, and are obviously ill adapted for crushing mollusk shells. An exhaustive food-habits study would probably show interesting differences in the diets of the sexes.

Common Map Turtle *Graptemys geographica* (Le Sueur)

(Plate 37. Figs. 13, 15. Map 14.)

Range: From Wisconsin southward through Illinois to Tennessee and westward through Missouri and Arkansas into eastern Kansas and Oklahoma. The northern limit of distribution extends across southern Michigan and southern Ontario to extreme southern Quebec and northern New

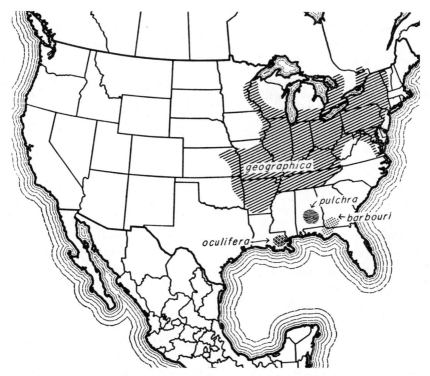

Map 14. Distribution of *Graptemys barbouri, G. geographica, G. oculifera,* and *G. pulchra.*

York and Vermont, whence the eastern edge of the range passes due south to the coast in the Susquehanna River drainage. The southern limit extends from coastal Virginia to northern West Virginia and thence southward into Tennessee.

Distinguishing features: This species is identified by the broad jaw surfaces, the detached, longitudinally directed spot on the side of the head, the large head of the female, the elongate precentral lamina, and the lack of a complex plastral figure. There is considerable variation, but if all these characters are considered identification is usually fairly certain.

Description: The shell is moderately low and flattened, with a central keel that may or may not be elevated to form a tubercle on some of the centrals. The maximum size is probably near 280 mm. (11 inches); shell length of the largest found by Lagler (1943b) in Michigan was 253 mm. (about 10 inches). Wood (1946) measured a male from eastern Tennessee that had the following dimensions: length of carapace, 156 mm. (6.15 inches); width of carapace, 127 mm. (5.00 inches); height of shell, 47.6 mm. (1.87 inches); length of plastron, 138 mm. (5.45 inches). The carapace is smooth, with a feebly serrate rear margin, elongate, wedge-shaped precentral, and reticulate pattern of light lines on an olive background. The keel on the second and third centrals, if present, is convex forward. The plastron is yellow to whitish, either unmarked or with a simple pattern made up of narrow black seam borders. The head is large, especially in the female, in which it may be tremendously swollen; alveolar surfaces of both jaws are very broad, those above nearly meeting on the median line. Ground color of head and legs is olive to brown with light stripes; the head pattern nearly always involves a somewhat elongate (longitudinally) spot behind the eye, and very frequently the fore end of a lateral stripe turns upward across the tympanum.

There is marked sexual dimorphism in this species, the males being much smaller (maximum size less than 140 mm. or 5½ inches), with relatively smaller head, very much longer and thicker tail, comparatively more extensive hind paddles, greater retention of the juvenile keel and coloration, a more wary disposition, and a more angular rear margin of the carapace.

The shell of the young individual is wider and deeper relative to its width, one specimen having the following dimensions: length of carapace, 40 mm. (1.57 inches); width of shell, 39 mm. (1.53 inches); depth of shell, 20 mm. (0.788 inches). The dorsal keel is much more pronounced and there is nearly always a plastral pattern formed by a simple dark line bordering each seam, but this is never conspicuous and double-lined as in *pseudogeographica* and is usually lost before maturity.

Habitat: This turtle shows a decided preference for rivers, where, however it may inhabit quiet backwaters and sloughs. It is more attracted by larger bodies of water than by small ones, being found in lakes but not ponds and but rarely in smaller swiftly flowing streams. Although fond

Plate 37. *Graptemys geographica. Upper and center:* Females; unknown locality. *Lower:* Female; Cumberland Gap, Tennessee.

of basking on logs, rocks, and sandy or grassy beaches, it is strictly aquatic and does not wander on land.

Habits: Of all the turtles studied by Newman (1906a) at Lake Maxinkuckee, Indiana, the map turtle reappeared earliest after its period of hibernation, an occasional individual showing up as early as April 10. At first the sunny portion of the day is spent basking in the sun; this turtle despite its extreme wariness is a confirmed basker and is one of the few species that habitually congregates on sand or grassy beaches for this purpose, as well as on rocks and logs. On the approach of warm weather the map turtles leave the shallower muddy lagoons for deeper water, perhaps in search of food but possibly for thermal reasons. Upon the approach of another winter they seek out the lagoons once more, where the soft mud affords desirable winter quarters. According to Newman they show a decided reluctance to retire into the mud, and the cold weather often catches them unawares; young specimens, especially, often become benumbed, and so many may be left stranded on shore when the ice has formed that it is possible to pick them up by the dozens. On the other hand it is a familiar sight to see map turtles swimming about or resting on open bottom even after their habitat has frozen over.

Newman regarded this turtle as neither remarkably timid, as stated by most observers, nor aggressively courageous. If pursued while in the water where there is plenty of bottom vegetation, map turtles quickly seek concealment in the maze of water weeds; when in the open they depend entirely on their speed in swimming, which is considerable, although as a rule if the water is shallow pursued individuals stop after a brief chase and retire into their shells. When captured they hiss and struggle, but never snap or thrust out the head to bite, and soon quiet down. In captivity they do not thrive, several writers having remarked that it is impossible to keep them alive for more than a year at the most. Mating has been observed in April and is thought to take place also in the fall.

Breeding: The laying season of the map turtle lasts from late May until July, although most individuals deposit their eggs during June. At this time the females often wander long distances inland in search of suitable nesting places, apparently preferring soft ploughed soil or clear, dry sand away from the water, but sometimes laying in heavy clay. The nest is a more or less symmetrical flask-shaped cavity excavated to the depth to which the hind foot will conveniently reach. Two layers of eggs are placed in the expanded portion, and the last two or three eggs are left in the narrow neck, where the uppermost may be within two inches of the surface. The eggs are ellipsoidal in shape, of dull white color and with a soft, easily indented shell. Cahn (1937) found the average dimensions of the eggs to be 32 by 21 mm. (1.26 by 0.827 inches). The normal complement appears to include from 10 to 16 eggs. Hatching occurs in Indiana and Illinois in late August and early September, and the young that emerge may have to travel almost incredible distances before reaching suitable water. Apparently an occasional nest, probably one made in the latter part of July, remains intact throughout the winter, since newly hatched map turtles are often found in May and June.

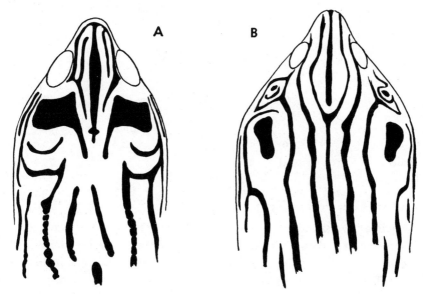

Figure 15. Dorsal head patterns of two forms of *Graptemys*. A: *Graptemys p. pseudogeographica*. B: *G. geographica*. The black bands in the diagrams represent light head stripes.

Feeding: As Garman remarked (1890c), the diet of the map turtle throws light on the curious modifications of the jaws and head, both of which are markedly broadened, sometimes spectacularly so. In all specimens examined by Garman in Illinois, the food was exclusively molluscan, in the young consisting of *Valvatra tricarinata* and other thin-shelled species and in the adult, of larger, thicker-shelled forms. Hay (1892a) related that at Lake Maxinkuckee a number of map turtles were kept in a tub in which, after a few days, numerous opercles of aquatic snails were found. Several other workers have found crayfish in the stomachs of map turtles, these in several cases constituting almost the only food present. Conant (1938a) found the diet to consist wholly of animal food, with snails, crayfish, aquatic insects, fish, and carrion predominant. The most recent study of

the food of this species is that of Lagler (1943b), who examined the digestive tracks of 27 Michigan specimens and summarized his findings as
shown in Table 5.

Economic importance: Apparently this species is of little interest in the

Table 5. Contents of stomachs and colons of map turtles.

Food item	Stomachs		Colons	
	Composition by volume (%)	Frequency of occurrence (%)	Composition by volume (%)	Frequency of occurrence (%)
Game fishes	1.8	8.3	trace	4.2
Forage fishes	11.3	8.3	—	—
Fish remains	2.4	8.3	0.6	8.3
Carrion	5.0	8.3	—	—
Crayfishes	52.4	8.3	13.0	25.0
Water mites	—	—	traces	8.3
Insects	8.6	41.7	12.0	50.0
Snails	17.3	83.3	57.7	79.2
Clams	1.3	8.3	12.5	37.5
Plants	—	—	4.2	16.7

turtle market, although all who have eaten it say that it is in no way except
size inferior to other species that are much esteemed. Clark and Southall
(1920) did not find it in any market, and Cahn (1937) said that in Illinois
it is caught in numbers by commercial fishermen but is either discarded or
only used locally. Lagler regarded it as of negligible importance as a fish
predator, although possibly offering some competition to game fish for
their food supply; and the same writer suggested a possibly beneficial role
in the destruction by map turtles of molluscan intermediate hosts of parasites of both fishes and man.

Ringed Sawback Turtle *Graptemys oculifera* (Baur)

(Map 14.)

Until recently I have been inclined to regard this name as representing
nothing more than a variant in the population of *G. p. kohnii*. The late
Dr. Leonhard Stejneger of the United States National Museum believed it
to be a valid species, however; and in a letter Dr. Fred Cagle of Tulane
University told me of his opinion that there are two forms of *Graptemys*

in southern Louisiana. I have recently re-examined one of the cotypes of *oculifera,* a well-preserved adult female sent to the Museum of Comparative Zoology by Baur in 1895. This specimen has the following dimensions: length of carapace, 202 mm. (7.95 inches); width of carapace, 167 mm. (6.58 inches); height of carapace, 88 mm. (3.46 inches); length of plastron, 186 mm. (7.33 inches).

The most conspicuous feature of this turtle is, as Baur (1890c) pointed out in the original description (and emphasized by proposing for the new form the name *"oculifera"*), a large light ring completely contained within each lateral lamina, with no interconnections between adjacent rings. Each marginal is marked by a light oblong figure formed by a line that passes from the upper section of the intermarginal seam backward to the upper hind corner of the lamina, where it does not touch the posterior seam but turns downward and extends to the margin to continue forward as a light marginal border.

The most striking element in the head pattern is an oblong, transversely oriented light blotch behind the eye, this being isolated and with no connections with head or neck stripes. There is a broad, transverse light bar on the chin.

The keels are obtuse on the second and third centrals, and are darker in color than the surrounding areas and only very slightly concave in lateral outline.

On the anterior part of the plastron there is a well-defined broad, dusky pattern, roughly following the trend of the seams. This is lacking posteriorly.

The alveolar surfaces of the upper jaw are narrow, not nearly meeting anteriorly. The tomium of the lower jaw is blunt and spoon-shaped terminally.

If this specimen actually represents a stock separate from *kohnii,* as would seem probable, then it may well be a very distinctive one. Besides the type, however, I have seen only two discolored and shriveled specimens and several eroded shells, all labeled "southern Louisiana," and can offer no valid opinion as to the status of the form. In his original description of the species, Baur stated that it was known from Pensacola, Florida, as well as from Louisiana, but this assertion has not been substantiated. I have not seen the specimens on the basis of which Stejneger and Barbour (1943) extended the range into Texas. It is probable that the studies of Dr. Cagle and his associates in Louisiana will clear up the uncertainty surrounding this turtle.

No information on the natural history of the ringed turtle is available.

Midland Sawback Turtle *Graptemys pseudogeographica pseudogeographica* (Gray)

(Plate 38. Figs. 13, 15, 16. Map 15.)

Range: Northern part of the Mississippi basin from west-central Wisconsin and northeastern Iowa southward to western Kentucky and northern Arkansas, where it intergrades with *G. p. kohnii;* an occasional inter-

Map 15. Distribution of the subspecies of *Graptemys pseudogeographica.*

grade may be found as far north as central Illinois. The western limit of distribution passes through eastern Nebraska, southeastern Kansas, and central Oklahoma, where intergradation with *G. p. versa* occurs. The easternmost limit of its range appears to be the extreme southeastern corner of Indiana and the western part of Kentucky.

Distinguishing features: This turtle may be known by its usually high, nodose dorsal keel with concave anterior profile; its usually small head and narrow alveolar ridges which do not tend to meet in the median line

Plate 38. Graptemys pseudogeographica pseudogeographica. Female; Hickman, Kentucky.

above; by the presence behind the eye of a transversely directed, crescentic extension of a dorsal head stripe, which terminates about halfway down the orbit, allowing other neck stripes to reach the orbital rim; and by the complex dusky plastral figure of the young.

Description: The shell is somewhat wedge-shaped, widest posteriorly and serrate at the hind margin. The largest specimen on record is a female with a shell 248 mm. (9.75 inches) in length. Cahn (1937) gave the following measurements for a male and a female from Illinois:

Sex	Carapace length	Carapace width	Plastron length	Shell depth
Female	195 mm. (7.67 inches)	149 mm. (5.87 inches)	181 mm. (7.14 inches)	63 mm. (2.48 inches)
Male	185 mm. (7.3 inches)	153 mm. (6.03 inches)	179 mm. (7.05 inches)	67 mm. (2.64 inches)

The carapace is smooth, with a prominent keel on which a more or less elongate tubercle extends backward toward the rear seam of each central, especially on the second and third of these; the anterior profile of this keel is concave as seen from the side. The precentral is often as broad as long. The ground color of the carapace is olive to brownish, with a very variable pattern that may include dark-brown smudges surrounded by greenish-yellow circular markings; the peak of each central tubercle is dark brown. The upper and lower marginals are usually decorated by ocellate spots, one at each seam. The plastron is yellow with or without a dark pattern following the general trends of the seams and at times quite extensive. The head and legs are greenish black with greenish-yellow stripes. The most conspicuous (and characteristic) mark on the head is the broadened anterior end of a neck stripe that crosses the nape near the mid-line, passes forward, and turns downward behind the upper edge of the orbit to extend a varying distance ventrally, but not to go so far down as to exclude other forward-trending stripes from the orbit. There is a yellow, rounded spot on the horny ventral surface at the tip of the lower jaw. The masticating surfaces of the jaws are not markedly broadened, those above not nearly meeting in the median line, those below not spoon like.

There is relatively little difference in the size of the sexes in the midland sawback, a most surprising feature in view of the striking disparity in size in males and females of some of its near relatives. The head of the female is conspicuously larger, however, and the tail of the male is larger and heavier and with the anus located farther back.

The young are about 32 mm. (1.26 inches) long at hatching, with a nearly circular shell, very high tubercles on the second and third centrals, and deeply serrate hind margin of the carapace. The ocellate markings of the carapace are quite distinct and the plastral pattern is usually extensive and intense.

Habitat: This turtle is found in lakes, rivers, sloughs, and backwaters, usually in the vicinity of abundant aquatic vegetation. It shuns strong currents and clear, sandy or gravelly bottoms with no plant life. It is very aquatic.

Habits: Despite the relative abundance of this turtle and the local gastronomic popularity that it enjoys, its habits in nature have been very infrequently observed. For most of the points in its life history we have to turn to the account of Cahn (1937), from which sections are quoted below:

In the water it is wary and secretive, disappearing from view at the least sign of danger, and remaining beneath the surface amid the protection of the dense vegetation for a long period before reappearing. These turtles are very gregarious and are often seen basking lazily in the hot sun. Basking places well away from the shore are best liked, and a protruding dead-head or a stranded log is ideal from their point of view. If such a site is not available, they will line up on half submerged logs along the shore, but under such conditions the turtles are more than ordinarily wary and on the alert, seeming to sense the added dangers which proximity to the shore brings. At the least sign of disturbance every turtle slides into the water, and this characteristic action has given to them the local name of "sliders." If caught on shore they behave much as do the turtles of the preceding species, withdrawing within their bulky shell and remaining stubbornly retracted. . . .

In the northern part of the state this species goes into hibernation during October and remains inactive until after the ice has gone out. The turtles bury themselves to a depth of from four inches to a foot or more in the soft mud of the bottom, or crawl into the underwater entrances of muskrat houses and there bury themselves. In the southern part of the state the turtles do not go into hibernation at all, but remain active, though but sluggishly so, throughout the coldest months of the winter. There is a hint that during this time they do not feed, or at least feed sparingly, for such turtles taken in the winter have invariably had the stomach empty. With the warming of the water in the spring, the turtles, as *G. geographica,* come out and sun themselves during the brightest hours of the day. The first chill of the afternoon sends them back into the water, yet occasionally an individual is found in which the reaction time to the change in temperature was too slow to get it back into the protection of the water before being paralyzed by the cold. Such an individual seems always to die with one night's exposure. Such incidents are perhaps more common in the fall than in the spring following it, though both have been observed.

Conant and Hudson gave 20 years and 6 months as the longevity record for this species in the Philadelphia zoo.

Breeding: Agassiz said that this turtle lays as early as June 1, but he did not distinguish between *geographica* and *kohnii*. Cahn's observations were to the contrary, the sawback having been the last of the several Illinois species observed by him to lay and the first nesting date having been July 6. On the latter date he made detailed notes on the laying activities of a female with a carapace length of 218 mm. (8.6 inches), which selected a nesting site in heavy black soil in the middle of an old road between the river and a ditch. Cahn came upon this turtle after excavation had begun, and the remaining part of the digging process watched by him lasted five minutes. She dug rapidly with alternating hind feet, pushing out accumulated soil with both feet and making a hole that sloped forward beneath her shell at a 60-degree angle. An unusual feature of the behavior of this turtle was the fact that she slowly rotated her body through a complete circle as the eggs were dropped into the nest, completion of the circle and deposition of the last egg coinciding almost exactly. During the filling process earth was raked forward with the anterior sides of the hind feet and then was patted a bit with the soles of the feet, after which the turtle left the site and headed for the ditch. The completing of the nest, deposition of the 11 eggs, and covering of the hole required seven minutes, and Cahn estimated that the turtle had been working perhaps five minutes when he encountered it.

The egg complement varies from 7 to 13, with 9 or 10 being the most usual number. The average dimensions of 71 eggs were 32.7 by 22.5 mm. (1.29 by 0.886 inches). Nothing is known concerning the incubation period.

Feeding: No detailed study of the feeding habits of this turtle has been made, but enough is known to indicate that the species is much more vegetarian in its diet than *G. geographica* (as the narrow masticating surface of its jaws would appear to indicate) once maturity has been reached. Like several other turtles the sawback is omnivorous, or perhaps principally carnivorous when young. Garman (1890c) apparently examined a number of stomachs and found them to contain mostly bulbs of a sedge

(*Cyperus*) with a few remains of mollusks and crayfish. Cahn found the young eating thin-shelled mollusks and an occasional worm and insect larva, while the stomachs of ten adults contained only plant material; he commented as follows concerning feeding by captive individuals: "Although dozens of specimens have been kept alive in captivity, they would never touch meat of any kind in any form, but ate lettuce and greens sparingly. They fed only when submerged and could never be induced to eat while out of water."

Economic importance: This turtle is highly regarded as food. The females are large enough to justify the labor spent in their preparation and they are sent to markets in various places, where they have sold for as much as a dollar each. Clark and Southall (1920) rated it as the best of the Mississippi Basin "terrapins" and said it had been used as a substitute for the diamondback.

Mississippi Sawback Turtle *Graptemys pseudogeographica kohnii* (Baur)

(Plate 39. Figs. 13, 16. Map 15.)

Range: Lower Mississippi Valley, from eastern Tennessee and northern Arkansas southwestward to eastern Texas and southward to the Gulf. Toward the east along the Gulf it is known from near Biloxi, Mississippi, and Baur (1890b) mentioned a specimen from Pensacola, Florida, but it is perhaps better to exclude this record until more material has corroborated it. Intergrades between *kohnii* and *pseudogeographica* are known from Arkansas and from northeastern Tennessee, and an occasional intermediate specimen turns up as far north as central Illinois. Intergradation between *kohnii* and *versa* probably takes places in east-central Texas, although I have seen no example of it. Mr. Glenn Gentry writes me that he found a form of *pseudogeographica,* together with *geographica* in the Tennessee River, in Humphreys County, Tennessee, and if this is *kohnii,* as seems likely, it would appear to represent the easternmost record for the form.

Distinguishing features: The following set of characters will serve to identify the Mississippi sawback in most cases: the high, dark tubercles of the central keel; the complex plastral figure of the young; the moderately large head with broad crushing surfaces in the jaws but with the upper jaw pavements not nearly meeting in the mid-line and those of the lower jaw not markedly wider in the middle than at the ends; and the deep

Plate 39. Graptemys pseudogeographica kohnii. A, D, and E: Male; Plaquemine, Louisiana. *B:* Female; New Orleans, Louisiana. *C:* Young specimen of unknown origin. *F:* Young; Plaquemine, Louisiana.

transverse postorbital crescent that excludes all other stripes from the rear rim of the orbit.

Description: The shape of the shell is similar to that of the midland sawback. Maximum size is perhaps slightly greater than that of the more northern subspecies. The largest specimen on record is the shell from Louisiana which Baur (1893c) referred to as *pulchra* but which very probably was an aberrantly colored *kohnii;* this measured 270 mm. (10.6

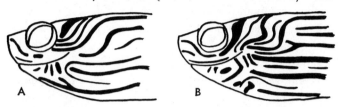

Figure 16. Lateral head patterns of two Mississippi Valley subspecies of *Graptemys pseudogeographica*. A: *G. p. pseudogeographica*. B: *G. p. kohnii*.

inches) in length. A male specimen from southeastern Texas has the following dimensions: length of shell, 137 mm. (5.34 inches); width of shell, 101 mm. (3.98 inches); height of shell, 59 mm. (2.32 inches): length of plastron, 125 mm. (4.93 inches). The ground color of the carapace is olive or brown, with an intensified brown on the keel and on its tubercles, of which those on the second and third centrals are very strong and convex anteriorly, that on the fourth somewhat weaker, and those of the first and fifth almost lacking. The pattern of the carapace is reticulate, each lamina having one or more circular markings that are interconnected by short bars; the marginals each bear a concentric light figure that is often open posteriorly. The plastron is greenish yellow with a very variable dusky pattern; this is usually a broad concentric set of dark lines suggesting the pattern of the plastron of *Chrysemys p. bellii,* but it may be greatly reduced or almost absent. Ground color of the head and soft parts is brownish to black. The crescent behind the eye connects with a dorsal head-and-neck stripe above, and below extends forward beneath the orbit. There is a rounded spot on the lower tomium of the lower jaw. The head is on the average broader than that of northern *pseudogeographica,* and with considerably broader alveolar surfaces, but is smaller than that of female *geographica* and the upper alveolar pavements do not tend to meet in the mid-line, while those below are not broadened mesially.

Sexual differences are similar to those in the northern race.

The young have short, high shells. A yearling from central Louisiana has the following shell dimensions: length, 35 mm. (1.38 inches); width, 34 mm. (1.34 inches); height, 18 mm. (0.71 inches). The central keels are probably a little less strong than those of young midland sawbacks, but are much more tuberculate than those of mature individuals of *kohnii*. The most arresting feature of the young is the very sharp, horny edge of the shell; the marginals are compressed and with keen outer borders of clear horn, which posteriorly are so deeply indented both at and midway between the seams that the rear margin of the shell may look as if set with a row of long, flattened, and very sharp transparent spines. This condition is still evident in juveniles of 65-mm. (2.56-inch) shell lengths but begins to disappear soon afterward. Something of the same condition is evident in young *pseudogeographica*, but I have never seen it so extreme as in the southern race.

Habitat: The Mississippi sawback occurs in rivers in their quieter reaches, lakes, and sloughs. Viosca (1931) described the habitat as "silt bearing streams and other waters in their flood plains." The availability of dense aquatic vegetation is apparently a desideratum.

Habits: The natural history of this race is almost entirely unknown. There is one record of laying at Natchez, Mississippi, on June 1, but the extent of the season has not been outlined, the eggs have not been described, and the size of the complement and the duration of incubation are not known.

With respect to the feeding habits of this turtle in Reelfoot Lake, Malcolm Parker (1939) said: "It feeds principally on such aquatic vegetation as *Ceratophyllum, Cabomba, Potamogeton* and *Lemnaceae*. The young are considerably less particular in feeding than the adults. The stomach of a young specimen contained about thirty damselfly nymphs and ten dragonfly nymphs."

A specimen collected in northern Louisiana disgorged a small clam, three snails, and the remains of a good-sized blue-tailed skink (*Eumeces*).

Economic importance. This sawback is eaten locally and may be found in the markets of cities of the lower Mississippi Valley. The young enter to some extent into the traffic in baby turtles for the pet shops, although the main burden of this trade is borne by other species.

Texas Sawback Turtle *Graptemys pseudogeographica versa* Stejneger
(Map 15.)

Range: This race is known only from four localities in the drainages of the Brazos and Colorado rivers, south-central Texas.

Distinguishing features: No adequate description of this turtle has been published, although the following diagnosis, quoted from Stejneger (1925), should make identification of it relatively easy:

Color pattern of postocular region that of typical *Graptemys pseudogeo-graphica*, fine light lines running obliquely upwards from tympanum to posterior edge of orbit, but postorbital spot extending backwards from its lower (outer) edge and not from its upper (inner) edge. . . . Seven paratypes from the same locality [Austin, Texas] are essentially like the type and differ markedly both from typical *pseudogeographica* and from *kohnii*.

A paratype in the Museum of Comparative Zoology does not agree with this description, in that the postorbital spot is the downward and backward extension of a supraorbital stripe; no stripe continues backward from the spot along the side of the head and neck.

In 1944 Charles Bogert lent me a series of five live sawbacks from Fort Sill, Oklahoma, caught in Medicine Creek by Lieutenant H. Epstein. While I am not acquainted with the subspecies *versa*, both the characters and the locality of these Fort Sill specimens lead me to believe that they represent intermediates between *versa* and *pseudogeographica*. If *versa* is as remarkably different a race as these intergrades imply, the following comments from my notebook will not be out of place:

The series comprises 5 adult males and one young female. Carapace light brown, the strong tubercles dark brown, and the laterals each with two or three reddish brown to orange ocelli; centrals with numerous finer and more irregular reticulations. Tubercle on second central the most prominent, the keel concave before it but straight on the third central. Marginals reticulated above and with broad-lined ocelli below. Bridge with broad-lined longitudinal figure similar to marginal ocelli. Plastron light-pinkish and greenish white, with a dusky figure which in those specimens that have not lost it is like that of *C. p. bellii*. Spots on head very intense lemon yellow. The side of the head is conspicuously three-spotted, besides being longitudinally striped. The uppermost spot is the postocular mark of typical *pseudogeographica* but here it is usually detached from all neck and head stripes. The second spot is beneath the eye and the third on the mandible beneath the second. The horny surface of the lower jaw is spotted below. The masticating surfaces of the jaws are narrow, in the males as well as in the female, not nearly meeting in the median line above. Stripes about ten on the fore surface of fore leg, that to second toe the broadest.

The heads of the males of this batch seem astonishingly small—in fact the whole front end of the animals seems dwarfed in contrast with the ponderous tails and broad, paddlelike hind legs. The nails of the fore toes of the males are much longer (up to twice as long) than those of the female, as in *Pseudemys*.

The young female has a carapace 140 mm. (5.52 inches) long, while that of the largest of the males is only 114 mm. (4.50 inches) long and he is obviously an old individual. This marked sex dimorphism suggests *geographica* instead of the races of *pseudogeographica* but the other characters all indicate the latter.

Nothing has been published concerning the habitat or habits of this turtle.

Alabama Sawback Turtle *Graptemys pulchra* Baur
(Figs. 13, 17. Map 14.)

Range: The Alabama River, south-central Alabama, and probably westward an undetermined distance along the coast. It is possible that this turtle intergrades with *G. p. kohnii* in Mississippi and Louisiana. It seems more likely, however, that intergradation with *oculifera* will be found, provided the latter proves to be valid. There is very close agreement between the type specimens of *pulchra* and *oculifera*.

Distinguishing features: The Alabama sawback turtle may apparently be recognized by the combination of a head pattern like that of *barbouri* with a narrow head and a reticulate instead of ringed pattern on the carapace, ocellate marginal pattern, narrow alveolar surfaces, and short, broad precentral lamina.

Figure 17. Dorsal head pattern of *Graptemys pulchra*. This style of pattern is closely similar to that of *G. barbouri*. Black areas in the diagram represent light markings.

Description: The original description of this turtle, by Baur (1893c), is as follows:

The coloration of the skull [head] and neck distinguishes this species at once from all others. The whole space between the orbits is characterized by a continuous yellow figure which sends backward on each side behind each orbit a strong process of the same color.

The head resembles that of *Graptemys kohnii*, but is more slender. The symphisis of the lower jaw is longer and the nose more projecting. In all the skulls examined the jugal is excluded from the orbit, a character not seen in the other species of *Graptemys* or *Malaclemmys*. The form of the carapace is very close to *Graptemys kohnii*; the dermal shields (laminae) are very thin. It is the largest form of *Graptemys*, the shell reaching a length of over 170 mm. in straight line. The color of the shell is light olive with yellow marks on the marginals; the plastron is yellow with some darker marks. Types: No. 8808, Smithsonian Institution, Washington, D.C. Two not full grown specimens collected by Dr. T. H. Bean at Montgomery, Alabama.

This account fails to mention certain features of importance. The pattern of the carapace is of light, black-bordered lines with reticulate arrangement, two or three lines meeting at right angles near the center of each lateral. The upper and lower marginals are marked with multiple-circle ocelli. The precentral is nearly square. The alveolar surfaces of the upper jaw are narrow as in *pseudogeographica*, not nearly meeting in the median line.

Dimensions of the cotypes, both females from Montgomery, Alabama, are as follows:

Head width	Shell length	Shell width	Shell height
32 mm.	177 mm.	139 mm.	67 mm.
(1.26 inches)	(6.97 inches)	(5.48 inches)	(2.64 inches)
26.8 mm.	179 mm.	135 mm.	67 mm.
(1.05 inches)	(7.05 inches)	(5.32 inches)	(2.64 inches)

In the same paper quoted above, Baur mentioned having earlier received a skull and a very large living specimen of this turtle from Mr. Gustav Kohn of New Orleans. He gave no more definite idea of the origin of the specimens. The skull would in any event have been unidentifiable, since I have found the character of the jugal and its relation to the orbit to be unreliable. As to what the live turtle might have been, my guess would be *oculifera*.

No data on the habitat preference or habits of this little-known turtle are available.

Genus *CHRYSEMYS*

The Painted Turtles

The genus *Chrysemys* is composed of turtles so similar in most ways to *Pseudemys* that the two have frequently been combined. *Chrysemys* comprises smaller species with shells that are relatively smooth, unserrated behind, and often middorsally striped at some stage. The upper jaw is notched in front and with the notch bordered by a toothlike projection on either side; the alveolar surface is narrow and with a feeble ridge. The genus in-cludes a single species and a number of intergrading races, all North Amer-ican. A fossil from the Pleistocene of Nebraska has been assigned to this genus, but in my opinion it might just as reasonably be placed in *Pseu-demys.*

KEY TO *CHRYSEMYS*

1   Plastron immaculate yellow, unmarked 2

1′   Plastron with a symmetrical dusky figure 3

2(1) Seams of second and third central laminae in nearly the same transverse line as those of adjacent laterals; eastern North America from New Brunswick to Georgia and probably north-ern Florida .... *picta picta* (p. 214)

2′   Seams of second and third centrals alternating with those of laterals; Mississippi Valley from Illinois southward to Louisi-ana .... *picta dorsalis* (p. 224)

3(1′) Dusky figure on plastron without marked transverse extensions along seams, the width of the figure about half the width of the plastron or less; Allegheny Mountains west to Wisconsin and southward to Tennessee .... *picta marginata* (p. 228)

3′   Dusky figure on plastron extensive, covering most of the surface of the plastron; transverse extensions along seams marked; west-ern Illinois northward to southern Ontario and northwestward to the Cascade Mountains and western British Columbia; south-westward to Texas and New Mexico .... *picta bellii* (p. 219)

Eastern Painted Turtle *Chrysemys picta picta* (Schneider)

(Plate 40. Map 16.)

Range: Eastern North America, from Nova Scotia, New Brunswick, and Quebec, south of the St. Lawrence River, southward through the coastal states to extreme northeastern Florida. Intergrades with *C. p. mar-ginata* are known from along the St. Lawrence River in southern Quebec, and from Maine, Vermont, New Hampshire, Massachusetts, Connecticut, New York, Maryland, West Virginia, and eastern Pennsylvania. Just where in the Gulf states the area of intergradation between this subspecies and *dorsalis* takes place has not been ascertained, but it may occur in Alabama and Mississippi.

Distinguishing features: Aside from the generic jaw characters men-tioned above, the eastern painted turtle is known by its aligned central and lateral laminae, the light fore margins of the lateral seams, the plain yellow plastron, and the low, unkeeled carapace.

Description: The shell of the eastern painted turtle is low, broad, and smoothly arched above. The size is small, maximum shell length being in the vicinity of 180 mm. (7.10 inches). Nichols and DeSola gave the following measurements of a very large female from Croton Point, New York:

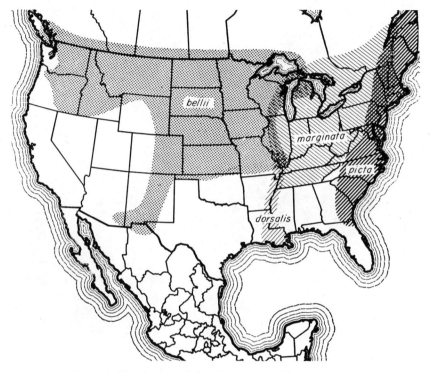

Map 16. Distribution of the subspecies of *Chrysemys picta*.

length of shell, 179 mm. (7.05 inches); width of shell, 140 mm. (5.52 inches); length of plastron, 171 mm. (6.75 inches); width of plastron, 114 mm. (4.50 inches). The carapace is oval as seen from above, somewhat elongate, depressed, unkeeled, and with its widest part toward the rear. It is smooth or with slight concentric striations. The second and third central laminae are aligned with the second and third laterals; that is, their anterior and posterior seams are nearly continuous across the shell. The carapace is olive to dark brown in color, with a narrow and sometimes incomplete middorsal light line. The dorsal laminae are broadly edged before with yellowish or reddish. The marginals are conspicuously marked

Plate 40. Chrysemys picta picta. A and F: Male; Dumfries, Virginia. *B, C, and E:* Male; Hammonton, New Jersey. *D:* Young; Westerly, Rhode Island.

above and below with red bands, bars, and crescents, more or less concentrically arranged. The plastron is plain yellow or, rarely, marked with black. The head is dark olive to dark brown, with a moderately constant arrangement of light lines running longitudinally and usually comprising a short bar from the upper orbit, a lesser mid-orbital line, a broad stripe from the lower orbit, and a very broad but usually short stripe from the angle of the jaws. The chin is marked with a pair of broad lines that enclose a median stripe and converge at the tip of the jaw, a very popular emydine decoration that turns up throughout *Chrysemys* and *Pseudemys* and in other forms as well. There are other subsidiary and variable lines in the pattern. The legs are dark, streaked with red.

The shell of the male painted turtle is more depressed than that of the female, the anus of the male is more distally placed, and the nails of the fore feet are longer than those of the female.

A hatchling measured by Wilcox (1933) had a shell 25.6 mm. (1.01 inches) long and 24.6 mm. (0.97 inches) wide. The same writer gave the dimensions of nine other hatchlings, all from Long Island and from a single litter; these varied in shell length from 23.8 mm. (0.94 inches) to 25.4 mm. (1.00 inches), and in shell width between 23.0 mm. (0.905 inches) and 25.4 mm. (1.00 inches). One of this litter had a carapace 25.4 mm. long and 25.4 mm. wide. Otherwise the young are similar to the adults except for the tail length, which is disproportionately great in young painted turtles, as in those of most species.

Habitat: This is a turtle of quiet waters—ponds, protected lake shores, ditches, and slow streams and marshes. It is extremely aquatic and, although an inveterate basker, it appears to leave the immediate vicinity of water only to lay or to make short migrations to more suitable neighboring habitats. I saw such a migration along a fill through a New Jersey marsh shortly after a June thundershower. Heavy traffic was passing over the road and I estimated that there were between 200 and 400 dead painted turtles within a distance of 300 yards.

Habits: The eastern painted turtle hibernates under water, usually well buried in mud or trash. It emerges in March and April in the northern part of its range and earlier farther south, while there is evidence that in the southernmost extremes of its territory it may in some years not hibernate at all. It is very fond of basking, and a dozen or more sometimes congregate to take the sun on a log or other partly submerged object. It occasionally wanders about on land, once in a while in such numbers that a concerted migration from one body of water to another is evidently under way. Even when no such large-scale hegira is in progress, individuals of both sexes may sometimes be found in the woods. Although published references to the voice of one turtle or another are frequently met, that which strikes me as the most extraordinary that I have come across is J. A. Allen's statement that the song of the painted turtle "is frequently heard in May and June, especially during intervals between showers on hot, sultry days" (1868). In this case it would seem that the observer had attributed to a turtle which he saw, the song of a *Hyla* that he could not see; but the only eligible *Hyla* in Massachusetts would have been the spring peeper, and the song of that frog is so universally recognized by anyone with the most elementary interest in natural history that I cannot believe that Allen would have mistaken it. This is just another of those inexplicable aberrations in observation, like the two published notes (one on a New England specimen, the other on one in Australia) that I have come across describing nest construction as a sort of drilling operation in which the female turtle planted one foot firmly as an auger and then gyrated about it to bore out the nesting hole.

Breeding: Copulation has been observed in April (Connecticut). The eggs are laid from May to July, usually late in June or July in the north, and usually in May or June in the south. They are from 3 to 11 in number, the most usual complement being 5 or 6, and are elliptical in shape, the average dimensions being about 33 by 23 mm. (1.3 by 0.905 inches). Data obtained by Finneran in Connecticut during June, 1942, give some idea of what is very likely the normal average incubation period:

Date (June, 1942)	Number of eggs	Date of hatching (1942)
3	7	8/18
3	6	8/23
5	5	8/20
25	5	9/2

The female digs the nest with her hind feet in almost any sort of soil in either woods or open country. She arranges each egg in the nest with a foot as it is dropped and covers the nest by raking in soil with the hind feet. An individual observed by Wilcox (1933) pulled in bits of straw to conceal the site and then walked away without inspecting it.

Nichols believed that in the majority of nests constructed on Long Island the young do not emerge until the April or May following oviposition, and there is evidence to support this view from other areas.

Feeding: The eastern painted turtle is an omnivorous feeder. It has been seen to take a number of different kinds of plant and animal foods, both in the natural state and in captivity, but no detailed study of its feeding habits has been made and the quantitative make-up of its diet is unknown. In a large percentage of the cases in which painted turtles are known to have eaten fish, the latter have probably been found dead or injured. Although this species prefers to swallow its food under water, Brimley (1943) saw one captive specimen eating on land.

Economic importance: Apparently this turtle is eaten in various places

from time to time, but it is very small and of no great consequence as food
for mankind. McCauley (1945) commented on this subject as follows:

Their flesh is palatable but they are considered too small by most people.
They are sold occasionally, nevertheless. At one time I interviewed a fisherman
on the Eastern Shore who was able to get $1 per dozen for them from individ-
uals in Pennsylvania. Whether the purchaser was under the impression he was
obtaining the famous diamondback at a bargain, or whether he desired a cheap
substitute was not ascertained.

Western Painted Turtle *Chrysemys picta bellii* (Gray)

(Plate 41. Map 16.)

Range: From southwestern Ontario southward through northern Michigan and western Wisconsin and Illinois; westward across the southern
edges of Manitoba, Saskatchewan, Alberta, and British Columbia, where it
is distributed mainly east of the Cascades except for scattered occurrences in
the coastal area; toward the south the western limit of its range lies in west-
central Washington, specimens from western Washington being regarded
as artificial introductions, and in Oregon it ranges "westward through
the Columbia Gorge as far as Vancouver in Washington and the Willamette River and Portland in Oregon" (Evenden, 1948). Still farther south,
the western limit traverses central Wyoming and Colorado, west-central
New Mexico, and extreme southeastern Arizona to enter Mexico in northern Chihuahua and the adjacent Texas frontier area. Elsewhere in Texas
it is apparently lacking, the southern limit of distribution passing from
northeastern New Mexico through the panhandle of Oklahoma and eastward to southern Missouri.

Distinguishing features: This subspecies is distinguished from the other
painted turtles by the following combination of characters: the central
and lateral laminae are staggered, their seams not continuous; a middorsal
light line is very inconspicuous or lacking, as are the light anterior borders
of the transverse seams; a light reticulated pattern may often be evident
on the carapace; the red and yellow markings of the marginals are less
striking than in *picta* and *marginata,* since there often is no red on the
upper marginals and only dull red on the lower; the plastral dark figure is
the most extensive in the genus, filling most of the plastral surface and
sending branches along the seams. This is the largest member of the genus.
The record shell length appears to be that of a specimen measured by

Plate 41. Chrysemys picta bellii. A and B: Male; Iowa. *C:* Female; same locality. *D and E:* Young; Colorado Springs, Colorado.

Van Denburgh (1922), which had a carapace 250 mm. (9.84 inches) long and 175 mm. (6.9 inches) wide, and a plastron 232 mm. (9.16 inches) long and 139 mm. (4.47 inches) wide. The dimensions of a male and a female from Illinois were found by Cahn (1937) to be as follows:

Sex	Carapace length	Carapace width	Plastron length	Plastron width	Shell depth	Weight
Male	114 mm. (4.5 in.)	103 mm. (4.06 in.)	130 mm. (5.12 in.)	87 mm. (3.42 in.)	44 mm. (1.73 in.)	359 grams
Female	145 mm. (5.7 in.)	115 mm. (4.54 in.)	142 mm. (5.6 in.)	94 mm. (3.7 in.)	52 mm. (2.05 in.)	429 grams

As far as is known, there are no secondary sexual characters that are not found also in the other subspecies. The female is the larger, the fore toenails of the male are longer, and the vent of the male is located behind the rear margin of the shell, while that of the female is beneath it.

Carl (1944) described the hatchling as almost exactly 25.4 mm. (1 inch) long and with a strong dorsal keel. The central stripe is pronounced but very thin. The plastron is more intensely pigmented and the tail is relatively longer than in the adult. At a shell length of 43 mm. (1.69 inches) the shell has begun to elongate faster than it grows laterally, a specimen of this length measured by Anderson (1942) having a shell width of 40 mm. (1.58 inches).

Habitat: This race is found in a variety of situations—streams, lakes, ditches, prairie and pasture ponds, and roadside pools—which would appear to have little in common. It is said, however, that *bellii* shows some preference for shallow, quiet water with soft bottom and plenty of aquatic vegetation. It is known from altitudes as great as 6,000 feet in the Rocky Mountains.

Habits: Because of the location and extent of the area of intergradation between *C. p. marginata* and *C. p. bellii*, it is difficult to decide which form is treated in many published accounts, and indeed some of the best of these studies have involved intergrade populations. Much of Cagle's work in Illinois was based on stocks not at present subspecifically identifiable, as was that of Pearse and his associates in Wisconsin, and a part of Lagler's food studies in Michigan. In his *Turtles of Illinois,* Cahn (1937) segregated his data under the three headings *marginata, bellii,* and *dorsalis,* and while this may have been the best course open to him, it tends to obscure the intimacy and complexity of the relation existing among all Illinois *Chrysemys.* By this system, for instance, it would be quite possible to treat as subspecific behavior the feeding or nesting procedures of two individuals from a single litter of turtles. On the other hand, there can be little doubt that Cahn's scheme has more to recommend it than the common tendency among compilers to throw off responsibility for the chore of separating into appropriate categories the existing published information with the remark that these forms are all merely races and probably do just about the same things. In 1942a Cagle remarked that he had a lot of ecological data on Illinois *Chrysemys* that he would publish if the taxonomic situation were ever cleared up—in other words, when he would be able to state definitely what he was writing about—and for this state of mind one can feel only sympathy.

Breeding: The courtship of this turtle was described by Taylor (1933) as a most extraordinary set of maneuvers by which the male swam backwards in front of the female, arms outstretched and palms out, and vibrated his long nails against her chin and lores. This was observed in several cases involving different pairs of individuals; it occurred on October 16 and 17, in a tank in the laboratory of the University of Kansas. Nesting begins in late May over most of the range of the race, but according to Carl (1944) it may occur as early as the first part of May on the coast of British Columbia. Carl found that the nesting site selected was consistently a southern slope exposed to the sun and near the water. According to him, the soil is moistened before excavation begins; the eggs may be from 6 to 20 in number, but usually are 12 or 13. Carl also said that the covered excavation is patted by the plastron of the female before she departs. Blanchard (1923) found a somewhat smaller egg complement to be usual in Iowa, egg numbers in 10 clutches examined by him having been as follows: 13, 12, 11, 5, 6, 6, 7, 7, 9, 12. Carl's data appear to add weight to Pope's suggestion (1939) that the western subspecies may lay more eggs than the other painted turtles. In general, however, clutches of from 5 to 8 eggs are most usual. The average size of the eggs is about 34.5 by 17.8 mm. (1.35 by 0.70 inches). Stromsten (1923) presented a detailed account of the activities of a female that nested in Iowa on June 30; the main points were as follows:

The site chosen was in hard clay soil on the side of a hill, and to reach it the turtle traveled about 45 feet from the water's edge. The hole was excavated by the hind feet, working alternately, the body shifting from side to side as first the right and then the left foot was put into operation. After some preliminary scraping by the hind claws the soil was moistened by cloacal bladder water which sometimes was squirted into the hole with some force. The wetting continued to take place as the work was shifted from one foot to the other. The hole was laterally enlarged as it deepened, imparting a flasklike shape. Dirt was lifted out and deposited at the rear edges of the hole, later to be pushed backwards out of the way. Six eggs were laid at average intervals of 30 seconds, and each in turn was arranged in the nest by an inserted hind foot. Immediately after the last egg fell, the covering process began, dirt being raked in from behind and pressed down by alternate hind feet. Since the wet dirt had been removed first, it had been pushed farthest away and was the last to be raked into the cavity, where it was thoroughly kneaded by the feet and compressed by the plastron. Dry dirt was then scratched in and it almost seemed to Stromsten that the turtle made an effort to reach far out for grass and other plant debris with which to conceal the site. The time schedule of the various operations was as follows: turtle appeared at water's edge, about 8:00 P.M.; began digging, 8:17 P.M.; began laying, 9:47 P.M.; finished laying, 9:52 P.M.; finished covering and left nest, 10:23 P.M.

The incubation period has not been determined, but in many cases the young do not emerge until the following spring, and Carl is of the opinion that in Canada this is a usual occurrence.

Feeding: Carl stated that in British Columbia the food of this race is largely of animal origin, although it takes some plant material. He listed as food items found: trout fry, tadpoles, fresh-water mussels, snails, insects, water lilies, water milfoil, *Elodea,* bulrush, and other plants.

Pearse's data from Wisconsin intergrades (*bellii* × *marginata*) showed a seasonal change in diet, as follows:

Month	Vertebrates (%)	Insects (%)	Plants (%)
June	2.3	60.7	32.0
September	10.0	36.2	55.0

Lagler (1943b) examined the contents of the digestive tracts of eight intergrades (*bellii* × *marginata*) from Michigan and compiled the results as shown in Table 6.

Lagler's comments on these data were as follows:

It is interesting to note the large numbers of hard-coated seeds of the white water-lily (*Nymphaea odorata*) ingested. These mature seeds, 813 in number, make up most of the volume of the plant materials eaten. They appear no different in the colons than in the stomachs, and probably remain viable in passing through the turtle. This observation and others suggest that these animals may be of ecological significance in the dispersal of water lilies.

On the basis of this material, it appears that the intergrades have food habits much like those of the midwestern painted turtle, and that these turtles compete with fish for food, but do not prey upon them.

Table 6. Contents of digestive tracts of intergrades from Michigan.

Food item	Stomachs		Colons	
	Composition by volume (%)	Frequency of occurrence (%)	Composition by volume (%)	Frequency of occurrence (%)
Frogs	42.7	14.3	—	—
Crustaceans	6.0	57.1	6.4	57.1
Insects	15.4	85.7	7.7	57.1
Lower plants	trace	14.3	3.4	28.6
Higher plants	24.8	57.1	79.9	85.7
Vegetable debris	2.6	57.1	1.9	57.1

Economic importance: Although the largest of the four races of painted turtles, this is still a small animal and while undoubtedly perfectly edible it has no commercial importance as a market turtle. Its possible significance as a factor in aquatic ecology is suggested by the remarks of Lagler quoted above.

Southern Painted Turtle *Chrysemys picta dorsalis* Agassiz

(Plate 42. Map 16.)

Range: Southern part of the Mississippi Valley, mostly toward its western side (but possibly reaching the Florida panhandle via the Chattahoochee River drainage); northward to extreme southern Illinois. As was mentioned earlier, material from Alabama and eastern Mississippi would help greatly in clearing up the status of this race.

Distinguishing features: The following combination of features identifies this subspecies: the lateral and central laminae alternate, their transverse seams not coinciding; the plastron is plain or with one or two small spots, not marked with a dark central blotch; the carapace is olive brown to nearly black, with a very broad and conspicuous middorsal light stripe, either red or yellow; the anterior margins of the laterals are bordered with similar color, to a greater extent than in *marginata* and *bellii;* the marginals show little red above, usually only a short bar at the margin, but have larger, median red areas below; the soft parts and head are marked much as in *picta.*

Plate 42. Chrysemys picta dorsalis. Female; southern Louisiana. *Upper right:* Young; Plaquemine, Louisiana.

The southern painted turtle is said to be the smallest member of the genus. The largest of 69 specimens measured at Reelfoot Lake by Marchand (1942), and apparently the largest on record, was 150 mm. (5.91 inches) in shell length, and of this group 41 out of 54 females had shell lengths between 125 mm. (4.94 inches) and 140 mm. (5.52 inches), while 11 out of the 15 male specimens had shell lengths of between 100 mm. (3.94 inches) and 115 mm. (4.54 inches). From these limited data it would appear that besides being considerably larger than the male, the female southern painted turtle outnumbers him by 3.6 to 1, a sex ratio not incompatible with the findings of investigators of other species of turtles. The fore toenails of *dorsalis* are elongate in the male, of the same size as hind toenails in the female; the tail of the male is nearly half as long as the plastron in the male and less than a third its length in the female. The usual more distally located vent is characteristic of the male in this race.

Like the adult southern painted turtle, the hatchling is smaller than that of the other subspecies. Of four hatchlings with egg teeth, from south-central Louisiana, the smallest has a shell length of 22 mm. (0.866 inches) and is 21 mm. (0.827 inches) wide and 10 mm. (0.394 inches) deep, with a plastron 22 mm. (0.866 inches) long and a tail 10 mm. (0.394 inches) long. The largest has the following dimensions: length of shell, 27 mm. (1.06 inches); width of shell, 25 mm. (0.985 inches); height of shell, 11 mm. (0.434 inches); length of plastron, 26 mm. (1.02 inches); length of tail, 13 mm. (0.513 inches). The coloration and head striping of these is similar to that of the adult, although the relative width of the very conspicuous middorsal stripe is greater.

Habitat: The southern painted turtle is even more confirmedly a pond dweller than its near relatives. It is found almost exclusively in quiet water, usually in sloughs, overflow ponds, and backwaters of the larger rivers or in shallow ponds with much vegetation, which need not be more than a few inches in depth and which may sporadically dry up completely. Due to the difference in feeding habits of the young and the adults mentioned below, there may be also some slight divergence in habitat preference in the two.

Habits: The following notes are quoted from Marchand (1942) and are based on observations made by him at Reelfoot Lake, Tennessee:

The *Chrysemys* population at Reelfoot Lake showed little departure from the general pattern of behavior of painted turtles as observed in other areas. The turtles were most often noticed when sunning or when resting at the surface of the water with their heads protruding above the duckweed. Sunning was indulged in more in the mid-morning and mid-afternoon hours than in the middle of the day when the heat was perhaps too intense. Isolated individuals were occasionally seen out on the logs as early as 6:30 A.M., but the practice did not become common until about 8:00 A.M. A few times turtles on the logs were found to be asleep, and in these cases could easily be approached to within a few feet before awakening. In most instances, however, they slid off the logs at the slightest sign of danger. When the turtles were frightened off the logs they could usually be expected to reappear in the same area in a relatively short time. The practice of resting at the surface with the head thrust through the mat of duckweed, while not as conspicuous as the sunning on logs, is probably even more common, since it is indulged in at all hours of the day. When the turtles are frightened in this position they rarely reappear in the immediate area.

Adult turtles were easily obtained in the late afternoon hours, at which time they appeared to have lost a great deal of their shyness. I also thought I could detect less shyness in the turtles in that portion of the Bayou bordered by dwellings. Despite the extensive inroads that have been made in the population of young turtles by the local residents it was fairly easy to catch them. They were most easily taken along the Bayou when they were sunning or resting at the surface.

Breeding: Courtship and mating have not been observed. Nesting occurs in June and July, the latest date on which Marchand found shelled eggs in dissected specimens having been July 8. The process of nest excavation has not been described, but Cagle (1937) found that the distances from water at which eight nests at Reelfoot Lake were located ranged from 45 to 50 feet. The egg complement usually numbers from 4 to 6, and it would appear that but one set is laid per season, since counts of shelled eggs in the oviducts of females dissected and in nests tend to be approximately equal. The nest is closed by a plug of mud almost certainly formed by wetting the soil with water from the cloacal bladder of the nesting turtle. The average size of the eggs is about 32 by 19 mm. (1.26 by 0.75 inches). They are bluntly oval, with the calcareous surface finely pitted, and are (according to Cagle) very similar in appearance to eggs of *Pseudemys scripta elegans*. In eleven of 32 eggs Cagle found small fly larvae, and in several nests red ants, which, however, had apparently entered after some of the eggs had dried out. The period of incubation has not been determined.

Feeding: Although the subspecies is omnivorous, Marchand (1942), who examined the stomach contents of 10 juveniles and 69 mature specimens, found an interesting reversal of food preference that comes with maturity. Plant material constituted 88 per cent of the food of the adults and only

13 per cent of that of the immature specimens, while the respective percentages of animal food were 10 and 85. Of the plants taken by the adults the most frequent were duckweeds and algae, although coontail was abundant in a few stomachs. The small amount of animal food eaten by the adults was mostly dragonfly larvae and crayfish. The most important animal foods in the diet of the young were dragonfly larvae and amphipods, with the larvae of caddis flies, beetles, and Diptera ranking next in order of volume. Duckweeds were the plants most abundantly eaten by the young. Concerning this dietary change from youth to maturity, which, incidentally, has been noted also in *Graptemys p. pseudogeographica* and probably is by no means rare among emydid turtles, Marchand commented as follows:

A possible explanation for the reversal in food habits in the *C. p. dorsalis* might be that these turtles have a preference for animal matter, and while immature are able to satisfy their requirements with a variety of small forms. When the adult size is reached this same food can no longer be obtained in sufficient quantity to meet the normal requirements and the turtles turn to the most available food that can be secured in proper abundance. The Reelfoot Lake area is unusual in the vast amount of aquatic vegetation that is present, and plant food can be obtained in unlimited quantity and with little effort. It is possible that if so many hydrophytes were not available the scavenger tendency so often noted in *Chrysemys* would be more apparent here.

The stomach of a young Reelfoot Lake specimen examined by Parker (1939) contained about 18 crane-fly larvae, the remains of several other insects, and considerable vegetable matter.

Economic importance: The young of this subspecies are often included with the enormous consignments of baby *Pseudemys* and *Graptemys* that are shipped to pet shops from the lower Mississippi Basin. Cagle mentioned a cent each as the price paid fishermen at Reelfoot Lake for all little turtles. This turtle is sometimes eaten and occasionally even sent to market. From Cagle's remarks it seems probable that the eggs are used as fish bait along with those of *Pseudemys s. elegans*.

Midland Painted Turtle *Chrysemys picta marginata* Agassiz

(Plate 43. Map 16.)

Range: Southeastern Wisconsin, eastern Illinois, and western Kentucky and Tennessee eastward to near the North Carolina line, and to western Virginia, eastern Pennsylvania, and New York, the northern limit of its

Plate 43. Chrysemys picta marginata. A and B: Female; near McLean, New York. *C:* Young; Ithaca, New York. *D:* Female; same locality.

range extending northward in Ontario to the Sault Ste. Marie area and into the western part of southern Quebec. Besides a broad area in the east where scattered intergrades with *C. p. picta* have been found, there is intergradation between *marginata* and the western *bellii* in the upper peninsula of Michigan, in south-central Ontario, in Wisconsin, and in eastern Illinois, and between these two and *C. p. dorsalis* in a vaguely defined area centering upon the mouth of the Ohio River. Little or nothing is known of the painted turtles of the territory in western North Carolina, northern Georgia, and Alabama and eastern Tennessee, where integrades between *picta, marginata,* and *dorsalis* might be expected.

Distinguishing features: The recognition characters of this painted turtle are, besides those necessary to establish its genus, the staggered laminae of the carapace, where the central series alternates with the laterals, rather than lining up with them; the lack or obsolescence of the light anterior borders of the transverse seams of the carapace; and the presence in most specimens of a dark central plastral figure which, however, is only half the width of the plastron or less and which does not send out extensions along the seams. The red markings on the shell margins and legs, and the head stripes, are quite similar to those of *C. p. picta.* The hind marginals are said to flare outward somewhat more than those of *picta*. Size range in this subspecies appears to be about the same as in *picta*. Of 586 Ohio specimens examined by Conant (1938a), a female with a shell 159 mm. (6.27 inches) long was the largest. Shell lengths of 90 adult specimens measured by Raney and Lachner in New York ranged from 106 mm. (4.18 inches) to 175 mm. (6.90 inches), with an average of 135.6 mm. (5.34 inches). The smallest female with eggs found by Cagle (1942) had a shell 95 mm. (3.47 inches) long.

Males average considerably smaller than females in this subspecies. The largest male on record is a Wisconsin specimen with a shell length of 178 mm. (7.02 inches), while the record female, also a Wisconsin individual, was 188 mm. (7.41 inches) in shell length; these specimens were from samples in which the average shell length for males was 117 mm. (4.61 inches) and for females 143 mm. (5.64 inches). The anus of the male opens beyond the hind shell margin, while that of the female lies just below it. The nails of the forefoot of the male are at least twice as long as those of his hind foot, whereas in the female fore and hind nails are of nearly equal length.

The young individual, as is usual, has a nearly circular shell and a much longer tail relative to the shell length than the adult. Hartweg (1944) gave the following measurements of 27 hatchlings, some of which, at least, had passed the winter in their nests (measurements in millimeters):

Size	Carapace length	Carapace width	Plastron length	Plastron width	Shell height
Largest	28.0	27.0	26.3	13.4	4.3
Smallest	21.0	20.6	20.1	10.5	2.0
Average	24.7	23.4	23.0	12.0	3.0

Pigmentation of the dark plastral figure is more intense in the young than in adult individuals.

Habitat: The midland painted turtle is found in nearly any kind of quiet water, and to some extent in streams, although it seeks the slack-water shores and coves. It is partial to shallows with dense vegetation and is remarkably tolerant of industrial pollution of its habitat.

Habits: The midland painted turtle appears to be unusually resistant to low temperatures, since it has been reported as more or less active during every month of the year and has several times been seen moving about beneath ice. Most frequently, however, it goes into hibernation in mud or debris, under water, in October and November (Evermann and Clark, 1916, said they sometimes bask until December 3) and remains until some time in March. In most of its territory this is an extremely abundant turtle, more than one writer having referred to group basking by "several dozen" individuals on a single log or fallen tree. In Lake Maxinkuckee, Indiana, Evermann and Clark caught no fewer than 280 specimens in one haul of a seine only 35 feet long (July 25, 1899). The disposition of this race, like that of *picta*, might be described as ineffectually irascible, on first being caught, but its attempts to bite are rarely successful and the worst injury usually suffered by a handler is a scratch from its claws. It is readily tamed and flourishes in captivity. While both sexes of this turtle may often be found moving about on land, this appears to be mostly aimless wandering, although a definite tendency to migrate from one place to another has once been demonstrated. In Lake Mendota, Wisconsin, Pearse (1923a) recovered 166 of a large number of painted turtles that he had marked, and found that after an average period between marking and recapture of 5 months and 19 days the average distance traveled was only 365 feet per turtle. This would appear to indicate a surprisingly restricted home range for the subspecies, although it is perfectly possible that populations living under different conditions, especially where there were strong seasonal or sporadic changes in environmental factors, might migrate more or less extensively. The population with which Pearse worked was intermediate in character between *bellii* and *marginata,* perhaps closer to the former.

Cagle (1944d), also working with an intermediate stock (Illinois), found that painted turtles sometimes move overland from one body of water to another, and that the home range of an individual may "include parts or all of several water bodies."

Breeding: Mating has been observed from immediately following the termination of hibernation in the spring until late summer and even early fall. It has been suggested, from a failure to find live spermatozoa in the genital tract of a female after copulation, that the late matings may be nonfunctional, but this has by no means been established as a fact, and, if true, it is a most noteworthy phenomenon—for a lower animal, at least. Nesting occurs most often in June but has been observed from May 31 through July. The female seeks an open, fairly high location and may wander distances of several hundred yards to find an acceptable site, or on the other hand she may lay very close to the water. A nesting female watched by Hartweg (1944) excavated a nest on a sandy knoll near Ann Arbor, Michigan, on June 8, 1941, and laid in it 6 eggs that were deposited at intervals of from 25 to 73 seconds, a hind foot being thrust into the nest after each egg fell. After this nest had been covered Hartweg found five small sticks on top of it. Hatchlings from this complement of eggs passed the winter in the nest and were found in mid-March. In this same area Hartweg found other nests in which baby turtles had overwintered, whether in the egg or as hatchlings in the nest is not known, but it seems likely that both occurred, since some of the little turtles had an egg tooth and some did not. Emergence of some of these hatchlings was deferred until as late as May 2. Hartweg reasonably suggested that in the cases in which hatchlings overwintered in the nest their failure to emerge was due not to late deposition of the eggs from which they came but rather to a necessity to await such softening of the confining soil as would occur only with the spring rains. In Florida I have observed several turtle nests in which hatchlings were hopelessly imprisoned by a hard-baked nest roof which only the hardest deluge could have softened. The normal incubation period is unknown. The eggs are 3 to 11 in number, usually 5 to 8, and they are elliptical in form, ranging from 28.6 to 31.8 mm. (1.13 to 1.25 inches) in longest diameter and from 17.5 to 20.6 mm. (0.690 to 0.812 inches) in shortest diameter.

Feeding: The midland painted turtle is omnivorous, the bulk of its diet being composed of aquatic invertebrates and plants. By far the most extensive study of the feeding habits of this form was made in Michigan by Lagler (1943b), who examined the contents of 394 stomachs that contained food. His results are summarized in Table 7. There is essential agreement between the findings of Lagler and those of other workers who have made less definitive analyses of stomach contents of this race in other places.

Table 7. Stomach contents of midland painted turtles.

Food item	Composition by volume (%)	Frequency of occurrence (%)
Game fishes	1.0	1.0
Forage fishes	0.3	1.5
Fish remains	0.1	0.8
Frog remains	0.4	0.3
Carrion	2.5	1.5
Spiders and water mites	traces	3.0
Leeches and "earthworms"	0.4	3.4
Crustaceans	5.0	15.5
Insects	19.5	55.1
Mollusks	5.5	15.7
Lower plants	30.7	58.9
Higher plants	30.8	48.2
Vegetable debris	3.7	35.5

Economic importance: Although these little turtles are eaten once in a while by the people who catch them, they are apparently not sent to market. Pearse, Lepkowsky, and Hintze (1925), referring to a Wisconsin intergrade stock, stated with astonishing finality that these turtles constitute a menace to fish management and should be destroyed. Neither anything that those writers found nor anything discovered since they wrote may be regarded as justification for such a remark, which is really a classic example of the sort of judgment the scientific practitioner of conservation tries to avoid. It contrasts sharply with the following more level-headed conclusions of Lagler, based upon a far more comprehensive body of data:

The number of game and pan fishes consumed by the painted turtles examined by me is insignificant for fish management. Regarding the oft-repeated accusation of egg predation by this and other aquatic turtles this series of specimens offers only negative evidence. Several of the individuals collected near centrarchid nests contained no trace of eggs or fry of these fishes. Members of this species have, however, been seen by Vern Winey of Kalamazoo on bluegill nests, supposedly foraging on the fry, in a position which made the turtles appear to be standing on their heads. No specimens were captured to verify these observations but Winey was certain that fry were present in the nests.

On the basis of the data assembled the mid-western painted turtle appears to have significance as a food competitor of game and other fishes but may hardly be considered a predator of these forms. Without additional quantitative work on kinds and amounts of food available and kinds and amounts actually consumed by associated fish species, I cannot estimate the seriousness of the food competition which they offer to fishes.

After a study made in Chautauqua Lake, New York, Raney and Lachner (1942) concluded that *C. p. marginata* is not important as either a predator or a competitor of game fishes.

Genus *PSEUDEMYS*

This group of medium-sized to large turtles, often referred to as "terrapins," is distributed from New England to Brazil and the West Indies. Fossil representatives have been found in Florida deposits dating back to the Pliocene, and the group was well represented in the Pleistocene.

In the species of this genus the alveolar surface of the upper jaw always has an elevation parallel with the edge; this may be a continuous ridge or a row of more or less separate conical tubercles. There is a bony temporal arch. The toes are fully webbed and are four and five on the fore and hind feet respectively. In all the forms the nails of the fore feet of the males are much longer than those of the females. The carapace of the adult may be keeled or not, and it is never marked by a median longitudinal stripe. There are always light stripes (usually some shade of yellow) on the head and neck except where lost in older melanistic individuals. The carapace is usually more or less wrinkled and has a serrated hind margin.

On numerous occasions herpetologists have proposed the splitting up of *Pseudemys* into two, three, or even four genera corresponding to varyingly distinct group trends that may be recognized. At least one such suggestion—that which would revive the old generic name *Trachemys* for the *scripta* section—seems fairly reasonable. The characters used in defining the subgeneric groups are subject to so much variation, however, that I can see little to be gained by subdividing the small genus and elevating to generic rank what is probably merely a subspecies complex.

KEY TO *PSEUDEMYS*

1   Alveolar surface of upper jaw usually not narrower anteriorly than posteriorly; alveolar ridge coarsely and irregularly tuberculate; general outline of tomium of upper beak, from both anterior and ventral aspects, rounded and obtuse, although sometimes with a median notch that may or may not be bounded on either side by a cusp; underside of mandible flat 2

1′   Alveolar surface of upper jaw usually narrower anteriorly than posteriorly and with the ridge smooth or finely serrate; general outline of fore part of cutting edge of upper jaws angular from anterior and ventral aspects; underside of mandible rounded 14

2(1)   Upper jaw smooth or notched at the symphysis, but without cusps bordering the notch; if such cusps are present, then the head pattern is of numerous stripes, many of which are subequal in breadth, broken, and interconnected laterally on the head and neck, and with the expanded portion of the supratemporal stripe detached 3

2′   Upper jaw with a terminal notch bordered by a strong cusp on either side; stripes of the head either greatly reduced in number or conspicuously of several orders of width and not extensively broken or laterally interconnected; expanded portion of the supratemporal stripe not detached 11

3(2)   Upper jaw smooth at the symphysis, lacking both notch and cusps 4

3′   Upper jaw notched at symphysis, with or without lateral cusps 9

4(3)   Plastron without dark markings 5

4′   Plastron with a rudimentary to extensive black or dusky figure, at least in juveniles and younger adults 6

5(4)   Supratemporal and paramedian head stripes confluent or nearly so behind the eye, continuing forward along the top of the head toward the snout as one line; lower marginals with solid, smudgelike blotches; plastron (in life) greenish white or greenish yellow; peninsular Florida . . . . *floridana peninsularis* (p. 289)

5′   Supratemporal and paramedian head stripes separately continuous along top of head; blotches on lower marginals enclosing light areas; plastron yellow or orange yellow in life;

coastal plain from North Carolina to Florida and Alabama
.... *floridana floridana* (p. 282)

6(4′) Length/height ratio, males 3.06–3.46, average 3.26, females
2.62–3.11, average 2.87; upper jaw not, or but slightly, serrate; ground color of soft parts light to dark brown; stripes on
head and limbs yellow or orange, occasionally reddish; ground
color of carapace brown, usually with conspicuous light markings 8

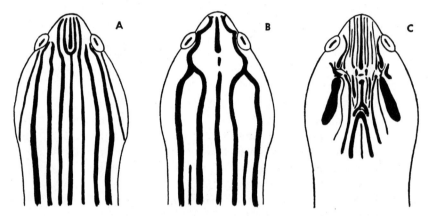

Figure 18. Three styles of dorsal head patterns in the *floridana* section of *Pseudemys. A:* The *floridana-mobiliensis-concinna-hieroglyphica* pattern. *B: P. f. peninsularis. C: P. f. texana.* The black bands in the diagrams represent light stripes.

6′ Length/height ratio, males 1.71–3.16, average 2.35, females
2.28–2.73, average 2.63; ground color of head, legs, and carapace black or brown and upper jaw serrate, often strongly so;
if upper jaw not serrate, then markings of carapace reduced,
the ground color greatly predominating 7

7(6′) Ground color of carapace, limbs, and head light to dark
brown, the stripes and reticulations yellow, orange-yellow,
or reddish; four or more lines on outer surface of fore limb;
outer surface of hind limb striped; usually seven or more lines
between the eyes; Gulf coast from extreme western Florida
to eastern Texas .... *floridana mobiliensis* (p. 301)

7′ Ground color of carapace, legs, and head lustrous, sooty black,
the stripes and reticulations light greenish-yellow; two or three
lines on outer surface of fore leg; outer surface of hind leg not
striped; five lines between the eyes; rivers of the Gulf drainage
and coastal waters in Florida from near Apalachicola to Citrus
and Sumter counties .... *floridana suwanniensis* (p. 296)

8(6) Alveolar surfaces of lower jaws relatively narrow and without high, isolated conical teeth on the median ridge; fore and hind feet mostly dark, unstriped above and below; tail not striped above; Maryland to eastern Tennessee and northern Alabama in the Piedmont .... *floridana concinna* (p. 286)

8′ Alveolar surfaces of lower jaws relatively broad and with a few high, isolated conical teeth on the median ridge, that nearest the symphysis especially strong; fore and hind feet marked with continuations of the light leg stripes; tail striped above; Illinois and Indiana south to northern Louisiana, between the Appalachians and the Mississippi River ....
.... *floridana hieroglyphica* (p. 304)

9(3′) Upper beak with the notch at the tip bordered by strong, sharp or rounded cusps; lower jaw with a strong, cusplike projection on either side of the median tooth; southwestern Texas and northeastern Mexico .... *floridana texana* (p. 312)

9′ Upper jaw without cusps at the symphysis; tomium of lower jaw also without cusplike projections other than the terminal tooth 10

10(9′) Lines on top of head between the supratemporals at a point above tympana numerous, usually eight or more, subequal in breadth, many of them discontinuous and interconnected by one or more straight or crescentic transverse bars behind the sagittal line; plastron very light, often almost white; west of the Mississippi River from southeastern Kansas and southern Missouri to northwestern Louisiana and central Texas ......
.... *floridana hoyi* (p. 307)

10′ Lines on top of head between supratemporals and above tympana usually fewer than eight, not extensively broken or interconnected by transverse bars; ground color of plastron not whitish but deeper yellow, sometimes orange or reddish; Illinois and Indiana south to northern Louisiana, between the Appalachians and the Mississippi River ....
.... *floridana hieroglyphica* (p. 304)

11(2′) Pattern of head and of carapace much reduced in adult; top of head often plain black; shell with one broad light vertical band across each lateral and another across each marginal, or irregularly mottled with red and black; carapace without a reticulate pattern of many thin lines 12

11′ Head with a few to many thin, well-defined lines always evident between the eyes; pattern of carapace a complex system of reticulate or ocellate thin lines; the *"alabamensis"* phase of

P. f. *suwanniensis* or of P. f. *mobiliensis;* return to couplet
number 7 ....

12(11) Centrals convex, the posterior ones slightly keeled; widest
        point of carapace usually anterior to middle; markings on
        lower marginals solid, smudgelike; peninsular Florida ....
        .... *nelsoni* (p. 274)

12′   Centrals concave or flattened; widest point of carapace usually
        posterior to middle; markings on lower marginals enclosing
        light areas 12

13(12′) Length/height ratio usually more than 2.5; Atlantic coast
        from North Carolina to Long Island ....
        .... *rubriventris rubriventris* (p. 267)

13′   Length/height ratio usually less than 2.5; Plymouth County,
        Massachusetts .... *rubriventris bangsi* (p. 273)

14(1′) Side of head usually with a conspicuous oblong or triangular
        yellow patch immediately behind eye; shell short, rough, and
        highly arched; Virginia to Florida in the coastal plain ......
        .... *scripta scripta* (p. 241)

14′   Side of head with lines extending from posterior margin of
        orbit, or with a detached, elliptical spot on temporal region;
        shell usually smooth, moderately elongated, and not highly
        arched 15

15(14) Pattern of carapace, if evident, with conspicuous transverse
        markings, lacking regularly placed black spots; pattern of
        plastron of ocellated or solid black spots that usually do not
        run together to form a linear, oblong figure 16

15′   Pattern of carapace obscure or reticulate, and usually includ-
        ing a black spot on the lower hind corner of some or all of the
        laterals and upper marginals; pattern of plastron a single,
        oblong dark figure 17

16(15) Light stripes on head and on beak broad and yellow; under
        surface of neck, legs, and tail light with dark stripes; ocelli
        on lower marginals single, bridge with more light than dark
        color; upper reaches of the Tennessee and Cumberland rivers,
        eastern Tennessee .... *scripta troostii* (p. 258)

16′   Light stripes on head and beak narrower, usually with some
        red; under surface of neck, legs, and tail dark with light
        stripes; ocellated spots on lower surface of marginals concen-
        trically double; bridge predominantly dark; Mississippi Valley
        from Illinois, Iowa, and Ohio to Alabama and Texas ......
        .... *scripta elegans* (p. 248)

17(15′) Black spots on carapace mostly ocellate, light-centered; tem-

poral spot very conspicuous, detached, nearly round, and intensely black-bordered; southwestern Texas and northeastern
Mexico ....*scripta gaigeae* (p. 262)

17′   Black spots on carapace usually not light-centered; temporal
spot elongated, continuous behind with a stripe on the neck,
not strikingly black-bordered; southern Baja California ....
................................*scripta nebulosa* (p. 264)

The *scripta* Section

This is the genus *Trachemys* of some authors. Of the three sections of
Pseudemys it is most trenchantly differentiated and, as well, the most
widely distributed, being known from Virginia to Brazil and the West
Indies. In all the forms of this group the beak is relatively smooth, not
cusped, and without strong serrations in most cases. The cutting edges of
the maxillae meet at the symphysis at a vertical angle, where an auxiliary
notch of varying depth and acuity is often present. The alveolar surfaces
of the upper jaw are relatively smooth, their median ridges low and but
slightly toothed, if at all. The supratemporal and the orbitomandibular
head stripes are the most conspicuous, the former often with a marked
expansion of its anterior portion which may or may not be detached; the
lines on the top of the head are often reduced or lacking. The pattern of
the plastron varies through a geographic gradient from the simple one of
single laminal spots (which sometimes are entirely lacking) of *scripta* to
the very extensive dark plastral figure of *P. s. ornata* and other forms in
the American tropics. The taxonomic confusion that for more than a hundred years has existed in this section of the genus has been due mostly
to the marked sexual dichromatism that the group displays and to an
added distractor contributed by the fact that this dichromatism does not
distinguish the sexes at all ages and sizes. Time and again the males have
been described as species distinct from the females; in Cuba the country
people still know the female of the local representative as *jicotea,* while for
the melanistic male they have the name *jarico,* and zoologists still find it
convenient to cling to a shred of the old delusion and refer to the discolored males as the *"rugosa-*phase." Likewise, in the United States, herpetologists who had reached their majorities fifteen years ago when Percy
Viosca (1933) confounded reptile students and turtlemongers alike by ex-
plaining *"troostii"* as the melanistic male of *"elegans,"* still tenaciously
hold to their notion of a *"troostii-*phase" of the red-eared turtle (or of the

yellow-bellied turtle) and the necessity for reapplication of the name *troostii* in an altogether different sense makes hard going for them.

The phenomenon of melanism in old males occurs to a marked degree in North American and West Indian forms of the *scripta* group, to a lesser extent in the tropical mainland subspecies of the same complex, and rarely and in a much more desultory and erratic way in the *rubriventris* and *floridana* sections of *Pseudemys*. Barbour and Carr (1940) commented as follows regarding the evolution of the melanistic pattern:

Melanism seems to be a condition that is visited upon the males, not at the age of puberty or of sexual maturity, or on attaining any preordained length, but at some quite indefinite and variable time during adult life. Thus, one finds some males colored like the females and others like *"troostii"* (and others in between) and all about the same size. Moreover, we have seen certain indications that an occasional male lives to a ripe old age without ever becoming melanistic.

The first indications of incipient melanism are rarely found in specimens under 150 mm. (5.9 inches) in length. There is some evidence that, once begun, the process may be fairly rapid. The thin lines along the bridge and those forming the marginal concentric figures darken and begin to break up. The black borders of the stripes on the head and limbs darken and broaden. Flecks of black pigment are deposited along and between the plastral seams, at first in the form of a somewhat vague but symmetrical figure, which progressively grows more conspicuous, in one phase the entire plastron becoming vermiculate with pigment, and in another the pigment remaining confined to the region of the seams. Deposition of pigment on the carapace begins on the anterior laminae and progresses posteriorly; in many specimens the anterior half of the carapace is heavily vermiculated and the posterior portion untouched. As the black flecks anastomose and come to occupy more area than the ground color, the latter changes from the original olive or brown to slate-blue, horn-color or light yellow, the result being a black shell speckled with yellow or bluish. Nearly the same process takes place on the plastron, except in certain individuals in which, as stated above, the black pigment apparently remains permanently confined to the seams. Meanwhile, the black lines bordering the yellow stripes on head and limbs have broadened and broken up, as have the yellow stripes themselves. With the fragmentation of the yellow markings the yellow pigment (sometimes) increases markedly in intensity and often in amount. Eventually all the original markings are obliterated, and the legs are vermiculate, entirely black, or entirely yellow.

Although the above account of melanistic changes is quoted from a paper treating the West Indian members of the *scripta* group, it is almost wholly applicable to the process as it occurs in the related forms in the

United States. The one phase that often turns up among the Antillean turtles and that I have never seen in *scripta* or *elegans* is the predominantly yellow individual with relatively few dark vermiculations, but this is merely the accentuation of a common trend toward diffusion of the light markings.

In addition to the changes in coloring associated with sex, there is in the turtles of this group a tendency to obliterate the juvenile shell pattern, usually largely one of ocellated rings of pigment in the bony plates, by secondary deposition of black pigment in the overlying horny laminae. The laying down of these smudges occurs most conspicuously on the plastron but may take place on the carapace as well.

Yellow-bellied Turtle *Pseudemys scripta scripta* (Schoepff)

(Plates 44, 45. Map 17.)

Range: Extreme southeastern Virginia southward across the eastern half of North Carolina and all but the northwestern tip of South Carolina

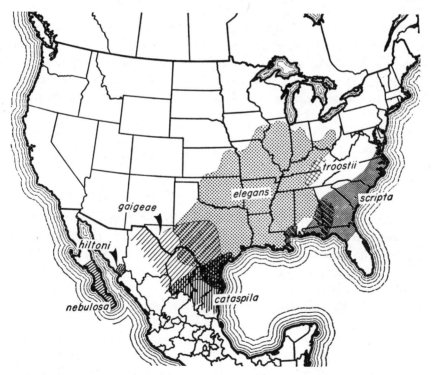

Map 17. Distribution of the subspecies of *Pseudemys scripta* in the United States and adjacent parts of Mexico.

to western Georgia and northern (Alachua and Levy counties) Florida. In the panhandle of Florida, especially in the Apalachicola drainage, and in western Georgia and southern Louisiana intergrades between *scripta* and *elegans* have been found, and in northern Georgia and Alabama and eastern Tennessee stocks intermediate between *troostii* and *scripta* should be looked for.

Distinguishing features: The large yellow patch behind the eye and the sparse, deep-black smudges on the plastral laminae are enough to separate this turtle from all others, except in the case of the old males, in which the color pattern is drastically modified and which often may be distinguished from similarly discolored males of *elegans* only with the greatest difficulty and uncertainty. The shell of *scripta* is nearly always shorter, heavier, broader, and considerably higher and more wrinkled longitudinally than that of the related subspecies, but in this respect as in several others *troostii* occupies a position intermediate between *scripta* and *elegans*.

Description: The shell of the yellow-bellied turtle is relatively short and broad, either straight-sided or, more frequently, oval in lateral outline, fairly deep, and with very thick, heavy bones. The size is moderate, the United States races of *scripta* being somewhat smaller than the average for the genus. Brimley (1907) mentioned a specimen with a shell 272 mm. (10.70 inches) long, but this is much larger than the usual mature size. A young adult male from Gainesville, Florida, and a female of medium size from Charleston, South Carolina, had the following shell dimensions:

Sex	Length	Width	Height
Male	142 mm.	111 mm.	61 mm.
	(5.60 inches)	(4.38 inches)	(2.40 inches)
Female	198 mm.	160 mm.	94 mm.
	(7.80 inches)	(6.30 inches)	(3.69 inches)

The carapace is more or less keeled above except in very old males, the precentral is long and narrow, and the hind margin is rather strikingly serrate. The ground color is dark brown to black with a varyingly evident pattern of transverse light bars on the laterals, one traversing the middle part of each lamina usually predominant. The lower surfaces of the shell are yellow with deep-black, finger-tip sized smudges at the posterior edge of each lower marginal, along the bridge, and on some of the plastral laminae; these smudges are actually composite markings, there being an ocellate ring in the bone beneath them which becomes hidden by heavy deposit of black pigment in the overlying horny lamina as the turtle grows.

Plate 44. Pseudemys scripta scripta. A: Nonmelanistic male. *B:* Hatchling, showing egg tooth. *C:* Young female. *D:* Old male. All from Gainesville, Florida.

Plate 45. Young specimens of *Pseudemys scripta* from Jackson County, Florida, showing variation in plastral pattern ranging from the scantiest "scripta" type (*upper left*) to a complex figure often seen in young *P. s. elegans* (*lower right*).

The spots are smudged in some hatchlings, but the majority begin life with only ocelli. On the plastrals the smudges occur one to a lamina, almost never more, and though their number is remarkably variable their distribution adheres to a constant plan. If only two spots are present they are almost invariably on the gular laminae; if three or four are present they occupy gulars and humerals, and so on, additional spots usually coming in from front to rear and never vice versa. Mr. C. S. Brimley of Raleigh, North Carolina, once counted the smudges on the plastra of 49 yellow-bellied turtles and found their number to vary as follows:

Number of spots	Number of turtles
1	2
2	36
3	2
4	3
5	1
6	2
7	1
10	1
12	1

In populations of *scripta* from areas near the range of *elegans,* which is a many-spotted form, a greater number of specimens have a greater number of plastral spots, as might be expected. The head and legs are brown to nearly black, with yellow stripes that in general are few and broad, especially on the head. The most distinctive marking of this turtle is a broad yellow vertical patch behind the orbit, formed by the fusion of dorsolateral and lateral neck stripes, but with the portion of the dorsolateral stripe immediately behind the yellow patch faded or obsolete, except in stocks from near the range of *elegans,* in which the connection is evident and the patch is often detached from the lateral stripe instead, and may be red instead of yellow.

Sexual differences in *scripta* include a somewhat smaller size for the male, which has also a longer tail, a lower, narrower shell, stronger, thicker, somewhat longer (but not greatly elongated, needlelike) and strongly curved nails on the fore toes, and a very radical modification of the juvenile color pattern which may occur in individuals of nearly any mature age but which usually begins to show up in specimens of 165 mm. (6.50 inches) or so, and which involves complete obliteration of the original yellow and black markings and the substitution of dusky and olive mottling. These changes are described in some detail in the paragraphs on the *scripta* section, in which the phenomenon is widespread.

The hatchling yellow-bellied turtle has a shell between 31 and 33 mm. in length and of nearly the same width. Its markings are those of the adult female and nonmelanistic male except that those of the shell show more of the pigment in the bone than of that in the laminae and thus are more evidently ocellated and less smudgelike. A hatchling from near Gainesville, Florida, showed the following changes in shell dimensions during a three weeks' period:

Date	Length	Width	Height
August 26, 1944 (about one week old; egg tooth and yolk sac present)	32.0 mm. (1.26 inches)	31.3 mm. (1.22 inches)	16.4 mm. (.650 inches)
September 12, 1944	32.5 mm. (1.28 inches)	33.0 mm. (1.30 inches)	16.0 mm. (.634 inches)

The decrease in shell height indicated and the rapid gain in width were evidently due to the fact that, even though a week old at the time the first measurements were made, the shell of the hatchling still showed the effects of having been cramped to conform with the eggshell. It is not rare to find young specimens of *scripta* that are wider than long, as at the second date above.

Habitat: This turtle is most typically a pond inhabitant. For instance, it is usually the most abundant chelonian in certain small, well-like sinkhole ponds in northern Florida. It is also found in small streams and sloughs and at the edges of lakes.

Habits: Although no systematic investigation of the habits of *P. s. scripta* has been made, scattered observations shed some light on its life history. In northern Florida it is frequently seen, owing to its partiality to small ponds and sinks which absence of aquatic vegetation may render unfit for occupancy by other emydid turtles. It may be this tolerance of relatively sterile environments that accounts for the fact that *scripta* is a more solitary turtle than its relatives, rarely being found in large sunning aggregations but more often occurring singly or in small groups of three or four individuals. It is sometimes seen sunning with other species of *Pseudemys* but more usually not, since a large percentage of the situations in which it lives are shared only with *Chelydra, Kinosternon,* and *Sternotherus,* and these are not particularly fond of basking on logs.

Further evidence of its ecologic tolerance is offered by Brimley (1920,

1943), who noted that during a period of progressive pollution of streams near Raleigh, North Carolina, the local population of *scripta* increased markedly in those streams while at the same time *P. f. floridana* almost disappeared.

Neill (1948a) learned that in Richmond County, Georgia, this turtle hibernates in mud and aquatic vegetation at lake margins, but attributed the finding in the woods of large numbers of its shells to poor weather forecasting on the part of belated or prematurely emerged hibernators. In northern Florida individuals may emerge to bask on nearly any warm winter day.

Breeding: I have found eggs or nesting females only during the months from May through July. Brimley (1909) collected a specimen that had just dug a nest cavity on May 24 in Craven County, North Carolina.

Although there is no published description of courtship in this subspecies, Dr. Lewis Marchand informs me that he watched captive individuals engage in the same bizarre maneuvers that have been reported for *P. s. elegans,* during which the male swims backward in front of the female as he strokes her face with his long nails. That this must probably be regarded as the normal and characteristic *liebespiel* of this section of the genus now seems evident, since Dr. Coleman Goin has recently informed me that he observed essentially the same antics performed by the closely related *P. terrapen* in a garden pool in Jamaica.

The nests are dug near water, rarely more than 200 yards distant, and in Florida the females do not wander so widely in prospecting as do those of *P. floridana peninsularis.* The size of the nest varies with the size of the turtle, the maximum depth of the bottom of the egg cavity being about five inches and the neck of the bottle-shaped cavity being barely wide enough to admit an egg. I have never found any auxiliary pockets containing single eggs, such as are common in other species. Eggs appear to be usually ten to twelve in number, although by chance the several females that I have found to contain shelled eggs or have discovered completing the nesting operation have all been small individuals, and it is possible that the complements of older females may be larger. Eleven eggs laid May 9, 1944, at Lake Alice, Alachua County, Florida, varied in length between 36.0 and 40.5 mm. (1.42–1.59 inches) and in width between 22.0 and 24.0 mm. (0.865–0.945 inches). These eggs, after being moved to the laboratory, hatched during August, 1944, and by the 26th of that month all the young turtles had broken their egg shells, although none had emerged from the confining sand. As in the case of other species the newly hatched young may remain interred in the nests for long periods of time awaiting heavy rainfall that is required to soften overlying encrusted soil.

Feeding: There is some evidence that this turtle is omnivorous in its feeding habits. Brimley found it common in polluted streams, and in the southern part of its range it often occurs where aquatic vegetation is meager or even entirely lacking. I have caught specimens on hooks baited with liver and watched an individual eating submerged plants. Marchand (1942) made the following notes regarding the food of *scripta:*

A male from Hart Springs, Gilchrist Co., Florida, length 130 mm.; Feb. 14, 1941. Stomach contents: 1 small crayfish; 2 Zygoptera larvae; 1 small shrimp (*Paleomonetes paludosa*); *Lemna sp.,* 50%; *Naias sp.* 50%.

A female from Hart Springs, length 90 mm., October 27, 1940. Stomach contents: dermal plate from small turtle with filamentous algae attached; other filamentous algae, trace; 2 Anisoptera larvae; *Ceratophyllum sp.,* trace; *Lemna sp.,* 90%; *Naias sp.,* 10%.

This meager evidence might indicate that *P. s. scripta* is more carnivorous than members of the *floridana* complex. The related subspecies, *P. s. troostii* [*P s. elegans*] is known to be largely carnivorous. On November 11, 1941, in the Chattahoochee River near Bascom, I caught a *P. s. scripta* on a trot line baited with meat.

Economic importance: Although sometimes included in market shipments of other more abundant cooters, *scripta* has little commercial value. However, it is not only caught for food by local fishermen and frog hunters, but in west Florida, where turtles are commonly speared in deep, clear springs, it is regularly recognized and is called "Baptist," with reference to the thick, hard shell which is more resistant to the gig points than that of the "Methodist," or softer-shelled Mobile turtle.

Red-eared Turtle *Pseudemys scripta elegans* (Wied)

(Plates 46, 47. Map 17.)

Range: Northern Illinois and Indiana southward to the Gulf and westward through southeastern Iowa, most of Missouri, eastern and southern Kansas, all but the panhandle of Oklahoma, and the eastern half of Texas. In central and southern Texas (and scattered through the border areas of eastern Mexico as well) intermediate specimens show its close relationship to *P. s. gaigeae*. To the east it is known from southern Ohio, the western halves of Kentucky and Tennessee to northeastern Alabama, and southern Louisiana. Intergrades between the red-eared turtle and the yellow-bellied

Plate 46. Pseudemys scripta elegans. Upper: Female; San Antonio, Texas. *Center:* Melanistic male; southern Louisiana. *Lower:* Female; Hickman, Kentucky. *Inset:* Young; Plaquemine, Louisiana.

turtle have been found in western Florida, western Georgia, and southern Louisiana. Geographically distributed intergradation between *elegans* and the Cumberland terrapin has not been demonstrated, but all the characters that differentiate the latter show up from time to time in nearly any Mississippi Basin population of *elegans*.

There is an interesting possibility that the red-eared turtle may have made material extensions of its range, or at least of its range-in-abundance, during relatively recent historic times. At the time of his death, Dr. Leonhard Stejneger of the United States National Museum had been gathering material for a monograph of the North American turtles, and from here and there among his notes I collected the following penciled comments on this possibility, and I quote them below almost as found:

Pseudemys elegans, of all our freshwater turtles the most easily and unmistakably identified from infancy up, because of the unique and insistently conspicuous red patch above the ear, is—with the possible exception of the painted turtle—at present the commonest and most widely distributed species west of the Alleghenies. Curiously enough, it was unknown to all our early travellers and herpetologists. Latreille, Daudin, Cuvier, Merrem, Bell, Say, Harlan, LeConte knew it not. Even LeSueur, who during his protracted stay at New Harmony paid special attention to the turtles and discovered *Graptemys geographica, G. pseudogeographica, Amyda spinifera,* and *A. mutica* in the Wabash River, had never seen nor heard of it there. Similarly, Troost, who collected assiduously for Holbrook, did not obtain it in Tennessee, and the latter consequently failed to describe and figure this remarkable turtle in his great *Herpetology of North America* (1836–1842).

However, about ten years after LeSueur's activity on the Wabash the prince Maximilian von Wied collected there the first specimen and described it in 1838 as *Emys elegans,* but it became not generally known until Agassiz in 1857 introduced it into North American herpetological literature as *Trachemys elegans.* Gray, it is true, in 1844, described a shell from Louisiana as *Emys holbrookii,* which has been identified as being that of a *P. elegans.* Agassiz knew the turtle from . . . [This sentence was not finished, but Agassiz had specimens from Texas; Iowa; Quincy, Illinois; Mississippi; Louisiana; one of the head waters of the Missouri; and from southern Texas and the border area of Mexico; he appears to have seen none from the region east of the Missouri River].

There are several specimens of *P. elegans* in the Paris Museum entered as having been [received] from LeSueur. However, they seem to have been acquired in 1847, and it seems probable that he may have obtained them after he had discovered the existence of Wied's new species. Wied, in 1864, wrote:

"M. LeSueur has assured me that during his long stay at New Harmony he never received this turtle."

The question thus naturally arises whether it has not greatly extended its range eastward, especially in the Ohio, Cumberland, and Tennessee rivers and their tributaries, during comparatively recent times.

Acceptance of this idea would make it somewhat easier to account for the presence in the upper reaches of the latter two rivers of a rather well-differentiated subspecies (*P. s. troostii*), since the two may have previously become isolated at some indefinite time in the past and have remained so until the present upriver extension of the range of *elegans*.

A range extension on a smaller scale has been noted by Edgren (1943, 1948) who found that a colony of *elegans* has increased the diameter of its range between 0.3 and 0.4 miles per year since the original introduction into a lake in Muskegon County, Michigan.

Distinguishing features: The red-eared turtle can usually be recognized among the members of its section of *Pseudemys* by its low, nearly or quite unkeeled and smooth shell, the several thin longitudinal stripes extending backward from the hind border of the orbit, and, most readily, by the long, oval expansion of the broad supratemporal stripe, which is usually bright red in color. There are thin, light lines on the beak and the plastron and bridge are extensively marked with black. The separation of old melanistic males of *elegans, scripta,* and *troostii* is very difficult, and in some cases impossible.

Description: The shell of this subspecies is a short ellipse, evenly convex before and at the sides, and rather low and flattened above. The size is medium; the maximum is said to be a shell length of about 280 mm., but the usual size is much smaller. Cagle (1944b and 1948a) showed that males become mature when their plastrons are between 90 and 100 mm. (3.54 and 3.95 inches) in length, while the females are ready to lay eggs at plastral lengths of about 160 mm. (6.31 inches). The same writer found that some old females with plastral lengths ranging from 210 to 220 mm. (8.27–8.67 inches) were sexually senile and apparently no longer capable of reproduction. Dimensions of a medium-sized male and female from Illinois were as follows:

Sex	Carapace length	Carapace width	Shell depth
Female	200 mm. (7.87 inches)	154 mm. (6.07 inches)	73 mm. (2.87 inches)
Male	156 mm. (6.15 inches)	122 mm. (4.80 inches)	55 mm. (2.16 inches)

The carapace is relatively smooth or somewhat rugose longitudinally and unkeeled, or with only a vestige of a keel, the posterior margin flaring outward and only slightly serrate. The precentral is long and narrow. The ground color of the carapace is brown or olive brown, with yellow and black lines and bands mostly longitudinally directed on the centrals, especially anteriorly, and transverse on the laterals, where there is usually one especially broad and conspicuous yellow band near the center of the lamina and a rather conspicuous black band or two and several less distinct yellow lines, all roughly parallel. The upper and lower marginals are usually black-spotted at the seams, the black spots often surrounded by concentric rings of black and yellow. The plastron is yellow, most often with a black smudge on each of the laminae overlying the juvenile ocellate spot on the bone beneath; this plastral pigment may extend over the greater part of the surface. The bridge and axial and inguinal laminae are usually heavily pigmented by black smudges or elongated bars, the bridge sometimes being almost entirely black. The head and legs are olive or brownish. The stripes on the head are mostly yellow, those on top very narrow and those on the sides of the head and neck broader. The most conspicuous of these stripes is a broad red or reddish-orange band that begins at the hind rim of the orbit and extends backward to the neck where it loses the reddish color and becomes quite narrow. A suborbital stripe passes back beneath the tympanum to be intercepted or not by another broad stripe, the mandibular, from the angle of the jaw. The legs and tail are striped with yellow.

The male red-eared turtle is smaller than the female, although not as markedly so as in other species in the family, and has longer, more curving fore nails (which are twice as long as those of the female or as those of its own hind feet). The vent of the male opens on the tail beyond the posterior edge of the shell, while that of the female does not extend beyond the edge. The most striking sexual difference is the melanism in old males by which a vague, dusky mottling and shading of shell and soft parts are substituted for the juvenile pattern of yellow stripes and black smudges or ocelli. Development of this condition is described in some detail in the remarks on the *scripta* section. The female may show a slight tendency toward the same secondary pigmentation but never to anything like the extent to which it occurs in the male.

The young are similar to the adult in most respects but have a higher, more strongly keeled shell and less pigment in the horny laminae of the shell, allowing the ocellated markings beneath to be seen more clearly.

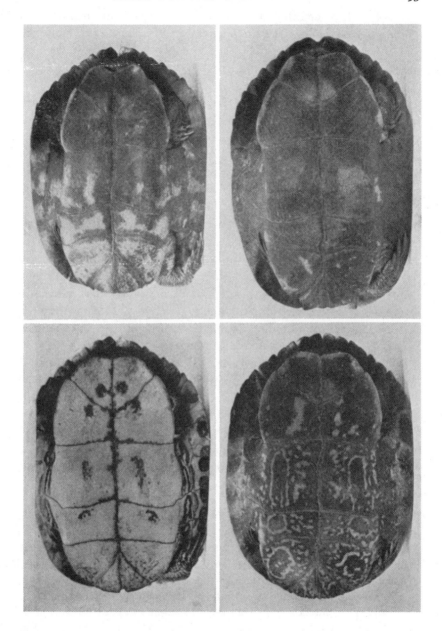

Plate 47. Pseudemys scripta elegans. Male specimens showing various degrees of melanistic change in coloration. All from Hickman, Kentucky.

Most of the light markings are more intense and vivid in small specimens. Young red-eared turtles are the universal pet-shop turtle and many thousands are handled by the trade each year. The shell of the hatchling is usually between 29 and 30 mm. long (1.14 and 1.18 inches) and of the same width or slightly less.

Habitat: Although typically an habitué of larger bodies of quiet water such as lakes and quiet coves and oxbows of the larger rivers, this turtle has also been found frequently in small, open, shallow prairie ponds, where at times the water level may be too low to cover its shell. Lakes and river coves with low, swampy shores and some aquatic vegetation appear to be the most favored habitats.

Habits: In his extensive studies of the biology of this turtle, Cagle was able to determine the extremes of temperature between which it is normally active in southern Illinois as 10° and 37° Centigrade. This temperature tolerance usually permitted continuous activity during the period from late April or early May to early October, but it was noted that during the winter there was temporary resumption of life processes at times when water temperatures were propitious. Also observed was a temporary recession of activity in midsummer and a subsequent increase in August and September. Feeding was found to cease after early October.

This is an extremely aquatic turtle that appears to take to land only for nesting purposes or to make migrations for one cause or another. Cahn regarded it as a "quick, active species," and it is of interest to note that this is not an appraisal that could be made of its southeastern relative, *scripta,* whose short, high, and very thick shell makes it a relatively slow and clumsy swimmer as compared with other species of the genus. Cahn made the following comments regarding the daily activities of *elegans* in Illinois:

If great beds of aquatic vegetation are available the turtles wander around, in and out, in the thickest of the growth, their perambulations often wearing open pathways or tunnels through it. As a result of their selection of quiet, vegetated regions they are, like the turtles of the genus *Chrysemys,* frequently found with dense growths of algae on the carapace, and little growths of the colonial protozoan, *Epistylis articulata* are often found upon the plastron. If logs are available out in mid stream, the turtles pile up on them two and three deep, and spend the hotter hours of the mid-summer days sunning themselves in the security of their island retreat. If logs are not available their heads may be seen poked out above the surface, just beneath which the turtles lie at rest for long periods at a time, apparently quite as relaxed as if they were on a more secure footing. Always watchful, the least sign of anything dangerous sends them beneath the surface like a flash, and once frightened, they are more cautious than ever. Refuge is sought at once in the heart of the densest vegetation, to the very bottom of which they burrow immediately, remaining motionless until the fright reaction wears off. The sloping bank of the water's edge is seldom utilized and one rarely sees a turtle of this species basking in such a location.

Cagle's studies of the growth of this turtle revealed the astonishing possibility that, although increase in size varies markedly with the individual and with the environment, under optimum conditions males may attain sexual maturity in "slightly more than one growing season," and females in three seasons.

Records of complete season's growth computed from measurements of growth rings permit a partial analysis of the influence of increasing size, attainment of maturity, seasons, and ecological factors on growth. The plastral length of *Pseudemys scripta troostii* [*elegans*] hatchlings computed from measurements of the birth plate on 32 juveniles 2–6 years of age ranged from 2.28 to 3.32 cm., mean 2.71. Individuals of this size group had a season's growth increment of 18–46 mm., mean 30 mm., these turtles thus average doubling their growth in the first season. This rapid growth decreased consistently in males with increase in size, but remained high until attainment of maturity (size group 10–11 cm.). Growth slowed rapidly beyond this point.

The females exhibit a similar but less rapid decrease in annual growth increments with increase in size [Cagle, 1948b].

The "birth plate" mentioned above by Cagle is the infantile lamina, which may persist in this genus for a few years as a smooth or granulated area arrounded by "annuli" or growth rings limiting the growth of subsequent seasons, but which is soon lost by periodic shedding and by wear. Chiefly because of this shedding of the laminae, occurring usually in late summer, the annuli are of relatively little value as age indicators after the first three years or so of life, but Cagle found them useful in growth studies involving short periods of time. His data from specimens released and later recovered indicated that the annuli are formed annually during the winter period of minimum growth and that the area between annuli is thus the growth of one season. Thus, if it were not that they are periodically shed and that the whole shell is subjected to abrasive wear, the number of annuli in this species would correspond fairly accurately to the age of the animal in years. Cagle supplemented his data obtained from actual measurements of recaptured specimens by calculating growth from the following proportion:

$$\frac{\text{Length of a given annulus of pectoral lamina}}{\text{Length of plastron at time annulus was formed}} = \frac{\text{Length of pectoral lamina}}{\text{Present length of plastron}}$$

Breeding: Several herpetologists have reported extraordinary courtship maneuvers in this turtle. Curiously similar to the *liebespiel* also described for *Chrysemys,* this activity involves the male's swimming backward before the female with his fore legs stretched forward and the palms outward. At repeated short intervals the toenails of the male are vibrated rapidly against the chin and sides of the face of the female for a second or so. To one who, like myself, has never witnessed this astonishing behavior, an earlier legend attributing to these long, needlelike male toenails a worm-spearing function seems almost more credible! Their use in courtship, however, has been well corroborated, for captive specimens at least. Such courtship has been observed in March, May, and October.

Nesting begins in early April in the southern part of the range and in May or June toward the northern extreme, and the females usually come ashore for this purpose in early morning or late afternoon. Details of the nesting procedure were recorded by Cagle (1937), who on June 20 and July 24 found a large number of nests on a road embankment eight feet above and 60 feet distant from a bayou at Reelfoot Lake, Tennessee. He found that the dry soil in which the turtles were nesting was moistened prior to excavation by large amounts of water from the cloacal bladder, resulting often in the wetting of an area two feet or so in diameter. The front feet of the digging animal were kept planted on the ground as braces to anchor the body while the hind feet scraped alternately, and, according to Cagle, awkwardly, at the hole being enlarged. A preliminary pit little more than an inch in depth was excavated by semicircular scraping movements of the hind feet. Then, abandoning the scraping motions and using a scooping technique, the turtle began to enlarge the nest proper in the bottom of the upper hole, making thus a globular cavity from 3 to 10 inches in diameter. Throughout the procedure the turtles appeared apprehensive and if disturbed abandoned the work permanently to return to the water. The eggs were deposited "within a few minutes," sometimes arranged by the hind feet, sometimes not, and usually lay in the nest in a variety of positions. The wet soil previously removed from the nest was remoistened and used to plug the cavity above the eggs. The plug was found to vary in thickness from half an inch to two and a half inches. Loose material from the surrounding area was pulled in and smoothed over and patted down by the lower shell and the hind feet. Cagle believed that the moist seal helped to maintain in the nest relatively stable temperature and humidity.

The egg complement varies from 5 to 22, with the average about 10. As in other species, the smaller individuals lay fewer eggs than the larger ones. Of 221 eggs measured by Cagle (1944c), the lengths ranged from 30.9–43.0 mm., with an average of 36.2 mm. (1.22–1.69 inches; average, 1.43 inches), and the widths from 19.4–24.8 mm., with an average of 21.6 mm. (0.765–0.977 inches; average, 0.850 inches).

Although there is available little information on the length of the incubation period, it seems evident that this is very variable, that it may be protracted by adverse conditions, and that in the northern populations the eggs may pass the winter in the nest, to hatch the following spring. Cagle found indications that the umbilical scar of the hatchling may close within three days after hatching and that the egg tooth may be retained for about ten days or somewhat longer.

Feeding: Most writers have regarded this as essentially a carnivorous turtle which on occasion takes vegetable food and which is not averse to scavenging. It seems probable, however, that the observations on which these conclusions have been based have been made largely on captive individuals which were eating merely what was made available to them, or what seemed most palatable in an unnatural diet. In Reelfoot Lake, Tennessee, at least, *elegans* appears to be largely herbivorous in maturity, the adult, as in several other species, showing a reversal of the predominantly carnivorous diet selection of the juvenile. Parker (1939) found mostly vegetable material in the stomachs he examined from that locality, although insect remains turned up in the stomachs of a number of juvenile specimens. Marchand (1942) examined the stomach contents of ten immature and twelve mature specimens and summarized his findings as follows:

Here again a reversal is noted between feeding habits of the immature and mature groups. In the latter, plant matter made up 89 percent by volume of the food consumed, while it comprised only 30 percent of the total in the immature group. In the group of larger turtles animal matter was found to compose 9 percent of the total, while 52 percent of the diet of the immature group was composed of such food. Of the plant matter eaten by the adult group the most important items were as follows: *Wolffia* 35 percent, *Lemna* and *Spirodela* 19 percent, *Cabomba* 8 percent, *Ceratophyllum* 8 percent. These were also the foods that were preferred by the painted turtle, except that *Cabomba* and *Ceratophyllum* seem to be relished more by *P. s. troostii* [*elegans*] than by *C. p. dorsalis*. Crayfish made up almost all of the animal food consumed by the adult group. Amphipods and Zygoptera larvae were the animals most frequently eaten by the young turtles, while the vegetable material consisted almost wholly of *Lemna* and *Spirodela*.

Economic importance: Since this turtle occurs where snappers and soft-shells are easily obtained, it is in most places not of great commercial interest for eating purposes, although frequently it finds a place on the tables of the people who catch it and sometimes is sent to market, most frequently in shipments that also include *Graptemys*. Cahn (1937) said that the meat is "of high quality and delicious flavor," and I find this quite easy to believe.

It is probable that the economic importance of the young in the pet trade far exceeds that of the adult in any of its relations to man. I have no definite figures on the numbers of baby *elegans* that annually are sent from the Mississippi Valley to pet stores all over the United States, but they must be tremendous. And even this mass cradle-robbing is not the end of the drain that man makes on the slider population. Cagle (1937) made the following remarks concerning the use of the eggs at Reelfoot Lake:

Undoubtedly man is the greatest enemy of the turtle. Although the slider turtles are an important food crop of the lake, the fishermen dig thousands of eggs annually. They are supposed to make excellent fish bait. These fishermen often use several hundred eggs in one morning to bait their trot lines. Several thousand pounds of the dressed meat of this turtle are shipped from the lake each year. The eggs of the worthless mud turtle cannot be used for bait because of their hard shell.

Children of all ages engage in prolonged turtle egg hunts during the summer months. The eggs collected are sold to the fishermen for fifteen to twenty cents per hundred. These enthusiastic egg hunters follow the turtles into the fields, break open the carapace with a hammer, remove the eggs and leave the turtles to die. One farmer said they had killed so many turtles in his field that his mule "refused to plow."

When one learns that sliders are still common in Reelfoot Lake, one can only marvel at the fecundity of the population and the fertility of the environment in which it lives.

Cumberland Turtle *Pseudemys scripta troostii* (Holbrook)

(Plate 48. Map 17.)

Range: Upper reaches of the Cumberland and Tennessee rivers in Tennessee and eastern Kentucky. Although geographic intergradation between this race and *elegans* toward the west and *scripta* toward the southeast has not been demonstrated, this is obviously due merely to a lack of

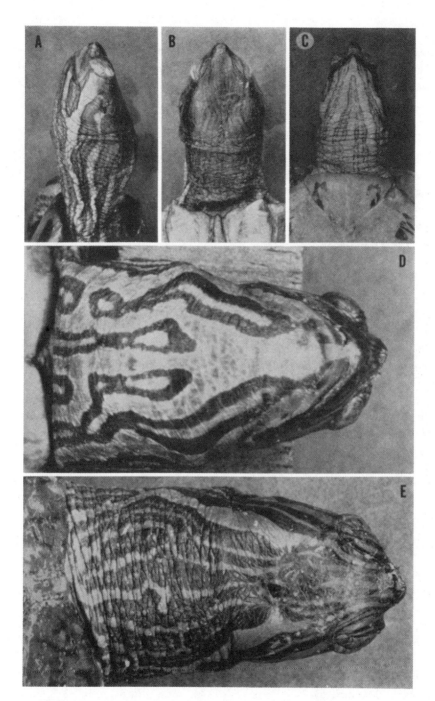

Plate 48. Pseudemys scripta troostii. Male specimens (preserved) from the Little Tennessee River, Kingston, Tennessee. *A and B:* Views of the same turtle. *C and E:* Another individual. *D:* A third individual.

specimens from the important intervening territories, since overlapping of the *troostii* characters occurs in many stocks of the other subspecies. Moreover, from one point of view *troostii* itself is merely a population of intergrades, in that its characters are almost precisely intermediate between those of *scripta* and those of *elegans*.

Distinguishing features: This turtle could be described as a *scripta* without the vertical postorbital patch, with a smoother lower shell, and with more, and more regularly ocellate, spots on the lower shell surfaces. Or on the other hand it might also be adequately defined as an *elegans* with a yellow bar extending from the upper hind rim of the orbit instead of a red bar, with fewer and much broader stripes on all the soft skin surfaces and on the horny beaks; simpler, ringlike ocelli on the lower marginals; less black on the bridge; smaller ocelli or smudges on the plastral laminae, or none at all; and lighter (yellow) nails and ground color of the carapace.

Sexual differences are essentially the same as in *elegans* except that the nonmelanistic male of *troostii* appears to have less black pigmentation of the lower shell surfaces (lower marginals, bridge, plastral laminae) than the female. The evolution of the melanistic pattern appears to proceed as in other members of the group.

Of fifteen mature specimens from eastern Tennessee (Hawkins County), the size of the nine males ranged from 140–210 mm. (5.52–8.28 inches) and that of the six females from 172–206 mm. (6.78–8.12 inches). The width of the shell was, on the average, 75 per cent of the length in the males and 76 per cent in the females; its height in males averaged 36.2 per cent of the length and in females 37.4 per cent.

The young show an accentuation of the subspecific characters of the adult. In a specimen 37 mm. (1.46 inches) long the stripes are all intensely bright yellow and very broad, especially those on the under surface of the throat; the bridge and first two pairs of plastrals are marked by simple ocelli, the other plastrals by small smudges, one to each lamina. The nails and upper surface of the tail are mostly yellow. The shell is remarkably smooth, with no sign of a keel. The hatchling is not known.

Note: Since there has been a good deal of confusion surrounding the use of the name *troostii*, it would perhaps not be amiss to review briefly the somewhat involved recent history of the term.

Prior to the appearance of the 1939 edition of the Stejneger and Barbour check list two "sliders" were recognized in the Mississippi drainage, *P. elegans,* described by Wied from New Harmony, Indiana, in 1838, and *P. troostii,* described by Holbrook two years earlier (1836) from the Cumberland River in Tennessee. *P. elegans* was the red-eared species and *P. troostii* the dun-colored, long-nailed, long-tailed form.

Then after nearly a century came Viosca's 1933 paper showing that the dun-colored turtles were nothing but a group of discolored males of the attractively decorated *elegans,* and one of the names had to be abandoned. Since *troostii* was the older name, the whole complex became known as *P. troostii.*

At this point I complicated the situation by demonstrating (to some people's satisfaction) that *troostii* intergrades with the yellow-bellied turtle, which bore a still older scientific name, and that the sliders would accordingly have to be called *P. scripta troostii.* Dr. Stejneger, who was the more-or-less final arbiter in such matters, never accepted my notion, and for this reason "*P. troostii*" remained the specific name of the Mississippi Valley sliders as far as he and the check list were concerned.

No sooner had herpetologists managed to stop thinking of the sliders as *elegans* and start calling them *troostii* than the 1939 check list was published, and in it two sliders were again listed, this time as *P. troostii troostii* and *P. troostii elegans.* This was too much for most people and they simply ignored it and continued to call all sliders *troostii.* The usage was clearly explained in a footnote in both the last two editions of the check list, but the type is small and apparently no one has ever bothered to read it.

Some time previously, Stejneger and Hartweg and, later but independently, I, had come to believe that the sliders of the eastern edge of the Appalachian Plateau in Tennessee and Kentucky, long ago given the name *cumberlandensis* by Holbrook, were a recognizable subspecies. Only to Stejneger, however, did it occur that the name "*troostii*" had been applied by Holbrook not to melanistic males of the wide-ranging Mississippi Valley form but to the male of *cumberlandensis.* Since *troostii* was described four years earlier than *cumberlandensis,* the former name had to be used for the highland subspecies, and this left the wide-ranging race to take the next available name, which was *elegans.* At present, there can be little doubt that both these turtles are races of *scripta.*

This is exactly the sort of mess that the layman finds it hard to forgive the systematic biologist, not realizing that such occasional minor disruptions inevitably attend the maneuverings of an orderly system of classification and that they are sought out and relished for their own sake only by the exceptional, underprivileged person.

There appear to be no published natural history data that can be regarded as applying to this upland subspecies.

Rio Grande Turtle *Pseudemys scripta gaigeae* Hartweg
(Plate 49. Map 17.)

Range: Northern Mexico from northern Chihuahua southward to Coahuila and eastern Durango; Brewster County, Texas. Toward the westward this subspecies is replaced by *P. s. hiltoni* of Sonora and in the coastal lowlands to the southward by *P. s. cataspila*. In Texas the region of intergradation with *P. s. elegans* was delimited by Hamilton (1947) as beginning "immediately north of the Rio Grande . . . to extend to a point somewhere between Shackleford County and the headwaters of the Trinity River." The same writer pointed out the fact that along the coast the population seems to be intermediate between *elegans* and *cataspila*, the *gaigeae* tendencies being little in evidence there. In a broad area extending from near the mouths of the Brazos and Colorado rivers westward and southward across the border into northern Tamaulipas and central Coahuila, there is a complex population of three-way intergrades that includes a surprisingly high incidence of individuals closely approaching *elegans* in appearance. Dr. Karl P. Schmidt of the Chicago Natural History Museum lent me for examination a lot of 25 specimens from a population of this sort from the region around Musquiz in central Coahuila, and of this series only one specimen showed a predominance of the *gaigeae* characters, all the rest being what I should call *elegans-cataspila* intergrades, but with a strong leaning toward eastern *elegans* in many cases. Shannon and Smith (1949) described a specimen from Hidalgo County in the southern Rio Grande Valley as intermediate between *gaigeae* and *elegans* but nearer the former. A series of specimens from this area should be very interesting.

Distinguishing features: The most conspicuous feature of this handsome and distinctive turtle is a large, rounded, isolated yellow spot on the side of the head at the level of, but well behind, the eye; this spot is black-bordered, and outside of the black margin there are usually two more or less complete white or yellow rings. Another smaller, similar spot touches the hind border of the orbit. The carapace is comparatively low, smooth, and unkeeled and is marked with an indefinitely reticulated pattern that characteristically includes one or more ocellate spots (a light ring around a black ring) toward the lower or posterior edges of the laterals and marginals; the lower marginals are marked with black-bordered ocelli. The plastral pattern varies from the *"ornata"* type, with the principal pigmented area a median one spreading outward along the transverse seams,

Plate 49. Pseudemys scripta gaigeae. Young; Boquillas, Brewster County, Texas. Preserved specimen.

to the *"elegans"* type in which longitudinally directed black blotches cross the transverse seams; the usual condition is somewhat intermediate.

Sexual differences are apparently as in *elegans,* the older males showing much the more marked inclination toward melanism and loss of the juvenile markings.

The young are keelless and have vivid and more strongly ocellated markings on the shell, both above and below, than the adult but are otherwise similar. The hatchling has not been described.

Habitat: This turtle has so far been recorded only from rivers, but it probably occurs in prairie ponds and tanks throughout its range where these lie reasonably near more permanent water; at any rate, this is the case with western *elegans.*

Habits: There is apparently no information available on the habits of this little-known subspecies. Judging from the character of the bodies of water from which it has been collected, it must be largely carnivorous.

Baja California Turtle *Pseudemys scripta nebulosa* (Van Denburgh)
(Map 17.)

Range: Southern Baja California, from San Ignacio south to San José del Cabo.

Distinguishing features: This subspecies may be recognized by the presence on the temporal region of a large yellow or orange spot, which is the terminal expansion of the uppermost line on the side of the neck [supratemporal] and thus represents the isolated spot of *gaigeae;* in *nebulosa,* however, the spot is attached behind and is not strikingly black-bordered, and there appear to be rarely any dark rings on the upper shell. The carapace is short and broad, obtusely keeled, and brown in ground color, with light markings but no black ocelli on the laterals; each upper marginal has a dark smudge roughly surrounded by a squarish light mark. The precentral is about as broad as long. There is a black smudge toward the hind seam of each lower marginal, and usually an elongate one on each section of the bridge. The plastral pattern is a dark figure located mesially and sending out broad lateral extensions roughly coinciding with the seams. There are four yellow stripes on the outer surface of the fore leg and no rounded spot touching the hind rim of the orbit.

Dimensions of a male and a female collected and measured in the San Ignacio River by Eric Waering of Petróleos Mexicanos were as follows:

Sex	Carapace length	Carapace width	Plastron length	Plastron width
Male	330 mm. (13.0 inches)	226 mm. (8.90 inches)	280 mm. (11 inches)	127 mm. (5.0 inches)
Female	370 mm. (14.6 inches)	264 mm. (10.4 inches)	330 mm. (13.0 inches)	162 mm. (6.38 inches)

In Mr. Waering's specimens the tails of the females were shorter and slimmer than those of the males and, as in most of the tropical races of *Pseudemys,* the males had longer, more pointed snouts.

Mr. Waering informed me in a letter (1943) that his series from San Ignacio differed from specimens described from the tip of the peninsula in having black instead of brown plastral markings, olive instead of brown skin color, and an orange temporal blotch instead of a yellow one.

Habitat: I quote again from Waering's correspondence (1943) to give his observations concerning the occurrence of *nebulosa* in the intermittently flowing Rio San Ignacio:

The San Ignacio River is surrounded by dry desert, making it impossible for the turtles to migrate except along the river. This stream normally consists of disconnected, small lagoons and lies in a canyon of up to 100 meters in depth. The river bed consists of dry desert with sand and boulders between the lagoons. The turtles therefore can migrate from one lagoon to another only after very heavy rains which occur very seldom. At the time of my visit no heavy rain had fallen since 1935.

The lagoons have very clear water, but it is slightly brackish, due to the outcropping sandstone in the river bed. These lagoons form small oases in which date palms, figs and other fruit trees flourish. I noted two small fish, less than three inches in length, abundantly represented in the lagoons. I also saw one small species of frog.

Natives believe that these turtles are a form of sea turtle that has been isolated from the Pacific Ocean!

Habits: For the only information on the life history of this race I must again quote from a letter from Mr. Waering:

The specimen . . . was found [near San Ignacio, Baja California] about 5:00 P.M., in June, 1943, laying eggs in a small hole located about 35 feet from the shore of the lagoon. The excavation, about the size and depth of her carapace, was dug in sand under the partial shade of a small tree, and was full of liquid, presumably urine. The turtle was inclined at an angle of about 20 degrees, with its posterior extremities in the hole filled with liquid. She exhibited no fear and remained motionless without trying to escape. Inasmuch as time was not available to watch her lay eggs the turtle was collected. This species is very vicious, exhibits an ugly temper and bites readily.

Economic importance: The same correspondent quoted above remarked that local inhabitants relished the meat of these turtles when cooked with chiles, and regularly caught them by diving into the lagoon and seizing them under water.

The *rubriventris* Section

In this group the upper jaw is deeply notched at the tip, with the notch flanked on either side by a strong cusp; and both jaws are characteristically strongly serrate, although this varies considerably. The median ridges on the crushing surfaces of the jaws are tuberculate, being similar in this respect to the races of *P. floridana* rather than to the *scripta* section, and the same relationship is indicated by the form of the lower surface of the mandible—flattened in *rubriventris* and *floridana,* rounded in *scripta.* The complex system of light lines on the head and neck shows marked similarity to that of the less specialized members of the *floridana* group, so much so, in fact, that the more conspicuous of these lines may be easily traced through most of the forms of both groups and, indeed, are to a certain extent evident in the *scripta* section and in *Chrysemys* and *Graptemys* as well. The most persistent of these lines are those that constitute the reduced head pattern of *Pseudemys nelsoni,* the southernmost representative of the *rubriventris* section, and are as follows:

An *orbitomandibular* stripe, forking at the angle of the jaw to send one branch to the lower margin of the orbit and another along the mandible almost to the symphisis; a *maxillary* stripe, extending obliquely from the nostril about half way to the angle of the jaw; a *tympano-orbital* stripe from the posteriodorsal corner of the orbit to the upper margin of the tympanum, sometimes continuing along the neck; a *supratemporal,* running obliquely upward along the dorsolateral edge of the neck and head, across the temporal region to the inner border of the orbit (the portion of this between the orbit and the anterior margin of the temporal fossa is often obliterated). Less frequently present and usually faint is the paramedian, which may either extend along the top of the head (just off center) to the tip of the snout or may join the supratemporal a short distance behind the orbit. A conspicuous *sagittal* stripe of varying length is usually present midway between the orbits [Carr, 1938a].

The fully expressed plastral pattern is the dendritic, seam-following type, most frequently present in complete form in *concinna* and *suwanniensis* of the *floridana* group. This pattern is usually absent in adult *nelsoni* and probably in most *bangsi*, but is often present in some populations of *rubriventris*.

The nails of the fore feet of the males are elongate and are straight, not curved as in the *scripta* group.

There is a consistent tendency in all the populations toward a very fancy erythrism of the plastron and of the light figures on the carapace. In a small percentage of old specimens (of both sexes) the juvenile pattern degenerates, as in old males of the *scripta* group, and is replaced by irregular red mottling of the shell and sometimes of the skin of the head and legs.

This complex of turtles is distributed from Massachusetts to extreme southern Florida. It is, however, a curiously fragmented stock, comprising at present three separate colonies, two of which overlap widely in their characters and one of which remains distinct.

Besides a partly shared range and abundant evidence of close relationship with the *floridana* group, as seen in their similarities of structure and pattern, there is a most extraordinary tendency in the two groups to show the same graded variations from the northern to the southern sections of their ranges, as I have mentioned elsewhere.

Red-bellied Turtle *Pseudemys rubriventris rubriventris* (Le Conte)
(Plate 50. Map 18.)

Range: The Atlantic coastal plain from New York to south-central Virginia. In the northern part of its range, especially, demands of the market have made great inroads in the populations, and the species apparently does not now occur in New York, although previously it was abundant there in the coastal streams. A few old records would extend the range southward to North Carolina and even to South Carolina, but I believe the South Carolina record to have been based on an erythristic variant of *concinna;* regarding the alleged North Carolina specimens Brimley (1943) commented as follows:

Yarrow's check list of North American Batrachia and Reptilia, published in 1882, lists one from Kinston in the U.S. National Museum as taken in May, 1875, and another from Wilmington sent in by a Mr. Davis, but no date given. It is possible that they were sent in from those localities but not taken there. I have never seen or heard of a specimen being taken in North Carolina in my time.

I might add that the late Mr. Brimley was an extraordinarily good field observer whose rather casual studies of *Pseudemys,* although a side line with him, were among the soundest ever made by anyone.

Map 18. Distribution of *Pseudemys nelsoni* and the subspecies of *P. rubriventris.*

Distinguishing features: The red-bellied turtle may be recognized by its notched and cusped upper jaw, the frequent occurrence of a conspicuous red pigmentation of the carapace and plastron, the light-centered blotches on the lower marginals, and the form of the shell, in which the highest point is usually at the middle, the widest point posterior to the middle, and the ratio of the length divided by the height usually more than 2.5.

Description: The shell is elongate, oval in lateral outline, usually flattened or even middorsally concave in old specimens, and often constricted in the region of the sixth marginals. The size is large, the maximum shell length being close to 400 mm. (15.7 inches); there are even uncorroborated reports of specimens as much as 458 mm. (18 inches) long, which would make this the largest of American emydids. Very large specimens are not often seen today. The longest shell measured by McCauley (1945) was 281 mm. (11.0 inches); Conant and Bailey (1936) collected a New Jersey specimen with a shell 290 mm. (11.4 inches) long; and the largest I have measured was a specimen from the Potomac River 300 mm. (11.8 inches)

Plate 50. Pseudemys rubriventris rubriventris. A: Male. *B, C, and D:* Female. *E and F:* Young. All from Hammonton, New Jersey.

long. Shell dimensions of a large female from New Jersey and a large male from the Potomac River are as follows:

Sex	Length	Width	Height
Female	297 mm.	216 mm.	123 mm.
	(11.7 inches)	(8.50 inches)	(4.85 inches)
Male	212 mm.	148 mm.	72 mm.
	(8.35 inches)	(5.83 inches)	(2.84 inches)

The ratio of the length divided by the height varies between 2.85 and 3.27, average 3.01, in males; and between 2.41 and 2.83, average 2.60, in females. The ratio of the length divided by the width ranges from 1.36 to 1.50 in males, average 1.42; and from 1.21 to 1.43, average 1.33, in females. The carapace is often wrinkled lengthwise and is never keeled except in young. It is brownish with a reticulated pattern in young individuals, the light markings often red, and the most frequently occurring red mark being the broader transverse line on each lateral, which also persists longer than the other lines when the animal matures. There is a transverse light bar (usually red) on each upper marginal. The plastron is orange or red, with a dark figure in young and usually none in old individuals. The submarginals are more often reddish than the plastron, and if spots are present they usually have light centers. The head is dark brown in color, with light stripes as described under the remarks on the Rubriventris Group but with a few subsidiary, narrower lines usually present between the principal ones in younger adults.

The shell of the male is lower and very slightly narrower, on the average, than that of the female. The usual tail and toenail differences between the sexes exist. The females are a little larger, but not conspicuously so, although this may be due to the fact that the bigger females have mostly been caught and eaten. As was mentioned above, the pattern degeneration and spreading of red pigmentation are not confined to the male, as is a somewhat similar condition in the *scripta* group, but are found in both sexes, although perhaps more frequently and to a more marked extent in the male. Among the red-bellied turtles so beautifully illustrated by Agassiz' lithographs, the specimens in Plate 27 are males, but that represented in Plate 26, G, is a young female and this shows unmistakable evidence of incipient erythrism (Vol. 2, 1857).

The shell of the hatchling is between 30 and 32 mm. (1.18 and 1.26 inches) long and usually within one millimeter of this in width, being in some as wide as long. The markings are generally brighter and more definite than in the adult, but there is less (often no) red pigmentation and no mottling of the shell. The plastral figure is usually conspicuous and there is a keel. For a comparison of the young of this turtle with that of Nelson's turtle the reader is referred to Richmond's and Goin's table in the account of the latter form.

Habitat: As regards living conditions the red-bellied turtle shows remarkable tolerance. It occurs in ponds and lakes as well as in streams, and of the latter it inhabits both clear, swiftly moving creeks and lower silt-bearing rivers which it even descends into their brackish lower reaches. It has been taken in salt marshes.

Habits: Although the red-bellied turtle is regularly seen basking on logs, it is said by all who are acquainted with it to be extremely wary and difficult to approach, perhaps because of long-time persecution by man, who has been weeding out the unwary individuals for a century or more. Conant (1945) pointed out that in the Delmarva Peninsula its basking is nearly always done either in inaccessible spots or within easy reach of deep water where the turtles swim rapidly away if alarmed. McCauley (1945) noted a contrast in the reactions of the adults and the young in this respect, the latter being much less skittish and sometimes even allowing themselves to be caught in a dipnet.

Breeding: The courtship and mating activities of this species have not been observed. The laying season extends from early June through July at the northern extreme of the range and from late March through June in Virginia. The nesting process has been described in detail only by Smith (1904), whose account is worth quoting despite certain patent inaccuracies, including the statement that the nest is excavated with the fore feet.

The egg-laying season is in June and July, and the place where the eggs are laid is usually a cultivated tract, often a cornfield adjoining the water. It is probable that a field would always be selected, but when there is a high, steep bank the eggs are of necessity deposited on the shore. The terrapins visit the fields only during egg-laying time and only for this purpose, and sometimes make their nests more than one hundred feet from the water. It has often been observed that six or eight terrapins will lay on the same shore or in the same field, their tracks being easily discernible in the moist or soft sand or loam. The nest is made in sand, clay or loam, a sandy loam or sandy clay being most frequently chosen. The nest, which is shaped like a carafe, is dug by the female with the fore legs. Its size depends on the size of the animal, or, what amounts to the same thing, on the number of eggs to be laid; an average nest would be four inches deep and four inches wide at the bottom, the opening being somewhat smaller than a silver dollar. When on the shore the nest is always above the high-water line.

All the eggs are laid at one time, and when the laying is completed earth is scraped into and over the hole and packed tightly. The packing is accomplished by the terrapin raising herself as high as possible on all four legs and then dropping heavily, by the sudden relaxation of the extensor muscles. Immediately after covering the nest, the terrapin withdraws to the water.

The size of the eggs varies somewhat with the size of the terrapin, but averages one inch by three fourths of an inch. A six-inch terrapin lays ten or twelve eggs, while the largest terrapins, fourteen or sixteen inches long, lay as many as twenty-five to thirty-five eggs, possibly more. When a terrapin is disturbed while making a nest or laying she will abandon the nest. On one occasion, when a terrapin was discovered over a nest in a cornfield, removed to see whether any eggs had been deposited, and replaced over the hole in the ground, it was found when the place was visited two hours later that she had left without laying any eggs. The eggs probably hatch during the summer, but on this point there have been no personal observations. The young, however, remain in the nest until the following spring (April 10 in one case), and when they emerge . . . they go to the water at once.

Richmond and Goin mentioned a nest containing twelve eggs and located fifty feet from the edge of a pond in New Kent County, Virginia. These eggs were laid at 11 A.M. on May 18 during a heavy rain. On the following day the observers discovered another turtle filling a nest in which there were thirteen eggs. These were taken to the laboratory, where they hatched shortly before September 1.

Conant and Bailey (1936) referred to a female from New Jersey, 290 mm. (11.4 inches) in length of carapace, which laid six eggs on July 21, 1935, and twelve additional eggs on August 10, 1935. The eggs of the second complement, which were slightly larger than those of the first, varied in length between 34 and 37 mm. (1.34–1.46 inches).

Feeding: Although there has been no systematic study of the food habits of this turtle, there are data which indicate that it may be omnivorous at all ages. There appears little doubt that this is true of captive individuals, at least.

Economic importance: During the latter part of the nineteenth century the immensely popular diamondback terrapin began to decline in abundance and to be increasingly replaced in many markets by red-bellied turtles of the Chesapeake Bay region. Although never attaining the high position of the diamondback in public esteem, the red-belly formerly supported regular fisheries, and as early as 1884 True was recommending its introduction into waters to the north and to the south of its natural range. Today, although its numbers have been greatly depleted in many parts of its former range, this turtle still may be seen in the markets of Washington, Baltimore, and Philadelphia. McCauley (1945) noted prices of from seventy-five cents up, according to size, in a Washington market, and pointed out that the species is now protected by law in Maryland from spring to early fall.

Plymouth Turtle *Pseudemys rubriventris bangsi* Babcock

(Map 18.)

Range: This isolated subspecies is confined to Plymouth County, Massachusetts. It is obviously impossible to demonstrate, between it and typical *rubriventris*, the geographic merging of characters which constitutes the conventional basis for the use of trinomials in continental populations. Such a colony is more like a stock living on an island, partly differentiated, but showing in its variations an overlapping of the characters of related forms.

Distinguishing features: The Plymouth turtle has not been defined by the detailed statistical analysis that would be necessary to show its minor differences from the *rubriventris* populations to the south, and probably never will be. Such a study necessitates large series of specimens and I don't suppose there are a dozen preserved specimens of *bangsi* in existence. The ponds that it inhabits are located in a populous district where its survival probably depends wholly on the conservation conscience of the local citizens, and further collecting in the colony would seem decidedly ill-advised.

In his original description of *bangsi*, Babcock (1937a) differentiated the race only on the basis of greater relative shell height, giving the ratio (presumably the average of all specimens measured by him) of the length of the shell to its height as 2.4. I have measured only one specimen of each sex of *bangsi*, but have calculated ratios for large series of the other two forms, *rubriventris* and *nelsoni*, and my data are summarized in Table 8. In the columns the first number is the maximum, the second the minimum, and

Table 8. Ratios of length of shell to other measurements in *Pseudemys rubriventris rubriventris*, *P. r. bangsi*, and *P. nelsoni*.

Form	Length to width		Length to height		Length to depth of bridge	
	Males	Females	Males	Females	Males	Females
rubriventris	1.36–1.50, 1.42	1.21–1.43, 1.33	2.85–3.87, 3.01	2.41–2.83, 2.60	8.33–12.04, 9.99	6.60–7.86, 7.35
bangsi	1.39	1.39	2.53	2.46	8.81	9.27
nelsoni	1.31–1.47, 1.40	1.32–1.47, 1.38	2.18–2.53, 2.32	1.93–2.28, 2.16	5.56–7.60, 6.35	5.41–6.92, 5.78

the third the average. Although the data given for *bangsi*, based on a single specimen of either sex, are of little real value in a comparison of this sort, the trends that they indicate are apparently fairly characteristic. In general appearance *bangsi* is more like *nelsoni* than like the intervening *rubriventris,* but certain features of conformation and coloration do not intergrade.

Habitat: This turtle is known only from ponds. Babcock (1916b) recorded it from the following Plymouth County ponds: Hoyt, Island, Nigger, Hillfield, Boot, Upper West, and Micajah. By 1937 he was able to add to this list Gunner's Exchange and Great South. Lucas (1916) mentioned a specimen from near Billington Sea, found in 1869, and another from Crooked Pond. I suppose it is possible that these red-bellies have been cut off from the main body of the population farther south during historic times by the activities of man, but I am inclined to doubt it. As Babcock (1917b) pointed out, in the area on Cape Cod that they now inhabit several southern species of plants and insects occur.

Habits: Very little specific information on the habits of this subspecies of the red-bellied turtle is in print. Babcock (1938) mentioned that it shows an inclination to aestivate or become inactive during the summer months, when not a single individual may be seen for weeks. The same author made the general statement that the eggs, which average 31 mm. (1.22 inches) in length, are laid on the shores of ponds or in neighboring fields during late June. He said also that the diet includes snails, fishes, tadpoles, crayfishes, and aquatic vegetation.

Economic importance: If it were not for the interest shown by local inhabitants of Plymouth County in the conservation of this interesting population of turtles it would soon be wiped out. There can be little doubt that this race is as good to eat as its relatives to the south, but there is not, and perhaps has never been, any traffic in it.

Florida Red-bellied Turtle *Pseudemys nelsoni* Carr

(Plate 51. Map 18.)

Range: Peninsular Florida. Known from Alachua and Levy counties southward to Cape Sable.

Distinguishing features: This is a large turtle of the *rubriventris* section, with the shell more similar in shape to that of *P. f. peninsularis* than to that of *P. rubriventris,* being highly arched and elevated along the centrals, the highest point often anterior to the middle and the widest point at the middle. The bridge is much deeper than in the northern relatives, in which it is again similar to *peninsularis.* The markings of the lower marginals are solid and smudgelike, not concentric, and the bridge is either unmarked

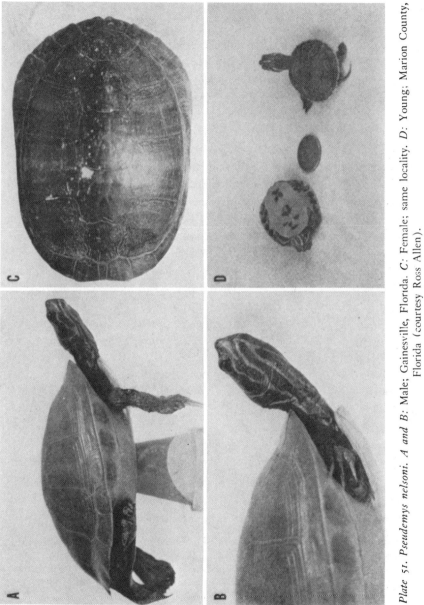

Plate 51. Pseudemys nelsoni. A and B: Male; Gainesville, Florida. *C:* Female; same locality. *D:* Young; Marion County, Florida (courtesy Ross Allen).

or with two or three dark smudges, with no linear or concentric figures. There are likewise no concentric figures on the upper marginals, and on the top of the head between the upper edges of the tympana there are in young adults four to six light stripes; these last two characters will distinguish *nelsoni* from the occasional aberrantly cusp-jawed individual of the *floridana* group that might turn up in its range. Coloration of the carapace is extremely variable but in the average example a dark, nearly or quite black ground color is conspicuously broken by a broad, red, wedge-shaped mark that traverses each of the first three laterals; there may be all kinds of subsidiary and more or less irregularly dispersed light markings of the laterals and centrals. The plastron is usually some shade of red, but not always, and is often without a dark pattern in adults. For further characters see the general account of the section, the table under the account of *P. rubriventris bangsi,* and the tabular comparison of the young of *nelsoni* and *rubriventris* below.

Besides the regular differences in the tail, the male *nelsoni* is somewhat smaller than the female and has much longer nails, which are very slightly curved, but not nearly so much so as in the races of *P. scripta.* The largest male *nelsoni* that I have measured had a shell 285 mm. (11.2 inches) long; that of the biggest female was 340 mm. (13.4 inches) long.

The following data comparing young of *nelsoni* and *rubriventris,* the latter egg-toothed hatchlings from New Kent County, Virginia, were taken from Richmond and Goin.

nelsoni	*rubriventris*
Bridge immaculate or with only a few spots	A heavy black band on bridge
Plastral markings absent or restricted to the region of the median seam	Plastral markings spread out over plastron, some usually present on each lamina
One to three stripes between the supratemporals behind the eye	Five to eight stripes between the supratemporals behind the eye
Postfemoral pattern mainly of two or three broad yellow bands	Postfemoral pattern of several faint horizontal bands and numerous spots
Ground color of soft parts dark black	Ground color of soft parts very dark green
Auxiliary stripes on the side of the head indistinct or absent	Auxiliary stripes on the side of the head distinct

Habitat: I have reported this turtle as found in "ditches, sloughs and marshes; lakes, ponds and streams; mangrove bordered creeks along the southwest coast" (1940). In Rainbow Springs Run, a broad, slow-moving, clear, spring-fed stream with heavy growth of *Sagittaria,* Marchand (1942) found the Florida red-bellied turtle to make up 2.1 per cent of a turtle population that included six other species, the most numerous of which were *P. f. suwanniensis* and *P. f. peninsularis,* basing his estimate on a collection of 1,022 turtles. I can detect no differences in the habitat preferences of *nelsoni* and *P. f. peninsularis,* the latter merely being the more abundant of the two in nearly every situation in which they are found together.

Habits: The life history of the Florida red-bellied turtle is very poorly known. In general its habits appear to be similar to those of *P. f. peninsularis* and *P. f. suwanniensis,* with which it is frequently found. It suns itself on the same logs with the other forms and feeds in the same beds of submerged vegetation. Females prospecting for nesting sites are often found in the same ploughed fields or lake-shore hammocks in which *peninsularis* is laying, and a percentage of the numerous varmint-ravaged nests that may be seen each summer in peninsular Florida are undoubtedly of this species, although the shells of its eggs are not to me distinguishable from those of *peninsularis.* There are indications that the nesting season may extend throughout the year, although the late spring and early summer months represent the peak of laying activity. A specimen that I kept in a tub laid several eggs over a period from June 14 to June 18, 1934, but these were dropped in the water and when subsequently buried in sand they failed to hatch.

Feeding: The only difference that can be noted in the feeding habits of *nelsoni* and those of *P. f. peninsularis* is what appears to be a slightly greater tendency of the former to scavenge. I have watched individuals bite at floating watermelon rind and eat dead fish. That the species is predominantly vegetarian, however, is indicated by several scattered observations and by a feeding experiment carried out by Allen (1938). In an attempt to gain some idea of the volume of food consumed by turtles Allen placed measured quantities of the plant *Sagittaria sinensis* in an enclosed pool in which 28 specimens of *nelsoni* were kept with 76 turtles of four other species of *Pseudemys.* On one occasion the 104 individuals ate no less than three and three-fourths bushels of the plants in one day. Allen noted little difference in the food preferences of the five species, finding all to be mainly herbivorous, occasionally taking a dead fish but never attempting to catch the large schools of small fish that were continuously present in the pen.

Another point, only casually mentioned in Allen's report but to me appearing to be extremely interesting, was the observation that although all the turtles appeared quite fond of water hyacinths they were able to eat them only when the buoyant plants had been overturned in the water and floated leaves downward. One can feel only sympathy for the thousands of Florida cooters that inhabit hyacinth-ridden lakes and streams and can only gaze upward at the acres of succulent but upright and unassailable salad, and await the providential upset of a plant!

Economic importance: The Florida red-bellied turtle is usually not recognized as distinct from *P. f. peninsularis* by the local people, who eat both but who only rarely try to market them. I should add that I also have been served stews made with both *nelsoni* and *peninsularis* on two occasions and in neither case could I distinguish two grades of meat.

The *floridana* Section

This group is an alliance of subspecies the extremes of which diverge widely but which are all interconnected by intergradation of one sort or another. Beginning in the Atlantic coastal plain with the lowland race, *floridana,* the populations extend westward through the southeastern United States in a double rank, an inland series (*concinna-hieroglyphica-hoyi*) and a coastal series (*peninsularis-suwanniensis-mobiliensis*), the two meeting in *P. f. texana* of the frontier areas of southern Texas and eastern Mexico.

The turtles of this section are on the whole larger than those of the other two divisions. The majority of the subspecies may be distinguished by the beak of the upper jaws, the two sides of which do not meet at a vertical angle in front as in the *scripta* section and which typically do not have a median notch bounded by toothlike cusps such as those of the *rubriventris* section. Instead, the symphysis is wide and nearly level or broadly concave as seen from the front, the only regular exceptions being in the cases of *hoyi* and occasional specimens of *hieroglyphica,* in which there is a median emargination and the very divergent *texana,* in which the notch is deep and is flanked by a strong cusp on either side. To this extent, there is overlapping between the present section and the *rubriventris* group discussed above. The ridges of the alveolar surfaces are feebly to very strongly dentate and the upper jaw is variable as to the degree of serration of its cutting edge, while that of the lower jaw is usually deeply serrate. A plastral pattern of the seam-following type is present in some forms and absent in others, appearing to be associated to some extent with the stream habitat. The pattern of stripes on the head is modified from one form to another but clings to a basic plan (discernible also in the *rubriventris* group and described under that heading) and affords valuable characters for defining the races. The group contains turtles highly rated for edibility and one or two that rank among the most succulent in American fresh waters.

These big "cooters" are notorious as a "difficult" group, by which the zoologist means that he finds it hard to force them into his standard taxonomic compartments. One main reason for this difficulty is that some of their more conspicuous characters vary to a bewildering extent and in apparently meaningless ways. From a single pond, for example, one may catch twenty turtles no two of which have the same shell pattern, and among which extreme examples may show differences from one another that are more striking than those that distinguish the subspecies to which they all belong. This happens so often that one gets the impression that the ancestors of these turtles must time and again have been split up into isolated colonies, allowing slight differences to arise, only to be reunited later on and to dump into the common pot all the accumulated differences and bequeath to the modern strains a polyglot heredity that has not yet settled down into relative stability.

Over and above the annoying inconsistencies that they exhibit, the pseudemyd cooters are a remarkable lot, offering the student of evolution-at-work field examples of nearly any situation he may have read about or may have conjured up in his own imagination. For besides the seemingly erratic variations one can find in these animals other tendencies that are quite clearly correlated with the way they are distributed geographically or with the special kind of environment they select to live in. Of such associative characters there are all kinds and orders, some obviously of value to the animals and others of no imaginable use. They divide and subdivide the complex through all its dimensions and, as I said before, strain our scheme of nomenclature but offer uncommon stimulation for the inquiring mind.

For one thing, the *floridana* group includes examples of the so-called "ecological subspecies," an evolutional entity that people talk about a lot but only rarely hold up for our admiration. The ecological subspecies is the more or less strongly differentiated stock that occupies the same geographic area as a closely related and intergrading population (in direct defiance of a strict interpretation of "Jordan's Law") but is able to hold on to its divergent traits because a different way of life keeps it partially isolated. This almost legendary situation seems to me to be illustrated by a group of turtle races that meet in west-central Florida near the base of the peninsula. Here there come together a northern and a southern inland subspecies (*floridana* and *peninsularis*) connected by intergrades that occupy a narrow intervening zone and which are, thus, ordinary geographic races. They have no plastral markings and they inhabit all kinds of bodies of water within their ranges except the rivers of the northern Gulf drainage, which are the exclusive freehold of a turtle that has a plastral pattern and that comprises not one race but two. The more southern of these, the locally famous "Suwannee chicken," ranges from just north of Tampa up to the Apalachicola delta, and the other, called "Mobilian" and also revered for its eating qualities, takes over there and goes on around the Gulf to Texas. The relationship between these two is evidently another straightforward example of geographic speciation, but what are the factors that keep them from mixing, socially as well as genetically, with the closely related stocks whose territory their own rivers traverse? I don't claim that the forefathers of the Suwannee chicken founded a new breed of cooter merely by moving into the Suwannee River and kicking out all nonconformers, but the facts remain that there *is* a Suwannee chicken and that it *does* predominate in the Suwannee while the lakes, ponds, and ditches of the same neighborhood are populated exclusively by *peninsularis*. While direct intergradation between these two is rare, that the relationship between them is a subspecific one is amply demonstrated by a merging of their traits via other races of the alliance. There is complete intergradation, for example, between *floridana* and the Mobilian.

To sum up this situation, we find in Florida a peninsular subspecies intergrading with a more northern coastal plain form, with each of these replaced in the rivers of the Gulf drainage by a fluviatile relative, and the two river subspecies merging at almost precisely the same latitude in which the inland cooters come together. Moreover, it seems nearly certain that another Gulf coast turtle known to herpetologists for the last fifty years as *P. alabamensis,* and characterized by having a deep notch and toothlike cusps at the tip of the upper jaw, is really just a variant form that occurs with apparently equal frequency among Suwannee chickens and among Mobilians.

This particular variation distresses the turtle student because a cusp-bounded notch has always been the one reliable feature by which he has been able to distinguish another section of the genus, the *rubriventris* group. The situation becomes even more puzzling when we come across

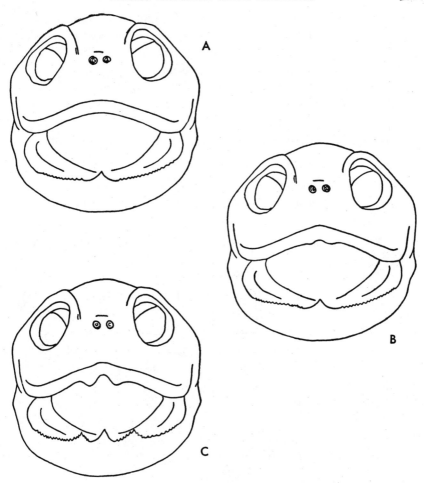

Figure 19. Jaws of three intergrading subspecies of *Pseudemys floridana. A: concinna,* North Carolina. *B: hoyi;* Arkansas. *C: texana;* Kerr County, Texas.

this same character once again in *P. floridana texana,* the extreme western representative of the *floridana* complex. Here, however, it does not mark merely an occasional freak, but constitutes the principal differential character of the subspecies, occurring in this restricted and intergrading population with the same stability and constancy as in the *rubriventris* group.

It should be emphasized that all this eccentricity occurs within an orthodox framework of geographic subspeciation that is almost diagrammatic in its arrangement. As a last distractor, however, one may notice that the slight differences which separate the northern and southern forms of *P. floridana* are almost exactly the same features that differentiate the northern and southern representatives of the *rubriventris* group. This parallel variation might seem to imply genetic contact, but in spite of it no one has ever been able to detect signs of interbreeding between the two in any of the territory that they share, until they reach southern Florida. In the Everglades, however, they appear to hybridize on occasion.

Coastal Plain Turtle *Pseudemys floridana floridana* (Le Conte)

(Plate 52. Fig. 18. Map 19.)

Range: Atlantic coastal plain from extreme southeastern Virginia to Alabama and northern Florida. It intergrades with *concinna* in the ex-

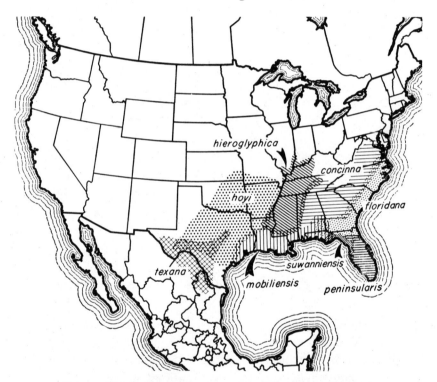

Map 19. Distribution of the subspecies of *Pseudemys floridana*.

treme northern part of its range and along the fall line, with *peninsularis* in the northern part of peninsular Florida, and with *mobiliensis* in Alabama and in the Florida panhandle.

Distinguishing features: The coastal plain turtle may be recognized by

Plate 52. Pseudemys floridana floridana; Billy's Island, Okefinokee Swamp, Georgia. In some respects these specimens vary slightly toward *P. f. peninsularis.* The individual shown in lower left was photographed while laying eggs on July 7, 1921. All are females except the turtle on the left in lower right.

its high shell, widest behind and highest in the middle, its separate supra-temporal and paramedian head stripes, and its smooth, unnotched upper jaw, and by the absence of a plastral pattern in a majority of specimens. The submarginal spots usually have light centers.

Description: The shell of this subspecies is highly elevated (although somewhat less so than that of the peninsular turtle) and with the highest point at the middle. This is a large race, the biggest specimens being on the average those from northern Florida but the maximum size (as far as my experience goes) being represented by a huge female from Charleston, South Carolina, with the following shell measurements: length, 397 mm. (15.6 inches); width, 270 mm. (10.6 inches); height, 175 mm. (6.90 inches).

In the specimens that I have measured, the ratio of the length divided by the width varied in males between 1.17 and 1.72 and in the females between 1.20 and 1.40; the ratio of the length divided by the height ranged in males from 2.19 to 3.40 and in females from 2.10 to 2.75. The ground color of the carapace is light to dark brown (usually lighter than in *peninsularis*), with complex, light-colored linear, reticulate, or concentric markings, in which transverse lines may predominate. The plastron is light yellow with no greenish tint, and there is usually no dark plastral pattern. The lower marginals are marked by a dark, light-centered spot on the rear section of each lamina, usually straddling the seam. The ground color of the head and legs is light to medium brown, not deep black, and the stripes are yellow with no greenish cast. On the head, the paramedian and supra-temporal stripes do not converge, but extend forward separately to a point between the eyes, at least, and there is usually a postorbital stripe extending from the posterior border of the orbit to the anterior edge of the tympanum. The portion of the supratemporal stripe on the neck is usually much broader than the corresponding part of the paramedian.

In a series of North Carolina specimens, fully mature males of between 228- and 256-mm. (8.97- and 10.01-inch) shell length average 3.6 per cent narrower and 6.7 per cent lower than mature females, and the average head width is 19 per cent less. The nails of the fore feet of the male are much longer than those of the female, and its tail is heavier and with much more distal vent.

The hatchling has not been described and I have seen none. A series of thirty specimens from North and South Carolina indicate that in growing from a shell length of 46 mm. (1.81 inches) to a shell length of 240 mm. (9.46 inches), the shell becomes 24 per cent narrower and 9 per cent lower, and the head becomes 11.2 per cent narrower, relative to the total length of the shell.

Habitat: This turtle shows little choice as regards its habitat. It is found in rivers, ponds, lakes, and swamps. It is partial to bodies of water in which hydrophytes are abundant.

Habits: Published data on the habits of this race are scarce. Students at the University of Florida have accumulated a good deal of information but the *floridana* population there, although including a fair percentage of individuals that must be identified as *P. f. floridana,* is actually an intergrading stock, and its habits cannot be regarded as necessarily typical of the cooter of the coastal plain farther north.

In southern Georgia and in coastal North Carolina this turtle is common, although it never reaches the abundance shown by the peninsular race or by the intermediate colonies in northern peninsular Florida. It is fond of taking the sun in small bands on snags or logs in quiet sloughs and backwaters of the lower streams. Under such conditions it is not approached easily, but on the whole it does not appear to me to be an unusually wary turtle. It disappears during the winter months, although as far north as the Ogeechee River drainage it may turn up again during Indian summer days or after short periods of unseasonably warm weather at any time of the winter.

Breeding: The only published information of the breeding habits of this subspecies is that of Wright and Funkhouser (1915), who found it to be the commonest turtle about the islands of the Okefinokee Swamp, Georgia. They gave May and June as the nesting season, with the height of activity coming in June. Egg complements included twelve to twenty eggs, which were ovate-elliptical, with soft shells and pinkish white, slightly granular surface. The eggs (103 were measured) varied in longest diameter between 33.0 and 40.5 mm., with an average of 34.1 mm. (1.30–1.59 inches, average 1.34 inches); in short diameter they ranged from 22 mm. to 27 mm., with an average of 24.15 mm. (0.866–1.06 inches, average 0.95 inches). Extensive depredations were made on the newly laid eggs by various animals in the swamp, mostly by bears, raccoons, and king snakes.

Feeding: Captive specimens are omnivorous in youth and mostly vegetarian when mature.

Economic importance: This turtle makes excellent soup and is eaten by rural people throughout its range. It was formerly sold in Charleston and Savannah in some numbers.

River Turtle *Pseudemys floridana concinna* (Le Conte)

(Plate 53. Figs. 18, 19. Map 19.)

Range: From Maryland southward to Alabama, mostly east of the mountains and above the fall line, and into eastern Tennessee. This subspecies has been found to intergrade with the coastal plain turtle and with the hieroglyphic turtle in various places where their ranges are contiguous.

Distinguishing features: This turtle is characterized by the long, relatively narrow, and depressed shell; the unjoined supratemporal and paramedian head stripes, the portion of the former on the neck noticeably broader than the latter; the presence of a more or less extensive dark figure

Plate 53. Pseudemys floridana concinna. Top, left and right: Male; Columbia, South Carolina. *Below:* Young female; Savannah River, McCormick County, South Carolina.

on the plastron; the smooth upper jaw, not notched at the tip; and the light-centered black spots on the submarginal seams. The size is somewhat less than that of *floridana* and *peninsularis;* shell length of the largest male that I have measured (Sumter County, S.C.) was 316 mm. (12.4 inches) and that of the largest female (from Columbus, Georgia) was 322 mm. (12.7 inches). Shell dimensions of a large male and a female of medium size are as follows:

Sex	Length	Width	Height
Male	274 mm.	180 mm.	79 mm.
(Congaree River, S.C.)	(10.8 inches)	(7.10 inches)	(3.11 inches)
Female	280 mm.	199 mm.	90 mm.
(Virginia)	(11.0 inches)	(7.85 inches)	(3.54 inches)

The ratio of the length of the shell divided by its height ranges in fully mature males from 3.06 to 3.46 (average, 3.26) and in females from 2.62 to 3.11 (average, 2.87). The highest point of the carapace is at the middle, and it is sometimes laterally restricted in the region of the sixth marginals. The ground color of the carapace is light to dark brown, with light markings that tend more to be concentrically arranged than in the case of *floridana*. The plastron is yellow, or occasionally red, and a dark figure more or less confined to the region near the seams is usually present, at least anteriorly. The ground color of the head and legs is light to dark brown; the stripes are yellow to reddish. There are 4 to 6 lines on the anterior surface of the front foot, but these tend to disappear on the feet, which are usually uniform dark above and below. The tail above is usually unstriped. On the top of the head between the greatly expanded supratemporal stripes there are 5 to 7 lines, all parallel and without transverse interconnections, which are also lacking between the stripes on the sides of the head. The upper jaw has neither notch nor cusps at the tip, and the ridge on the crushing surface of the lower jaw is tuberculate but does not have isolated, very high, conical teeth.

The sexual differences in this subspecies are much the same as in *floridana*. Besides the very long, straight fingernails and longer, thicker tail of the male, the carapace is conspicuously more depressed relative to its length than that of the female (comparative ratios given above); the width of the shell relative to the length remains virtually the same.

Comparing shell proportions of thirty young specimens (from west-central South Carolina) of both sexes, we find that in passing from the size group with shells between 39 and 58 mm. (1.53 and 2.28 inches) long to that in which the shells are 163 to 184 mm. (6.43 to 7.25 inches) long, the carapace increases in length over the width by 17 per cent, the depth of the shell relative to its length decreases by 9.4 per cent, and the width of the head relative to the length of the shell decreases by 7.7 per cent.

Habitat: In his original description of this handsome turtle, published more than a hundred years ago (1836), Le Conte said: "Inhabits in the rivers of Georgia and South Carolina, where the beds are rocky. I have never seen them below Augusta on the Savannah or Columbia on the Congaree." It appears to be by choice a river turtle, being taken usually in streams with considerable current; this of course might be attributed mainly to the paucity of slow-moving streams in piedmont areas such as that inhabited by *concinna*. It is interesting to note that here in the *concinna-floridana* relationship, as in that between *suwanniensis* and *peninsularis,* a form addicted to life in currents is distinguished from a more lacustrine relative by the presence of a dark plastral pattern in the former.

Habits: Almost no data on the habits of the river turtle are available. Brimley (1943) found the intestine of a large specimen to contain nothing but filamentous algae, which have been casually noted in other cases. Conant and Hudson gave five years as the longevity record of an individual in the Philadelphia Zoo. Neill (1948a) was unable to find this turtle during the winter months in Richmond County, Georgia, during periods when other turtles were fairly active, perhaps indicating a tendency toward deep hibernation.

It is the male of *P. f. concinna* to which the widespread and ludicrous myth concerning the worm-spearing function of the elongated nails of the fore feet has most frequently referred. Although usually cited as of doubtful authenticity, this story was accepted as gospel by at least one respectable turtle specialist and has found wide credence among laymen. I was interested to find what appears to be the first printed appearance of the legend, not in this case applied to *concinna* but apparently to the closely related *P. f. suwanniensis.* The yarn is that of the Reverend C. F. Knight, a New England clergyman and avowed eyewitness of the astounding operation:

It has upon its fore feet three claws of ordinary length and two of an enormous development; they being often found nearly three inches long. The reason of this elongation was not apparent until by close observation from a boat at the mouth of the river Wakulla [Florida], the speaker saw two turtles of this species, thrusting these long claws into the holes made by some worm, with which the hard clay bottom of the stream was everywhere perforated. The transfixed worms were probably the common food of this turtle [Knight, 1871].

Peninsular Turtle *Pseudemys floridana peninsularis* Carr

(Plate 54. Fig. 18. Map 19.)

Range: Peninsular Florida, from Alachua and Putnam counties southward to the tip of the peninsula. North of Alachua County it intergrades with *P. f. floridana* and in the rivers of the Gulf drainage north of Homosassa it is replaced by *P. f. suwanniensis*.

Distinguishing features: This turtle is easily identified by its highly arched shell and greenish-white, unmarked plastron, by the confluence of the paramedian and supratemporal stripes on the side of the head behind the eye and by the solid, smudgelike submarginal spots.

Description: The shell is very highly domed, elliptical in lateral outline, and with the highest point usually anterior to the middle of the long axis. This subspecies may reach a larger size than any other North American member of its family except *P. f. suwanniensis*. The maximum measured size appears to be that of a female from Lake Apopka (Orange County) with a shell 402 mm. (15.8 inches) long, 282 mm. (11.2 inches) wide, and 184 mm. (7.25 inches) high. The carapace is relatively smooth and typically with the pattern of reticulate or transverse light markings less conspicuous than in most other members of the group. The ground color is black or very dark brown. The bridge is extremely deep, usually easily visible from a side view of the shell. The under shell is very light yellowish or greenish yellow and not marked. On the lower marginals there are large, dark smudges, not light-centered, one bisected by each seam but often obsolescent anteriorly and posteriorly; one or two similar spots may mark the bridge, which, however, is more frequently immaculate. The ground color of the head and legs is lustrous, sooty black with greenish-yellow or greenish-white stripes. The supratemporal and paramedian head stripes meet behind the eye to continue forward as a single line, or if they barely fail to meet, then only one of them continues anteriorly. Otherwise the head stripes are much reduced as compared with those of other members of the complex. There is usually no postorbital stripe from the hind edge of the orbit to the tympanum; the portion of the supratemporal on the neck is not, or is but slightly, broader than the paramedian. The jaws are smooth or feebly serrate, the upper not notched at the tip. There are three (rarely five) stripes on the top of the head between the supratemporals at a point above the tympana and one to four stripes on the anterior surface of the fore legs.

Plate 54. Pseudemys floridana peninsularis. A, B, and E: Male; St. Petersburg, Florida. *C:* Female; same locality. *D:* Head of adult male; Hendry County, Florida.

The males are somewhat smaller than the females and with the shell lower, but there is little difference in relative width. Very large females tend to develop a distinct shoulder in the shell above the fore legs; this type of carapace is usually also strongly domed and often depressed along the centrals, imparting a curiously bilobed appearance. Shell measurements of a male and a female of large size from Orange County, Florida, are as follows:

Sex	Length	Width	Height
Female	366 mm.	210 mm.	132 mm.
	(14.4 inches)	(8.26 inches)	(5.2 inches)
Male	314 mm.	197 mm.	123 mm.
	(12.5 inches)	(7.76 inches)	(4.85 inches)

The nails of the fore feet of the males are longer than those of the females (the male whose dimensions were given above had nails 21 mm. long), but this difference is by no means as extreme as in some of the other subspecies. The tail of the male is longer, thicker, and blunter, with the anus located nearer the tip. The plastron of the male is not concave.

After losing its initial egg-shape the hatchling's shell remains comparatively broad for some time, losing about 10 per cent in relative width on reaching mature size in both sexes. The height of the shell relative to its length appears to remain fairly constant with growth. The hatchling is about 32 mm. (1.26 inches) in shell length and gains a couple of millimeters in width on expanding immediately after hatching. Of 16 hatchlings that I dug out of a nest in Alachua County, Florida, the largest and smallest had the following shell dimensions:

Length	Width	Height
33 mm.	31.2 mm.	18.2 mm.
(1.30 inches)	(1.24 inches)	(0.716 inches)
27 mm.	25 mm.	16.1 mm.
(1.06 inches)	(0.985 inches)	(0.635 inches)

The coloration of the young is more vivid, but the reduced head striping characteristic of the race is evident in it. The pattern of the carapace of the adult develops by the spreading, on a light background, of about nine rows of dark embryonic spots, still discernible as such in an occasional hatchling. What appear as light stripes and lines are really the remnants of the light ground color after expansion of the embryonic color areas. Week-old young have a strong egg tooth and a freely retractile yolk sac vestige.

Habitat: These turtles are found in almost any kind of aquatic situation, preferably where there is considerable vegetation. They are most abundant in lakes and sloughs and in certain streams of the St. Johns drainage.

Habits: Like most turtles that inhabit peninsular Florida, this species does not hibernate, in the true sense of the word. It may slow down considerably and even bury itself in the mud during cold spells, but the first warm sunshine brings it out to resume what appears to be normal activity. I have found these turtles sleeping among bonnet stems (or sometimes even feeding) in shallow ponds and have caught them by the sackful on nights when I had to wade to shore to thaw out by a huge fire every twenty minutes or so. During unseasonably warm winter weather they often wander about on land, and while this sometimes may be attributed to a migratory urge set up by a low water level in their home ponds, they do it at other times when no explanation is evident. At such times males as well as females may be encountered in the woods or on road shoulders, indicating that the wanderlust is not merely an ill-timed nesting instinct. Whatever the motive, it appears to be a disadvantageous one, since large numbers of cooters are mashed on the highways, killed and eaten by animals (what animals I don't know, but some creature that prowls the Florida woods can shuck out a peninsular turtle as easily as you or I can shell a peanut), or are wedged under fence wires, where they kick for a few days and then languish and die. I have several times seen a half-dozen or more shells under a few hundred yards of hog fence, and only recently counted eleven trapped in one fifty-yard stretch.

In certain places this turtle is very abundant and its sunning aggregations may include 20 or 30 individuals, or even more where there is a strategically located fallen treetop or a long cypress log lying at water level and providing ample room. Nowhere, however, does the peninsular turtle form assemblages as large as those of the Suwannee turtle in the middle reaches of the Suwannee River and in Suwannee Sound. In Rainbow Springs Run (Marion County), however, which is one of the very few places in which the peninsular and Suwannee turtles may be found together, Marchand (1942) found the turtle population to include 37.3 per cent *suwanniensis* as compared with 33.3 per cent for *peninsularis*.

The only data available on the growth of this species is that of Marchand, who marked and released a large number of specimens in spring runs of the peninsula. His scatter diagram indicates that the greatest growth shown by any of the 59 retaken specimens was that of a male which made an increase in its 160-mm. (6.3-inch) shell length of about 12 per cent per year; and the next greatest increase was that of a 240-mm. (9.46-inch) female, with a gain of about 8 per cent. In general the greatest gains were made, as one would expect, by the smaller adult individuals (juveniles were not included in the study). The same writer found that body weight often fluctuated without reference to shell dimensions.

Marchand saw strong evidence that the present species is mostly diurnal in its activities, and my own observations support this view, although on several occasions, in a colony of cooters most of which were asleep or at least inactive, I have come across two or three individuals that were feeding.

Information concerning the home range of the peninsular turtle was contributed by Marchand (1945b), who outlined the movements of 33 specimens recovered (out of 151 marked) in Crystal Springs, Pasco County, as follows:

Of the 33 *Pseudemys f. peninsularis* which were retaken once, 16 had not moved. Six of these returns came within a month, while the greatest length of time elapsed was 7 months (2 cases). Of the 17 individuals which did wander, 8 moved 100 yards or less, and the greatest distance covered was 300 yards (3 cases: 7 months, 1½ months and 7 days, respectively). In the group of eight specimens which were retaken twice, the greatest distance from the original point was 265 yards, this in a period of one month. Two individuals had not moved at all and two had returned to the site of initial capture. In the group of three specimens with three returns, two showed attachments for certain places, while the third wandered extensively, covering 400 yards in 19 days, although the total distance from the starting point was only 225 yards. The turtle that was retaken 4 times had not moved on one occasion, and the maximum distance covered on any movement was 175 yards, this in 18 days.

In another spring run (Rainbow), marked turtles showed greater tendency to wander, 14 of 20 retakes (two species, *suwanniensis* and *peninsularis*) having traveled 700 yards or more and some having moved five miles downstream. It is possible that in this case the increased movement was due to the fact that all the specimens were held in captivity for some time and then released in large numbers in one place on the run, thus building up there an abnormally large population.

In thermal experiments with this turtle Bogert and Cowles kept a 2,951-gram specimen for 40 hours in a chamber in which the temperature was maintained at 38° Centigrade and the relative humidity at 37 per cent. During this period the turtle maintained a body temperature of between 35° and 36.9° Centigrade, and at the end had lost 14.4 per cent of its original body weight with little harmful effect evident. When removed and placed in water it soon assumed a new weight that exceeded the original by 115 grams. Such data show that these turtles are able to resist desiccation to an extent that would permit protracted sojourns on land.

Breeding: Apparently nothing has been published concerning courtship and mating in this turtle. I have seen a male and female together at various times during the year, but never under conditions propitious for observation.

Laying has been observed from November to June, and probably occurs sporadically throughout the year. It is much easier to find nests in late spring and early summer than in the fall, however, and the peak is apparently reached in May throughout northern and central Florida. The nest may be made in nearly any kind of soil, the only apparent specification being a certain amount of friability. The nest is usually between 4 and 5 inches deep and the egg pocket about the same width at the widest point; the excavation is roughly flask-shaped, and there may or may not be one or more accessory side pockets, located 2 to 5 inches from the main hole and usually containing a single egg each. I have never watched the placing of these eggs in the side pockets, but I assume that they occupy the trenches that most nesting peninsular turtles kick out on either side of the nest and in which the feet rest during oviposition. The eggs fall singly into the nest, and, unlike some sea turtles, this cooter often pauses between eggs to arrange those in the bottom of the nest.

The number of eggs is quite variable, undoubtedly depending to some extent upon the number of times the individual has laid before during the season. Most females appear to lay at least twice a year, and the assortment of immature eggs to be seen in the ovaries of some specimens would seem to indicate that several layings may take place. However this may be, nearly every female that lays in May, if dissected, will prove to have ovarian eggs of about half an inch in diameter, and it appears likely that these eggs would mature within a matter of weeks. The usual number of eggs laid is about 20; nests have been found with from 12 to 29 eggs. The size of the eggs is also variable; they have been found to range between 29 and 38 mm. (1.14 and 1.49 inches) in length. An average-sized egg, 34 mm. (1.34 inches) long, is about 26 mm. (1.02 inches) in short diameter. The incubation period is markedly affected by temperature and moisture, but appears to average something over three months. Eggs laid during late winter or very early spring sometimes hatch no earlier than others laid two months later. In three nests that I marked after seeing the eggs laid on May 7,

1944 (at Bivins Arm of Paynes Prairie, Alachua County), hatching occurred sometime prior to August 26, when I dug out the nest. The weather had been unusually dry and it appeared that the hatchlings would have remained in the nests until a heavy rain came to soak and soften the overlying soil. A clutch of eggs watched by Goff and Goff (1935) hatched after 150 days' incubation, having been laid before the end of the cool weather (February 8, 1931).

A female may wander widely before choosing a site and beginning excavation of her nest, or she may head purposefully up the shore for a few feet and begin digging without preliminary. I saw one individual cross more than a hundred yards of newly ploughed and perfectly homogeneous ground, obviously exercising the most fastidious sort of discriminatory faculty, and stop at last to spoon out her nest in soil which to my rude eye was exactly like all the scorned earth behind her. A female usually consumes about one hour or slightly less in combined digging, laying, and meditation, and then usually heads directly for the water. In every newly made nest that I have seen, the soil appeared moister than that of the environs and I have no doubt that it is wet by cloacal water, as is customary with many species, but I have never actually observed the wetting process. A nesting *Pseudemys* is not characterized by the bemused singleness of purpose shown by sea turtles, and any obtrusive behavior on the part of onlookers interrupts her neuro-endocrine train and she may either retire into her shell for several hours or claw herself frenziedly back toward the water.

I have never been able to understand the value of the single eggs in the side pockets. The ostensible aim is to drag a red herring before the ravening skunk and coon, but if this is really the idea the gesture is a pathetic one and so hopelessly ingenuous that I wonder how the poor cooters get by at all. Instead of being a proper red herring, the scarcely covered, sometimes completely exposed, egg in the accessory hole is more like a beacon to the prowling oöphage and would appear almost certain to decrease, rather than increase, the chances of a nest's escaping notice. From the sad appearance of a popular nesting ground during the height of the season I would judge that very few nests indeed do escape notice, and I imagine that most of the peninsular turtles alive today owe their existence to the fortuitous coming of a heavy shower soon after their mother laid her eggs. Allen (1938) believed that 95 per cent of all eggs laid are destroyed by skunks, coons, opossums, king snakes, and hogs. Of 287 freshly made nests found by him, no fewer than 267 had recently been destroyed.

Feeding: The Florida turtle is almost completely herbivorous. Allen offered various kinds of food to a large number of captive individuals and found that neither live nor cut fish was acceptable, while such aquatic plants as *Sagittaria, Ceratophyllum,* and *Myriophyllum* were eaten readily and in great quantity. In the stomach of a female specimen from Rainbow Springs Run, Marchand found the following plant food: *Naias sp.,* 99 per cent; *Sagittaria lorata,* 1 per cent; *Lemna sp.,* unmeasured trace. The stomach of a male from the same place contained *Naias sp.* (90%) and *Sagittaria lorata* (10%). When vegetables from truck gardens in central Florida are placed in shallow lake margins to keep them fresh, these turtles sometimes congregate to eat them. They are very fond of watermelon.

Economic importance: Once in a while someone gathers a barrel or two of peninsular turtles and sends them to market, occasionally even shipping them north, but the total monetary value represented is negligible. This cooter is eaten locally, but not to the extent that it would be if people knew how good a stew it makes. It has always mystified me how often the most penurious cracker will court pellagra rather than relax his standards of dietetic refinement and eat cooter. The eggs are quite edible, a little more "sandy" than those of sea turtles, however, and this offends some palates.

Suwannee Turtle *Pseudemys floridana suwanniensis* Carr

(Plate 55. Fig. 18. Map 19.)

Range: Gulf coastal waters and rivers and runs of the Gulf drainage of northern and central Florida from the vicinity of Cape San Blas, Franklin County, southward to Pinellas County. Toward the southern end of its range the "Suwannee chicken" shows tendencies toward *P. f. peninsularis,* although the two subspecies mix in a few of the streams there without apparent interbreeding. In the headwaters of other west coast rivers there is sparing intergradation with *peninsularis* and in the area about the Apalachicola delta and Cape San Blas there is complete intergradation with *P. f. mobiliensis.*

Distinguishing features: Like all the cooters of the deep South, this one has a deep, highly and smoothly vaulted shell with a streamlined effect imparted by the distribution of the greatest elevations anterior to the middle of the long axis. The size is very large, a female of 416-mm. (16.3-inch) shell length apparently being the largest emydid on definite record. The shell of the largest male known is 359 mm. (14.1 inches) long.

The closest relative of this turtle is the Mobilian, which has similarly

Plate 55. Pseudemys floridana suwanniensis. A: Young individual; Dun-
ellon, Florida. (Courtesy Ross Allen.) *B and D:* Female; Withlacoochee River,
Marion County, Florida. *C:* Female; Levy County, Florida.

specialized and unusual habitat and ways of life. Aside from numerous slight but consistent differences in shell contours, the two may be distinguished at a glance by the difference in coloration, the ground color of the carapace and of the head and legs being much darker in *suwanniensis* and with reduced, greenish-white markings instead of orange-yellow or reddish stripes and lines as in *mobiliensis*. There are only two or three stripes on the fore surface of the fore leg, and the outer surface of the hind foot is not striped. The plastral figure is dendritic and symmetrical, except in old, faded specimens, and does not include unpaired spots or blotches placed off center. There are about five greenish stripes between the eyes on the sooty black skin on the head, and the arrangement of these stripes is like that in typical *floridana,* with the supratemporal expanded on the neck and not confluent with the paramedian as in *peninsularis,* from which the presence of a plastral pattern also distinguishes *suwanniensis.*

The male is smaller than the female. Of 440 specimens measured by Marchand (1942), 232 were females ranging in shell length from 120–400 mm. (4.73–15.7 inches), and of these the greater part had shells 240–360 mm. (9.45–14.2 inches) long. Shell length in his males varied between 130–330 mm. (5.13–13.0 inches), most of them falling between 150 mm. (5.92 inches) and 270 mm. (10.6 inches). The male is also flatter than the female. Of 50 specimens of each sex Marchand found that the ratio of the length divided by the height of the shell was from 2.5 to 2.8 in the great majority of females and from 2.85 to 3.1 in nearly all the males. There is very slight overlapping in this respect, and perhaps none if the ratios were segregated into size groups. The male has, in addition to a very large tail, the ridiculously elongate straight nails on the fore feet that are so common in its section of the genus. Proportional to the depth of the shell, the paddle surfaces of the hind feet are much more extensive than in the female, and Marchand regarded these turtles as easily the fastest swimmers of all *Pseudemys,* a judgment with which my own experience is in complete agreement.

For some curious reason I have never been able to obtain hatchlings of the Suwannee turtle, and no account of the very young has been published.

Habitat: This turtle is found in rivers, spring runs, and salt-water flats in bays and offshore lagoons where there is abundant vegetation. In Rainbow Springs Run, Marchand found that "the most favorable sections of the run were those in which there was a heavy growth of *Naias sp.* and of *Sagittaria lorata,* and where the bottom was fairly flat and away from the main current of the run." I have seen the heads of literally hundreds of

Suwannee chickens thrust above the surface in shallow salt water covering grassy flats in Suwannee Sound. The extraordinary abundance attained by this turtle in the lower, brackish streams of the Gulf drainage was long ago remarked by Knight (1871), who commented as follows:

The *Emys serrata* [actually *P. f. suwanniensis*] in the early summer congregated in great numbers in the warm and still bayous near the mouths of those streams [along the coast near Tallahassee] which empty into the Gulf. On one occasion the speaker, floating quietly down stream, came upon one of these gatherings where there seemed to be many thousands within the space of two or three acres, covering every log and stump and hummock almost as thickly as shingles lie upon a roof.

In the original description of this subspecies (Carr, 1937a), I noted the marked preference shown by it for two adjacent but strikingly different habitats—the clear, spring-fed rivers and runs of the Gulf coast and the turtle-grass flats off the mouths of these streams. It is of interest to speculate on the history of this almost uniquely euryhaline turtle population and to note that its apparent tolerance of strong variations in salinity involves a passing between the sea and the same calcareous spring waters that are periodically or sporadically invaded by numerous marine fishes with what appears to be complete impunity. The Suwannee chicken sometimes wanders far offshore and is occasionally caught and taken to the Havana market by visiting Cuban trawlers. Specimens with heavy infestations of barnacles are often seen.

Habits: Nothing is known concerning the relative seasonal activities of these turtles, beyond the observation that they are less in evidence in the streams during cold weather. As might be expected, the bands that graze on the salt-water flats are much less susceptible to weather changes and may often be seen feeding when the air temperature is near freezing.

There is little information available on the home range and migrations of this turtle.

Marchand (1942) commented as follows regarding results of his marking experiments at Rainbow Springs Run (Marion County):

Some 20 returns, about equally distributed between *P. f. peninsularis* and *P. f. suwanniensis,* were obtained, and of these 14 had wandered 700 yards or more. Furthermore, throughout the entire period of work here, marked turtles were common even in the lower parts of the run, and since all specimens were released at the camp, a distance of several miles had been covered. Mr. E. Ross Allen informed me that he had captured one of my marked *P. f. suwanniensis* (a female) at the junction of Rainbow Run and the Withlacoochee River, a distance of about 3½ miles below the camp. Some of this wandering should probably be attributed to abnormal population pressure set up by releasing large numbers of specimens at the camp.

More extensive marking experiments are much to be desired, since they would probably furnish data on the degree of intercourse that occurs between the fluviatile and marine populations. It should be mentioned that of the 1,022 turtles collected in Rainbow Run by Marchand, *P. f. peninsularis* constituted 33.3 per cent and *P. f. suwanniensis* 37.3 per cent, with no evidence of intergradation or hybridization. This is a most intriguing observation in view of the fact that the divergence of the two forms is elsewhere bridged extensively by the characters of intermediate stocks.

In relatively undisturbed waters the sunning aggregations of this race are the largest I have ever seen. Although it would be hard to give a reliable estimate of the numbers represented in some of these gatherings, they must occasionally number in the hundreds, since they sometimes cover almost completely several huge cypress logs in a jam. Turtles in such groups are usually exceedingly shy, although this may be due to some extent to the fact that in few if any of the most popular sunning sites are they free from at least occasional persecution by man.

During a two-year period of work at Rainbow Springs Run, Marchand retook 101 marked specimens of *suwanniensis*. His graphs of growth rates calculated from these data show the greatest increase in shell length to have been made by a young female (length, 115 mm. or 4.54 inches), which increased its length by about 23 per cent per year. The next most rapid growth was that of a young male 140 mm. (5.52 inches) long, which increased its length by 17 per cent. The average growth rate for all retaken specimens was between 5 and 10 per cent per year, with the smaller specimens showing the more rapid rates.

Breeding: The only published information on courtship in this subspecies (or, for that matter, in the species) is that of Marchand (1944), who observed it at Blue Springs, Gilchrist County, and Rainbow Run, Marion County. The procedure was relatively simple, and by no means as bizarre as the antics of the related *scripta* group. The female swam slowly along under three or four feet of water while the male, after approaching from behind, swam just above her, his neck curved downward toward and to within a half inch or so of hers. Meanwhile, the long tail of the male was bent forward, and as he sank to a point just above the shell of the female the tip of the tail was thrust under the hind margin of her shell and the pair went to the bottom, where copulation was presumably effected.

Except for the preceding notes nothing is known of the reproductive habits of this turtle.

Feeding: Mature Suwannee turtles appear to be almost wholly vegetarian. Although Allen (1938) noted a slight tendency to scavenge and to take meat or dead fish, he found no predacious inclinations and a marked preference for plant food, both aquatic and terrestrial. Marchand (1942) examined the stomachs of 10 adult specimens taken in Hart Springs, Gilchrist County, between October, 1940, and February, 1941; the food present in these was as follows: *Naias sp.*, 82.3 per cent; *Lemna sp.*, 7.2 per cent; *Ceratophyllum sp.*, 5.5 per cent; *Sagittaria sp.*, 2.5 per cent; filamentous algae, 2.5 per cent. In the salt-water habitats the turtles appear to feed largely on turtle grass. Very young individuals in captivity are at least as carnivorous as herbivorous, if not even more so, as appears to be the case with the young of most plant-eating turtles.

Economic importance: The local esteem for the "Suwannee chicken" is profound and reverent and by no means unjustified, although in my judgment the disparity in flavor between this and the other hard-shelled cooters is not so great as the Suwannee River people would have one believe. The principal point of superiority shown by *suwanniensis* seems to me to be the absence of any sign of the "cottony" texture that detracts from the appeal of the meat of the others unless skillfully cooked. Moreover, there can be no more succulent soup than that made from these turtles.

There used to be a man on an island off the mouth of the Suwannee River who contributed much to the fame of the Suwannee chicken through the extraordinary artistry of the stews he made of it. People came from long distances for the privilege of eating turtle here, even though the last part of the trip involved a twenty-mile jaunt over a wheel-track woods road and a long haul by boat. In view of this type of enthusiasm it seems strange that the turtles have never found their way to an organized market.

Mobile Turtle *Pseudemys floridana mobiliensis* (Holbrook)

(Plate 56. Fig. 18. Map 19.)

Range: Coastal waters and lower streams of the Gulf drainage from the Florida panhandle in the vicinity of Cape San Blas and extreme southwestern Georgia westward to Texas.

Distinguishing features: This is a high-shelled version of *concinna* and *hieroglyphica*, or, if one prefers, a *floridana* with plastral markings. Per-

Plate 56. Pseudemys floridana mobiliensis. A and B: Male; southern Louisiana.
C: Young; same locality. *D:* Female; Mobile, Alabama.

haps the latter figure is the better, because in areas around the type locality
mobiliensis appears to differ from *floridana* in almost no other way. The
upper and lower jaws in specimens of comparable age and sex are usually
much more strongly serrate in *mobiliensis* than in *floridana,* but not always;
the shell shows slight average differences in markings and conformation,
and the Mobilian is much the larger. This race also intergrades with *P. f.
suwanniensis,* but typically is easily distinguished by its lighter ground
color of carapace and soft parts (brown in *mobiliensis,* sooty black in
suwanniensis); by the orange or reddish cast to the yellow markings
(greenish yellow and very light in the Suwannee chicken); by the presence of four or more lines on the outer surface of the fore limb and seven
or more between the eyes, while the corresponding numbers are respectively

two to three and five or more in *suwanniensis*, from which *mobiliensis* also differs in having the outer surface of the hind foot striped.

The maximum size is larger than that of most members of the genus (but apparently smaller than that reached by *suwanniensis* and *peninsularis*), being represented among specimens I have measured by a shell length of 280 mm. (11.0 inches) for males and 350 mm. (13.75 inches) for females. Much larger top sizes have been reported, usually without specific detail. True (1884), for example, gave 16 inches as maximum shell length, and DeKay (1842) said that the shell may be 15 inches long. Shell dimensions of a male and female of good size from Mobile are as follows:

Sex	Length	Width	Height
Male	255 mm.	172 mm.	96 mm.
	(10.0 inches)	(6.77 inches)	(3.78 inches)
Female	315 mm.	236 mm.	121 mm.
	(12.4 inches)	(9.30 inches)	(4.76 inches)

The ratio of the length divided by the height varies in specimens I have measured from 2.66 to 3.19 (average, 2.95) for males, and from 2.33 to 2.93 (average, 2.75) in females. The ratio of the length divided by the width ranged from 1.24 to 1.44 (average, 1.34) in males, and from 1.26 to 1.44 (average, 1.35) in females.

The pattern of the plastron is not altogether confined to the seams between the laminae, being often less symmetrical than in either *concinna* or *suwanniensis*. Frequently it comprises only a spot or two, placed off center, but in some specimens the typical dendritic figure so widespread in the genus (and in related genera) is to be seen.

Aberrant specimens of the Mobilian, like similar ones of the Suwannee turtle, may have the upper jaw terminally notched and with lateral cusps as in *texana* (and also as in the *rubriventris* section). This form was for years dignified by a name, *P. alabamensis* (Baur). Another common variation, especially in populations toward the western extreme of the range, is a conspicuous vertical bar through the tympanum, which is merely a sign of the genetic influence of the neighboring Texas turtle, in which such a mark is common.

The smallest specimen of which any measurements are available has a shell 40 mm. (1.57 inches) long, 39 mm. (1.53 inches) wide, and 17 mm. (0.67 inches) high. The plastral pattern is complete on all seams and almost symmetrical, and that of the carapace is well defined, with transverse light lines predominating on the laterals and longitudinally directed lines on the centrals. A keel is conspicuous on the first, second, and third centrals but absent on the fourth and fifth.

Habitat: This turtle occurs in lower coastal streams, ideally those with abundant submerged vegetation; in brackish estuaries; and occasionally on salt-water flats behind the offshore islands.

Habits: Almost nothing is known of the life history of the Mobilian. Reports of local inhabitants seem to indicate that its habits are similar to those of *suwanniensis*. There is no question but that it ventures regularly into salt water, if it does not in some cases actually remain there permanently. Regarding the finding of a specimen on Horn Island, an offshore bar in Mississippi Sound, Allen (1932) noted the following·

A large individual was on the landward beach . . . , June 28, 1930. It was very slow in its movements and seemed to be near death. It had probably been carried out to sea and become stranded on the island. The carapace and plastron of another fresh water turtle were found nearby, but the species could not be determined with accuracy.

It seems probable that Allen's turtle was suffering from some malady other than the effect of high salinity, since shrimp trawlers in former days caught considerable numbers of normal individuals in this same sound. That this race may be less strictly vegetarian in its diet than *suwanniensis* is suggested by Allen's remark that he caught a large individual on hook and line.

A specimen lived twelve years and six months at the Philadelphia Zoo.

The only breeding record for this turtle is Agassiz' statement that he found twelve eggs in a specimen from Natchez, Mississippi, dissected on July 12.

Economic importance: This turtle represents a fairly valuable and well-regarded food resource in the Gulf coastal area. Although the demand in fashionable restaurants is not nearly so great as for diamondbacks and green turtles, the Mobilian is found with far greater frequency than those turtles in the markets that cater to the small restaurant and to the housewife. There is some opinion among Gulf coast turtle connoisseurs that the Mobilian is somewhat inferior in flavor to its near relative, the "Suwannee chicken."

Hieroglyphic Turtle *Pseudemys floridana hieroglyphica* (Holbrook)
(Plate 57. Fig. 18. Map 19.)

Range: Mississippi Valley from southern Illinois and Indiana southward through western Kentucky and western Tennessee to central Mississippi,

Plate 57. Pseudemys floridana hieroglyphica. Upper and center: Male; Hickman, Kentucky. *Lower:* Male; Reelfoot Lake, Tennessee.

305

northwestern Alabama, and northeastern Louisiana. West of the Mississippi River this subspecies is replaced by *P. f. hoyi*, from which it is incompletely differentiated. Intergrades with *concinna* have been found in central Tennessee and with *mobiliensis* in southern Alabama and Louisiana.

Distinguishing features: This turtle is intermediate between *P. f. concinna* and trans-Mississippi *hoyi*, and between these two and coastal *mobiliensis* as well, both with respect to its characters and in its distribution. Typical specimens, constituting a majority in the area outlined above, are fairly easily identified by a combination of minor features. The shell is long, fairly narrow, and relatively very low, although some specimens vary toward the shorter oblong shell of *hoyi*. The shell is sometimes slightly constricted at the region of the sixth marginals. The size is apparently somewhat larger than that of *concinna*, maximum shell length being about 370 mm. (about 14.6 inches). The shell dimensions of the type specimen, a large male from the Cumberland River, Tennessee, are as follows: length, 265 mm. (10.4 inches); width, 178 mm. (7.02 inches); depth of shell, 74 mm. (2.91 inches); width of head, 29.3 mm. (1.15 inches). The carapace is olive to dark brown, with a conspicuous, rather broad-lined, network enclosing repeatedly concentric rings; these light markings are on the average composed of more and narrower lines disposed in less transverse and right-angular arrangement than in typical *concinna*. Each upper marginal is split by a T-shaped crossbar and each marginal seam is straddled by a concentric light figure. The plastron is yellow, with a more or less complete dark pattern along the seams; this pattern when fully developed is indistinguishable from those of *mobiliensis*, *concinna*, and *suwanniensis*. The head and legs are brown with yellow stripes; and the arrangement of the stripes on the head is as in *concinna*, the supratemporal enlarged on the neck and not joined to the paramedian; the other lines are of various widths and without lateral interconnections. The feet are marked by continuations of the leg stripes. The alveolar ridges of the jaws are strongly tuberculate, that of the lower jaw with a few high, conical teeth of which that near the angle of the jaws is the longest. The upper jaw is usually slightly notched at the tip but never with lateral cusps.

The males are somewhat smaller and distinctly lower than the females, and their fore toenails are very much longer, apparently averaging longer than in any other form except *hoyi*. Average ratio of length divided by height is about 3.26 in males and about 2.87 in females.

Very young individuals appear to differ from adults in the same respects as in the other related races. The hatchling has not been described.

Habitat: This turtle lives closely associated with rivers, although usually choosing the quieter and more protected coves, sloughs, oxbows, and pond-like bays. It is not often taken in the swift water of higher streams. It is occasionally found in isolated lakes.

Habits: Almost nothing has been published on the life history of this subspecies. From the oviduct of a female sent in late May from Reelfoot Lake, Tennessee, Cagle (1937) took nine eggs, the average dimensions of which were 37.6 mm. by 26.1 mm. (1.48 by 1.03 inches). The same writer mentioned "early June" as the season of nesting, and stated that of a number of nests of which he had records all were located within a hundred feet of the edge of the water.

Cahn (1937) named a long list of animals found in stomachs of specimens of this turtle to support his suggestion that it is largely carnivorous. Parker (1939), on the other hand, found plants to constitute 99 per cent of the contents of the stomach of a large individual dissected by him, and remarked that "small ones in captivity will eat bits of meat and insects, as well as vegetable matter." It seems probable that *hieroglyphica* is not markedly different in its food preferences from eastern subspecies of the complex, and that the carnivorous tendency noted by Cahn is actually characteristic only of the juvenile groups.

Economic importance: This turtle is used as food locally, but apparently is not sent to market with any frequency. Its eggs undoubtedly are hunted indiscriminately, along with those of *P. s. elegans,* where large populations of the latter in the waters occupied by the former make egg hunting popular.

Hoy's Turtle *Pseudemys floridana hoyi* (Agassiz)

(Plate 58. Figs. 19, 20. Map 19.)

Range: Southern Missouri and southeastern Kansas southward through Arkansas to northwestern Louisiana and southwestward across Oklahoma into south-central Texas, where intergradation with *P. f. texana* occurs.

Distinguishing features: This is a rather easily identified subspecies throughout interior parts of its range. As its most conspicuous character it possesses a median notch at the tip of the upper jaw. This notch is not flanked by sharp cusps, although the cutting edge (which is not serrate) may curve downward slightly on either side of the notch. While such a notch may occur in *hieroglyphica,* it is much less frequent there and less pronounced, and in the intergrading *P. f. texana* the notch is bounded by

Plate 58. Pseudemys floridana hoyi. A: Male; Imboden, Arkansas. *B and D:* Females; same locality. *C:* Female; Black River, Powhatan, Arkansas.

lateral cusps. The shell is relatively short and broad (as compared with *hieroglyphica* and *concinna*) in some females, especially, being so strongly oblong as to be nearly rectangular. Although a most arresting character, this squarish shell shape is by no means constant, but a tendency toward it, at least, may be seen in most populations of *hoyi*. The shell is brownish or olive above with numerous comparatively narrow, light, concentrically arranged lines, the transverse elements most conspicuous; below, it is a very light shade of yellow or even nearly white, with a more or less complete dark plastral pattern. The ground color of the plastron, from many places in the range of the subspecies, is so unusually light as to be one of the more striking features of the race; in other places, however, it is deeper yellow or even slightly orange, as in *hieroglyphica*. Another characteristic feature is the nature and arrangement of the light stripes on the side, and especially on the top, of the head and neck. These are numerous and, instead of being of various orders of breadth, are much more nearly equal in breadth and intensity than in any other race; there are nearly always as many as eight and up to nineteen of these subequal stripes on the neck between the supratemporals. Instead of being mostly continuously horizontal, the head stripes are often arranged as median U-shaped figures, or are extensively broken and interrupted or interconnected laterally by bars or crescents. Such departures from the basic plan of head striping occur in the other subspecies as aberrancies, but only in the present form and in

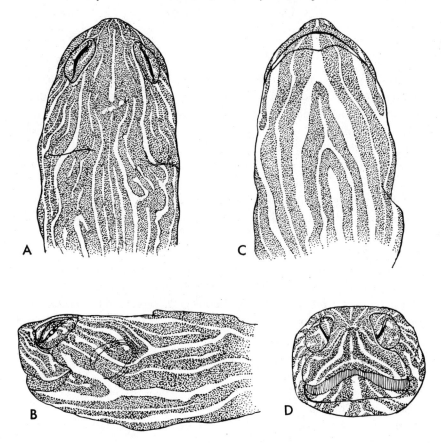

Figure 20. Details of the head pattern of *Pseudemys floridana hoyi.*

texana do they characterize a majority of specimens throughout most of the range. Burger, Smith, and Smith suggested that this race may be distinguished from *hieroglyphica* on the basis of the color patterns of the chin and lower surface of the hind foot, but I have been unable to investigate this point. The alveolar ridge in the lower jaw is strongly tuberculate, with the "teeth" few, isolated, and very high, especially posteriorly.

The size is evidently somewhat smaller than that of the eastern races, the maximum probably being represented by female specimens with shell lengths of about 300 mm. (11.8 inches). Dimensions of the shells of a mature male and female of fair size from Lawrence County, Arkansas, are as follows:

Sex	Length	Width	Height
Male	182 mm.	140 mm.	61 mm.
	(7.17 inches)	(5.51 inches)	(2.40 inches)
Female	231 mm.	177 mm.	85 mm.
	(9.1 inches)	(6.98 inches)	(3.35 inches)

The shell of the male tends to be lower than that of the female, very slightly narrower, and with less tendency toward the extreme oblong in outline. The fore toenails of the male are as much as three times as long as those of females of equal or greater size, and the disparity in tail size and conformation is as in other races.

I never saw a hatchling of this subspecies, nor even a very young specimen, and there is no published information on them.

Habitat: This is primarily a stream-inhabiting turtle, preferring rivers with silted bottom; it has also been taken, however, in lakes, small ponds, and drainage ditches.

Habits: In view of the broad territory occupied by this turtle, it is surprising to find an almost complete lack of published information on its habits. Strecker (1927c) stated that in nature it feeds almost entirely on Mollusca, principally snails of the genera *Sphaerium, Planorbis,* and *Limnaea,* and the character of the streams that it inhabits, flood-swept and often devoid of vegetation, may indicate that the subspecies is at least capable of subsisting on a diet composed largely of animal food. The thin, delicate jaws, however, are not those of a shell crusher, and there have been traces of vegetable matter in the intestinal tracts of specimens that I have received.

Economic importance: Hoy's turtle is eaten locally and is occasionally shipped to large markets along with consignments of some other, more abundant turtle.

Plate 59. Intergrade male specimen, between *Pseudemys f. hoyi* and *P. f. texana* (nearer the latter); Waco, Texas.

Texas Turtle *Pseudemys floridana texana* Baur

(Plate 60. Figs. 18, 19. Map 19.)

Range: Western Texas from Culberson and Medina counties into north-eastern Mexico in Coahuila and Nuevo Leon. The concurrence of distribution of this race of *P. floridana* and *P. s. gaigeae,* plus their "fortuitous" agreement in certain peculiar characters, is another of the eccentricities of this curious genus of turtles.

Distinguishing features: This is the most strikingly differentiated member of the *floridana* group. It is so completely different in so many ways from eastern representatives that nearly everyone who has seen it, including the describer himself, has thought of it as representing the *rubriventris* section. Besides this, until I disinterred the type specimen from underneath half a cord of sawfish saws and related plunder in a storeroom in the basement of the Philadelphia Academy of Sciences, this distinctive cooter really languished nameless, since its rightful name, *texana,* had been for years misapplied to the composite *hoyi-hieroglyphica* population. It is really difficult to look at typical specimens of true *texana* and of *concinna,* say, and persuade oneself that one of these could merge with the other if the vicissitudes of all its characters were traced through the intervening turtle populations, but such is the case. In south-central Texas all turtles representing the *floridana* section turn out to be some sort of intermediates between *hoyi* and *texana.*

The more obvious distinguishing characters of this turtle are the notched and cusped upper jaw, the very strong "teeth" on the alveolar ridges, and the much-broken head pattern, involving isolated spots and short, thick, often vertical bars. Other features are given in the description below.

Description: The shell of the Texas turtle is a short ellipse in lateral outline, and as seen from the side it curves evenly up to a highest point at the middle. The ratio of the length divided by the height ranges in mature specimens measured by me from 2.66 to 2.77, with an average of 2.7. There are insufficient data on which to base a statement of the maximum size. The biggest female I have seen is the type specimen, in which the shell is 273 mm. (10.75 inches) long, and the biggest male had a shell 244 mm. (9.62 inches) long. Shell dimensions of a small female and a medium-sized male from north-central Coahuila, Mexico, are as follows:

Plate 60. Pseudemys floridana texana. A, C, and E: Females; near San Antonio, Texas. B and D: Young specimen (preserved); Guadelupe Mountains, Texas.

Sex	Length	Width	Height
Male	195 mm. (7.67 inches)	145 mm. (5.72 inches)	70 mm. (2.76 inches)
Female	120 mm. (4.73 inches)	100 mm. (3.94 inches)	46 mm. (1.81 inches)

In the juvenile pattern of the carapace there is a network of fine-lined reticulations, the meshes of which are in large part filled with light-lined ocelli, and three to six of these may occur on a lateral or central lamina; the upper marginals are ocellated, each figure bisected by a seam. The ground color of the carapace is olive brown and the markings are whitish to reddish orange. The plastron is marked by a pattern of rather narrow dark lines following the seams; this tends to fade in the middle area of the plastron. There are an ocellated dark spot on each submarginal seam and a linear figure on the bridge. The ground color of the under shell is dilute yellow, but darker than in *hoyi*. The lines on the head are much broken, giving rise to a number of more or less isolated expansions or spots that are usually black-bordered (cf. *P. s. gaigeae!*).

The supratemporal stripe is usually markedly expanded on the temporal region and often breaks again behind the eye to leave there a small round spot. The stripe from mid orbit to mid tympanum is usually discontinuous also, leaving on the tympanum a bar, often vertically directed and sometimes connected with the orbitomandibular stripe. Many of the other stripes may be interrupted and cross-connected, making the most bizarre set of head markings imaginable, considering that all are derived from the comparatively simple *"concinna-rubriventris"* plan. The head markings are much more strongly suggestive of those of *gaigeae,* only remotely related but co-extensive in range; and further remarkable and wholly unexplained parallelism between these two groups in the Rio Grande area is seen in the melanistic degeneration of the juvenile shell pattern, which occurs to more marked extent in *texana* than in any form outside of the *scripta* group.

The upper jaw has a smooth cutting edge except for a notch at the tip, flanked by more or less pronounced cusplike projections of the tomium; these cusps are the termination of a pair of low, vertical ridges extending downward from the region of each nostril and giving a curiously angular appearance to the end of the snout, and in old males this may be markedly swollen and broadened. The lower jaw is coarsely serrate, with a long, sharp terminal tooth flanked by cusplike elevations of the cutting edge.

The alveolar surfaces of both jaws have very strongly tuberculate median ridges, that of the upper jaw usually with a very high tubercle anteriorly, followed by a wide gap, then another tooth, a short gap, and a final shorter tooth. In the lower jaw the median ridge has tubercles which are so high that they are easily seen projecting above the cutting edge from the anterior aspect.

There are striking secondary sexual differences in this subspecies. Besides a heavier tail and the more swollen snout mentioned above, there is a tendency toward longer nails on the front feet of the male, although this is less marked than in *hieroglyphica* and *hoyi* and the nails are sometimes more curved than in those races. Here again we note an unexplained parallelism with *P. s. gaigeae*. In the longest-nailed male I have seen, the animal was only 135 mm. (5.33 inches) long and the longest nail was 14 mm. (0.55 inches) long; the order of length of the nails was as follows: 3, 4, 2, 5, 1. The most remarkable sexual difference is the series of color changes undergone by older males, to be seen also in old females but to no such pronounced degree. On the head and limbs the light stripes tend to widen and later to break up and scatter, producing a homogeneously mottled effect; a similar mottling of the under shell is produced when the plastral figure and the markings of the bridge and submarginals disintegrate and scatter. Likewise, the linear pattern of the carapace fades and breaks up and is replaced by an intricate scattering of short lines and bars having little resemblance to the markings of young specimens.

The smallest specimen I have seen had a shell 38 mm. (1.49 inches) long and 17 mm. (0.67 inches) high; damaged marginals made measurement of the width pointless. In it, all the markings were exceptionally bright and well defined and the pattern of the carapace was very complex. The hatchling has not been described.

Habitat: This turtle is usually reported in rivers but is found also in tanks and ditches.

Habits: The scattered life history notes on *"texana"* that have been published have all actually referred to *hoyi* or to *hieroglyphica*. The fact that *texana* is often found in bodies of water in which no vegetation occurs may indicate that it can live on animal food, if necessary.

Economic importance: I know nothing of the relation of this turtle to man.

Genus *DEIROCHELYS*

The Chicken Turtles

This is a curious genus with only one species, and this is confined to the southeastern United States. It is characterized by an exceptionally long neck and by the length and slimness of the free upper ends of the ribs, which bend around an enormous bundle of retractile muscles for the neck. The shell is elongated and its surface is usually sculptured.

Of living relatives the genus *Chrysemys* is evidently most closely allied to *Deirochelys*, the chief differences between the two being the above-mentioned characters of the chicken turtles. There is considerable similarity between *Deirochelys* and *Emys*, but this is probably purely fortuitous.

This genus is known from the Pliocene of Florida.

Chicken Turtle *Deirochelys reticularia* (Latreille)

(Plate 61. Map 20.)

Range: Southeastern North Carolina southward in the coastal plain of South Carolina and Georgia to Dade and Collier counties, Florida; westward through southern Alabama and Mississippi to Louisiana and eastern Texas, whence it extends northward to southeastern Oklahoma and northeastern Arkansas.

Distinguishing features: This turtle, although superficially similar in appearance to *Chrysemys*, may be identified by its extraordinarily long neck, its elongated, finely wrinkled, and reticulated carapace, and by the vertical striping of the rump.

Description: The shell of the chicken turtle is long, narrow, and somewhat depressed. The size is moderate, a near-maximum shell length probably being 250 mm. (9.85 inches), and the more usual adult length not much more than 125 mm. (4.93 inches); shell width is usually about 65 per cent of the length, shell height, about 40 per cent, and head-and-neck length, 75 to 80 per cent. The carapace is usually sculptured with narrow longitudinal wrinkles, and is almost or completely keelless in adults and not serrate behind. It is olive to brown in ground color, with a very distinctive large-mesh reticulate light pattern. The plastron is large; it and the

Map 20. Distribution of *Deirochelys reticularia.*

lower marginals are yellow, usually with one or two black smudges on the bridge and perhaps on some of the neighboring marginals. The head and neck are olive brown with conspicuous but variable light stripes. The legs are marbled with whitish or yellow—somewhat irregularly so except on the rump, where there is a characteristic pattern of vertical stripes. The alveolar surfaces of the jaws are straight and narrow, and the jaws meet at an angle at the symphysis.

Although no detailed study of sexual characteristics in this species has been made, they appear not to be pronounced. The tail of the male is the longer. The females that I have seen have averaged larger and with broader shell, but I have no statistics to demonstrate this and none have been published.

A hatchling from southern Louisiana, with a sharp, brownish, thornlike egg tooth midway between the angle at the jaw tip and the nostrils, has the following dimensions: length of carapace, 28 mm. (1.10 inches); width of carapace, 25.5 mm. (1.0 inch); depth of shell, 14 mm. (0.55 inches); length of plastron, 26 mm. (1.02 inches); length of tail behind rear edge of cara-

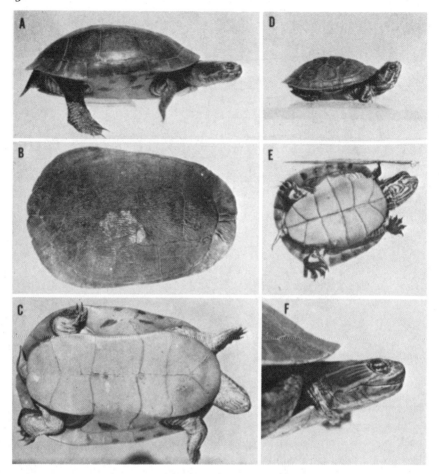

Plate 61. *Deirochelys reticularia.* *A and C:* Female; southern Louisiana. *B:* Male; Leesburg, Florida. *D and E:* Young; Okefinokee Swamp, Georgia. *F:* Female; St. Petersburg, Florida.

pace, 7 mm. (0.276 inches); length of extended neck and head beyond anterior edge of plastron, 21 mm. (0.826 inches); width of head, 8 mm. (0.315 inches). Markings are as in the adult except that more head stripes are evident and there is a dark linear pattern following the seams of the plastron and black spots or bars on the seams of all the lower marginals.

Habitat: The chicken turtle is essentially an inhabitant of quiet water—ponds, marshes, sloughs, and ditches. It is only rarely found in streams, and then usually in the quieter reaches or arms where pond conditions are approached.

Habits: Both sexes of this turtle are given to walking about on road shoulders and in flatwoods, and in the Everglades large numbers of individuals may at times be found mashed on the roads. Neill (1948a) recently suggested that the abundance of shells of this species to be found in the woods in Richmond County, Georgia, may be due to the turtles' having been caught by sudden cold weather while indulging this inclination to wander on land.

Breeding: Apparently, the chicken turtle lays in Florida at any time of the year; at any rate, eggs have been found from September to January, and females often wander about ploughed fields among nesting *Pseudemys* in June, although no nests have been observed at this time. Shelled eggs have been taken from the oviducts of females at widely spaced intervals throughout the year, but Cagle and Tihen showed that this may not necessarily always be indicative of the nesting season, since they found evidence that if suitable conditions for nesting are lacking, the eggs may be retained for months. In the case they described, shelled eggs were retained from June 30, 1946, until January 30, 1947, and some of them had been freed into the abdominal cavity by erosive destruction of areas of the wall of the oviduct. The eggs vary in length from 37 to 40 mm. (1.46 to 1.57 inches) and in width from 20.1 to 23.1 mm. (0.79 to 0.91 inches). The number of eggs per clutch may vary from 7 to 15. I have seen nests in heavy soil in ploughed lakeside fields, but have not witnessed the nesting process, and it has not been described.

Feeding: Few details of the feeding habits of the chicken turtle are known. Captive specimens accept nearly anything offered them, and I have seen them under natural conditions eating tadpoles, crayfish, and what appeared to be the bud of a bonnet (*Nuphar*). I believe it probable that the species is really omnivorous, and am certain that its natural diet includes more animal food than that of *Pseudemys*.

Economic importance: This turtle is nearly everywhere regarded as worth eating, and justly so. Formerly it was commonly sold in the markets in southern cities, but this occurs much less frequently today. The chicken turtle is recognized by rural people as distinct from the other cooters in many places in the south and is generally rated by them as even more succulent than the rest, although with this judgment I cannot agree.

Family TESTUDINIDAE

The Land Tortoises

THE tortoises of this family have curious feet. The hind legs are cylindrical and columnar like those of an elephant, and the fore legs have the anterior surface covered by thick, often bony scales. The toes are short, two-jointed, and without a trace of web. The neck and legs are retractile, although in some forms the fore limbs are merely drawn together in front of the withdrawn head. The plastron is firmly joined to the carapace by sutures. The head is covered by scales.

The living land tortoises are a widely scattered but numerically feeble remnant of a herbivorous tortoise fauna which, along with the ancestral horses, camels, elephants, and many other stocks, left the lowlands and invaded the vast open prairies that existed during early Cenozoic. It is thought that the group may have originated in North America, and during Oligocene and Miocene time some fifty species existed here in amazing abundance. During Pliocene time the formerly numerous giant species became extinct throughout most of their range and today are represented only by island forms of which there are two main groups, one in the Galápagos off the coast of Peru and another in the islands of the western Indian Ocean. In North America only three closely related small tortoises have survived.

My brother and I once found a giant tortoise swimming off a key in Florida Bay and were no end excited until we learned that it was one of C. H. Townsend's specimens introduced on Lignum Vitae Key. I have recently learned that members of this captive colony have, elsewhere, subsequently bred.

Genus *GOPHERUS*

The Gopher Tortoises

These tortoises are inconspicuously set off from *Testudo* by their more flattened fore limbs, which are adapted for digging. The fore toenails are heavy, broad, and short. There is one postcentral lamina. The top of the head is covered by scales. The alveolar surface of the upper jaw has a ridge running parallel with the cutting edge and a longitudinal elevation at the symphysis.

The genus includes three species distributed in the southeastern and southwestern United States and northern Mexico. The group is so slightly differentiated from *Testudo* that the positive identification of fossil ma-terial as representing *Gopherus* would be next to impossible.

KEY TO *GOPHERUS*

1   Hind foot diameter divided by head width less than 0.85; alveolar ridges of upper jaws making together an angle of more than 65 degrees (up to 80 degrees or more) 2

1'   Hind foot diameter divided by head width more than 0.85; alveolar ridges of upper jaws together making an angle of less than 60 degrees; southern Nevada southward into northern Sonora . . . . *agassizii* (p. 321)

2(1)   Average ratio of shell depth to plastral length about 0.40; angle between upper alveolar ridges more than 73 degrees; southern South Carolina westward near the Gulf to extreme eastern Texas . . . . *polyphemus* (p. 332)

2'   Average ratio of shell depth to plastral length about 0.50; angle between upper alveolar ridges less than 73 degrees; southern Texas and northeastern Mexico . . . . *berlandieri* (p. 328)

Desert Tortoise *Gopherus agassizii* (Cooper)

(Plate 62. Map 21.)

Range: Southern Nevada and extreme southwestern Utah southward through southern California and southwestern Arizona into southern Sonora, Mexico; Tiburón Island in the Gulf of California. Miller (1942), who found bones from the McKittrick Asphalt (California, Pleistocene) indistinguishable from those of this species, suggested that the desert tor-toise evolved east of the Sierran divide, and that the McKittrick population represented only a "far-flung western outpost of the species during a post-Tertiary expansion of its natal area."

Distinguishing features: This tortoise may be distinguished from the other members of its genus by its narrower head, sharp-angled intersection of the alveolar ridges of the upper jaw, and its large hind foot. These char-acters are expressed in more definite terms in the description below.

Map 21. Distribution of the species of *Gopherus*.

Description: The shell is oblong, with the sides either straight, slightly convex, or slightly concave as seen from above. The largest specimen on record was a male 337 mm. (13.2 inches) in shell length. The dimensions (in millimeters) of four specimens, two males and two females from Sonora, Mexico, are given in the following tabulation from Bogert and Oliver:

Sex	Carapace length	Carapace width	Shell depth	Head width	Diameter of hind foot
Male	228	157	95	35	34
Male	217	140	79	32	30
Female	202	153	79	30	27
Female	223	147	81	31	29

The carapace is more or less serrate, especially behind, and with well-marked growth furrows in most specimens. It is horn color or brown, often with yellowish centers in the laminae. The anterior projection of the

Plate 62. Gopherus agassizii. Upper and center: Male. *Lower:* Female. Both from Perris, California.

323

plastron is not, or is but slightly, bent upward; the plastron is yellow, shaded with brown along the edges of the laminae. The head is relatively small, rounded in front, and with the alveolar ridges of the upper jaws forming a relatively sharp angle with each other (less than 65 degrees). The hind foot is very large, the ratio of head width to foot diameter being less than 0.85. Table 9, condensed with slight modifications from Bogert and Oliver, affords a useful comparison of certain proportions of the species of the genus *Gopherus*.

Table 9. Comparison of certain proportions of the species of *Gopherus*. For each ratio the extremes are given first and are followed by the mean.

Species	Plastron width to plastron length	Shell depth to plastron length	Hind foot diameter to head width
G. polyphemus	0.71–0.94, 0.76	0.36–0.46, 0.41	0.53–0.78, 0.63
G. berlandieri	0.75–0.94, 0.80	0.45–0.57, 0.48	0.62–0.81, 0.69
G. agassizii	0.65–0.82, 0.73	0.36–0.46, 0.42	0.85–1.15, 0.97

The male desert tortoise is larger than the female and has a longer tail, longer gular projections at the fore end of the plastron, more thickened nails, and a more concave plastron. According to Grant (1936), these secondary sexual differences appear when the turtles reach shell lengths of between 120 and 140 mm. (4.73 and 5.52 inches).

Grant (1936b) gave the dimensions of a hatchling of this species as 36 mm. (1.42 inch) shell length and 39 mm. (1.54 inch) shell width. He found that the little tortoise underwent an "unrolling" process during the first five days after hatching, and that 10 months later it was 48 mm. (1.89 inch) long by 42.5 mm. (1.67 inch) wide. The egg tooth is not shed, according to Grant, but flattens out after several months.

Habitat: As its common name implies, this tortoise is partial to desert terrain where, according to Linsdale (1940), it may be widely scattered over dry, flat, sandy or gravelly ground. While Miller (1932) and other writers have regarded this as characteristically a burrowing species, Bogert and Oliver offered the following observations in this regard:

We did not observe any burrows in the vicinity of Alamos, nor can we find any records of burrows in the vicinity of Tucson. A single specimen was secured by the senior author in Sabino Canyon in the foothills of the Santa Catalina Mountains north of Tucson, and another specimen was taken near the Agua Caliente Mountains, 15 miles to the east, in August, 1942. But in neither locality were burrows discovered. Whether in the southern portion of its range the desert tortoise resorts to crevices rather than to burrowing can only be speculated. True considered it doubtful whether Berlandier's tortoise dug burrows, and so far no observations appear to have been reported which belie this supposition.

Woodbury and Hardy (1940, 1948) made a detailed study of the habits of a colony of some 300 desert tortoises occupying a territory of about 1,200 acres in southwestern Utah, where they marked and released 281 individuals and recaptured tortoises 812 times over a period of ten years. These observers found dens of two distinct types (see below) to be in regular use by the animals and to be of the utmost importance in the somewhat complex seasonal and diurnal behavior rhythms that survival in the harsh environment demands of them. The burrows are shared with a list of guest animals that at least approaches the roster of such commensals to be found in the dens of the southeastern gopher tortoise.

The desert tortoise has been taken at altitudes as high as 3,500 feet.

Habits: The desert tortoise is forced to contend with an extremely rigorous environment and as a result shows some unusually interesting adaptive features. Woodbury and Hardy (1940) found that in southwestern Utah tortoises tend to congregate in large winter dens during cold weather, spreading out again over nearby areas during milder weather, and descending into short individual burrows to escape extreme heat or drought. They found that the hibernating dens, which except for occasional brief winter emergencies were occupied from October to April inclusive, were excavated to depths of from 5 to 30 feet into gravel banks, whereas the summer burrows were sharply inclined and from 3 to 4 feet deep.

In a later paper Woodbury and Hardy (1948) summarized the roles of the two types of burrows in the annual and diurnal behavior rhythms of this species in Utah as follows:

The annual behavior pattern involves a short two-way migration, in which the tortoises inhabit the dens for winter hibernation where they utilize stored fat for nutrition, migrate to nearby ranges for summer, where they utilize the summer holes to avoid heat stroke and desiccation and then move back to the dens for winter. This is similar to a two-way bird migration, but the distances covered are approximately in proportion to their powers of locomotion.

The daily behavior pattern also follows a two-way movement, but on a much smaller scale even than the annual pattern. The daily pattern is inoperative during the winter except in the cases of tortoises that occasionally move about, but when they leave the dens during the mild spring temperatures the daily pattern complicates the seasonal pattern until they return again to the dens.

In the early spring they find comfortable conditions abroad during the day-time and seek protection from the cool or cold nights. A little later, when temperatures are higher, they are able to find comfort both day and night either in sunshine or shade and shift from one to the other as the daily rhythm demands. Still later, when it gets hot on the surface in the sunshine, pro-tection is sought in the summer holes in the daytime and activity abroad occurs at night or early morning. In the fall, the springtime march of the pattern is reversed.

The above writers found that the home range of individual tortoises was relatively circumscribed, most individuals remaining within areas of no more than 10 to 100 acres, with considerable overlapping between neigh-boring freeholds. Bogert (1937) found even fewer indications of wandering by three Mojave Desert specimens that he marked, released, and recovered after intervals of from 619 days to 4 years. That there must be considerable individual variation in habits, however, and that populations of desert tor-toises must diverge markedly in their habits according to the demands of the environment are indicated by conflicting reports on the migratory movements of the animal and by the discrepancies in the remarks of Bogert and Oliver (quoted above under "Habitat"), Woodbury and Hardy, and various other observers concerning the burrowing tendencies.

The desert tortoise appears not to use its burrow as a haven into which to scramble for safety in an emergency, as the gopher turtle does, but relies on it chiefly as an air-conditioned retreat from intolerable, or uncom-fortable, extremes of temperature or humidity. Woodbury and Hardy showed that the temperatures at a distance of 17.5 feet within a den are remarkably stable, and while they provide no data on evaporation rates that obtain there, these would undoubtedly be more propitious in the dens than outside.

The desert turtle has been observed to dig by alternately scraping with the fore feet and pushing with the shoulders.

Several writers have referred to the storing of water by desert tortoises in accessory bladders, as if to indicate that the animals hold in reserve a supply of water for use in future periods of drought. Actually, it seems highly improbable that such water is ever reabsorbed to any appreciable extent. Moreover, Woodbury and Hardy found that the accessory bladders of other turtles are in *agassizii* represented by mere vestigial pouches, and that the water referred to by others is actually nothing but urine in which the nitrogen compounds, as in some other reptiles and in birds, are in the form of insoluble uric acid instead of soluble urea, and which can thus be held in the bladder for much longer periods than mammal urine without becoming a dangerously poisonous solution. Although the body of a tortoise is well insulated against evaporation, there is constant water loss through respiration, and since most tortoises probably depend largely on water produced in their tissues by oxidation of food, the retention of urine is an important device for water conservation.

On aesthetic grounds Woodbury and Hardy do not take kindly to the idea that this water carried about by desert tortoises might constitute an emergency water ration for a lost and thirsty desert traveler, but agree with other writers that it might well be a vital factor in the survival of such an individual.

Young desert tortoises were found by Woodbury and Hardy to grow to a shell length of about 100 mm. (3.94 inches) in five years. They calculated that sexual maturity is reached in fifteen to twenty years when the females are between 230 and 265 mm. (9.05–10.4 inches) in shell length and the males between 250 and 316 mm. (9.85–12.4 inches). The most rapid growth noted by these writers was that of a tortoise that increased its length from 206 to 302 mm. (8.10–12.6 inches) between November 5, 1938, and March 19, 1943.

The apparent record for the survival of a desert tortoise in captivity is the two years and nine months cited by Conant and Hudson as the time an individual has lived in the Philadelphia Zoo.

Breeding: The breeding habits have been only little observed under natural conditions. Mating occurs in May in captive specimens. The males become unwontedly belligerent, and attack one another on sight, the long gular extension of the plastron being used, according to Grant (1936b), as a horn for "goring" and overturning an opponent. During courtship the male has been heard to make a grunting sound. Nesting occurs in June. It has been thought that the eggs are deposited singly, but Grant gave the size of the complements of five females, one of which included six eggs; one, two eggs; and three, five eggs each. The nesting holes were dug by the hind feet (which is a point worthy of note, in view of the extreme adaptation of the front feet to another type of digging) and the eggs were individually arranged in the bottom of the cavity by the foot as they fell. In one case Grant found the incubation period to be about 118 days (for eggs laid in captivity). The dimensions of two eggs were: 41.6 by 36.7 by 34.9 mm. (1.64 by 1.44 by 1.37 inches) and 48.7 by 39.6 by 38.2 mm. (1.92 by 1.56 by 1.50 inches). Bogert and Oliver referred to the eggs as spherical, and it is possible that the above measurements were made on aberrant eggs, although a few other data indicate that an elliptical and slightly flattened form may be usual.

Feeding: No study of the feeding of representative numbers of desert tortoises in nature has been made. In captivity they eat nearly anything in the vegetable line and occasionally take insects or other animal food, although apparently with no great enthusiasm. One kept by Clifford Pope bit at his little son's bare toes, but perhaps through sentimental rather than gastronomic motives. A perusal of the list of things that have been eaten by captive tortoises gives one more to wonder at human eccentricity than at the flexibility of tortoises.

Cox said that cactus is a natural food of tortoises. A. I. and R. D. Ortenburger found only grasses in stomachs, and Woodbury and Hardy also regarded grasses as the mainstay in the diet of the Utah population that they studied.

Economic importance: Although always a favorite with the Indians, who not only liked to eat tortoises but used the shell as receptacles and in other ways, this species is not spoken of with any great respect as an article of diet by those who have written of its eating qualities. Despite the repugnance of the thought to some, the potential value to the desert traveler of the water that it stores in its bladder cannot be overlooked. It seems most regrettable that the populations are being greatly reduced by capture of the animals for sale to tourists, who in most cases would be hard put to show any need at all for a desert tortoise.

Berlandier's Tortoise *Gopherus berlandieri* (Agassiz)

(Plate 63. Fig. 21. Map 21.)

Range: Southern Texas and northeastern Mexico; from Aransas, Refugio, Bexar, Medina, and Uvalde counties, Texas, across the frontier into the eastern parts of Coahuila and Nuevo Leon and southward to the southern part of Tamaulipas. Gunter (1945) recently expressed the opinion that this tortoise is spreading "northward and eastward from the lower Texas Border region."

Distinguishing features: Although quite similar in general appearance to the other American gopher tortoises, this species may be distinguished in most cases by its shorter shell (see table of comparative proportions under *G. agassizii*); the snout is usually more pointed than in the others and the angle made by the alveolar ridges, according to Bogert and Oliver (1945), is fairly characteristic and constant. The head is smaller than in *polyphemus*

Plate 63. Gopherus berlandieri. Upper and center: Specimens of undetermined sex; near San Antonio, Texas. *Lower:* Male (*left*) and female (*right*); Brownsville, Texas.

329

and the diameter of the hind foot is less (see same table). Smith and Brown called attention to the integumentary gland located beneath the angle of each ramus of the lower jaw, but this occurs also in *polyphemus*.

Description: The maximum size of this tortoise appears to be about 215 mm. (8.48 inches). The carapace is brown in color, with or without light yellow centers in the laminae. The head and legs and the plastron are all some shade of dull yellowish. The angle between the upper alveolar ridges is usually more than 65 degrees and less than 70 degrees. In a series of specimens from Coahuila, Schmidt and Owens found the plastral width to be from 78 per cent to 87 per cent of its length. Mittleman and Brown calculated the following ratios for the species: diameter of hind foot divided by width of head, 0.57–0.89, average, 0.75; width of carapace divided by length of carapace, 0.73–1.01, average, 0.85; shell depth divided by length of carapace, 0.49–0.72, average, 0.55.

There appear to be slight differences in the body proportions of the sexes, in addition to the usual concave plastron of the male. Mittleman and Brown gave additional ratios to illustrate some of these differences, taking their measurements from fourteen males and thirteen females from Texas and Nuevo Leon. They found the ratio of the diameter of the hind foot to the head width to vary from 0.69 to 0.89, with an average of 0.79, in males, and from 0.57 to 0.81, average 0.71, in females. Only two of the males in this series had a ratio of less than 0.74, the remaining twelve specimens ranging from 0.74 to 0.89; nine of the females had ratios of less than 0.74, while in four specimens the ratio varied from 0.78 to 0.81. Further

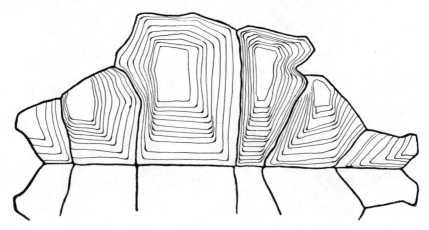

Figure 21. Plastron of *Gopherus berlandieri,* showing growth rings and evidence of radial displacement of the juvenile part of each lamina.

sexual dimorphism was noted in the ratio of the carapace width to carapace length, the males varying in this respect from 0.73 to 0.90, with an average of 0.82, and the females from 0.79 to 1.01, average, 0.96. In nine of the males the ratio was 0.82 or lower, while in only one female was it lower than 0.83. These writers, like Bogert and Oliver, failed to find any significant sexual divergence in the ratio of shell depth to carapace length, males varying from 0.51 to 0.60, average, 0.56, and females between 0.49 and 0.72, average, 0.55.

The shell of the hatchling may be as wide as long, the specimen figured by Agassiz (1857, Vol. 2, pl. 3) being 41 by 41 mm. (1.6 by 1.6 inches). Schmidt and Owens found that in a small series of small specimens 84–102 mm. (3.3 to 4.0 inches) in shell length, the ratio of the width of the plastron to the length of the shell varied between 0.85 and 0.87, while in adults it ranged from 0.78 to 0.84.

Habitat: This turtle is found in well-drained, sandy soils generally. Although perhaps partial to open woods, it is found also in chaparral and mesquite, often under near-desert conditions, and while not so tolerant of arid range as its relative, the desert tortoise, it appears nevertheless to thrive even where drought is quite severe. While Berlandier's tortoise often digs long burrows, in certain sections of Tamaulipas where the hard-baked soil had a high content of clay we never came across a burrow that did more than barely hide the body of the occupant, and we found some tortoises taking refuge under stumps and heaps of rubble. This species has been taken at altitudes between 2,600 and 2,900 feet in Tamaulipas.

Habits: Very few observations on the habits of Berlandier's tortoise have been published. In Tamaulipas in December the species appears to go into a state of semihibernation during the occasional periods of chilly weather. Since it seems to find the adobe soils common in that area too heavy for effective burrowing, it spends a lot of time passively awaiting better times, often with no more than half of its shell under cover. Of six individuals observed, only one was actively moving about, and this was in the middle of a bright, warm day. Grant (1936b) referred to what appeared to be a migration of these tortoises in southern Texas; and Hamilton (1944) made the following notes on a similar observation:

On August 3, 1938, while driving along U.S. Highway 96 between Skidmore, Bee County, and Mathis, San Patricio County, in the vicinity of Corpus Christi Lake, an apparent migration of this species was observed. Sixteen of these tortoises were counted on the road alone in a distance of 2 or 3 miles, and at least as many more were observed in the sandy, brushy country beside the road. This was just after the let-up of the disastrous flood rains which devastated North Texas that year, this part of Texas receiving a part of those rains, and the surrounding country was quite damp.

Hamilton mentioned the fact that large numbers of these tortoises are killed on the highways by cars.

Breeding: Hamilton saw one individual chase down and mount another (locality and date given above) and reasonably assumed the affair to be a straightforward sort of courtship, although the sexes were not determined. Otherwise nothing is known of mating, and the nesting and laying processes have not been described. Strecker (1929b) said that the eggs vary from globular to elongate. Three eggs were found beneath a tortoise which had been unable to construct a proper nest because of the rocky character of the soil, and which was accordingly in a state of peevishness not often seen in these turtles. She tried to conceal the eggs with her shell and bit a proffered stick.

Feeding: Mittleman and Brown gave the only data on the feeding habits of this tortoise under natural conditions that I can find in print. They remarked that it "feeds heavily on the ripe fruit of the common cactus [*Opuntia leptocaulis*] during the summer, while at other times it appears to subsist largely on young shoots and highly colored blossoms of a variety of plants." Captive individuals prefer plant food but have been known to eat meat.

Gopher Tortoise *Gopherus polyphemus* (Daudin)
(Plate 64. Map 21.)

Range: Southwestern South Carolina southward to near the tip of the Florida peninsula; westward through southern Georgia, Alabama, and Mississippi to Louisiana and extreme southeastern Texas. From Louisiana it barely extends into southeastern Arkansas.

Distinguishing features: This tortoise has a relatively elongate shell, as compared with Berlandier's tortoise. Its hind feet are comparatively small and the angle between the upper alveolar ridges is very wide. Details of these characters may be found in the comparative table under *G. agassizii* and in the descriptions below.

Description: It is difficult to decide what is myth and what is gospel among reports of gopher sizes. There is no doubt that individuals often reach a shell length of 300 mm. (11.8 inches), or at least that they used to attain this size before Florida became so cluttered with people. They prob-

Plate 64. Gopherus polyphemus. A: Specimen of undetermined sex; Lake County, Florida. *B:* Female; Lake Alfred, Florida. *C:* Young; Marion County, Florida. (Courtesy Ross Allen.) *D:* Female; Sanibel Island, Florida. *E:* Sex undetermined; Kissimmee, Florida.

333

ably became considerably bigger. I measured a large male specimen from Alachua County, Florida, that had the following dimensions: length of carapace, 343 mm. (13.5 inches); width of carapace, 263 mm. (10.3 inches); length of plastron, including gulars, 345 mm. (13.6 inches); projection of the gulars beyond anterior margin of carapace, 30 mm. (1.18 inches). Theodore Roosevelt (1917) made the statement that gophers reach a shell length of 18 inches (458 mm.); but when he said in the next breath that a 13-inch individual "easily walked away" with him standing on its back, having some vague recollection of the size of the President I lost confidence in the whole yarn. There may have been an occasional 18-inch gopher in the old days, but as far as I know, the specimen whose measurements I give above is the largest on definite record. A 12-inch gopher weighs nine or ten pounds.

The carapace of the gopher is brown or tan, sometimes with light centers in the laminae and sometimes without; growth rings are very evident in younger individuals but worn away in older adults. The plastron is yellowish and the soft parts are grayish brown. The head is large and blunt, and the angle between the upper alveolar ridges is wide—73 degrees or even more. Some body proportions are expressed as ratios in the table under *G. agassizii,* which compares these ratios for the three species of the genus. From this table it is evident that the head of *polyphemus* is comparatively broad, the hind foot small, and the shell rather elongate, relative to its height. Thirty-one specimens measured by Goin and Goff, ranging in shell length from 107 mm. to 240 mm. (4.22 inches to 9.46 inches) varied in the ratio of the shell width to shell length between 0.69 and 0.79, with an average of 0.74.

There was no significant correlation with sex in the above ratios. Like other species of the genus the male gopher has a concave plastron. There appears to be no important difference in size, but Goin and Goff found that "in all of the measured specimens in which sex could be determined [55 specimens], the plastron averaged 4.1 mm. (0.16 inches) longer than the carapace in the male and 2.1 mm. (0.083 inch) shorter in the female."

In very young gophers the width of the shell more nearly approaches the length, although this condition changes more rapidly than in *G. berlandieri.* Goin and Goff found that the ratio of width to length in a specimen 56 mm. (2.20 inches) long was 0.86, but that after this individual had reached a shell length of 63 mm. (2.48 inches), ten months later, the ratio was 0.79. I have lived with gophers most of my life and yet never saw a hatchling, nor can I find any published reference to one.

Habitat: The gopher tortoise is found most abundantly in sandy ridge and dune-sand areas where the water table never comes near the surface. It is characteristic in high pine woods in peninsular Florida.

Habits: Of all the habits of this interesting turtle, the most distinctive is its excavation of extensive, well-made, and permanently occupied burrows, which afford it protection from enemies and from incompatible meteorological conditions, and provide a home into which it may retire to pass its nights or to go into whatever partial hibernation the season may demand. In sandy areas of the deep South these burrows are an integral part of the landscape, and their capaciousness and frequence of occurrence have led to the development of a number of commensal guests which to varying degrees are dependent on the gopher's home for their own shelter or food supply. Some of these, like the opossum and the fox, may merely drop in casually to pass a rainy night or may take to a gopher burrow to escape pursuing dogs (I should say that a good 50 per cent of Florida fox hunts wind up when the fox seeks earth in a gopher hole). Some others, such as the indigo snake, the diamond-backed rattler, and the gopher frog, are merely strongly addicted to cavernicolous life and take advantage of gopher holes as the most abundant and most comfortably constructed caves in their home areas. There is a long list of others, mostly arthropods, that are to varying degrees what are called "obligate inquilines" and that would find life sorely rigorous or even impossible if deprived of this particular kind of refuge. A number of zoologists have interested themselves in these guests of the gopher tortoise, and of these Young and Goff have provided the most recent and one of the best accounts of the method of construction of the burrows, as well as of their inhabitants. Their description is quoted below:

It is interesting to note that a distinctive association of commensals has developed in such a simple habitat as the gopher burrow. But when we consider the fact that the gopher is a member of an ancient race which probably reached Florida soon after the elevation of the Pleistocene, and that the burrowing habits of the genus were probably established long before, it would be more noteworthy if no animals had taken advantage of the available protection and food supply. The open mouth of the burrow is a blanket invitation to any animals with cavernicolous tendencies, and the supply of dung at the bottom is an added attraction to coprophages. None of the true obligates, as far as we know, has been taken outside gopher burrows. Most of them are apparently found throughout the greater part of the range of the tortoise, yet no explanation of how they are transferred from one burrow to another can be offered.

Usually the burrow of the gopher is comparatively simple, varying in size and depth with the size and age of the individual inhabiting it. The young tortoise excavates for itself a burrow just a little larger than the greatest width and depth of its carapace, and from one to three feet deep. As the carapace grows the burrow is gradually enlarged until it may reach a maximum width of a foot or more, and a length varying from ten to 35 feet. There frequently are turns in the course of descent. These do not seem to be a protective or drainage device, but rather depend on the chance obstruction of the course by rocks, roots, harder material, etc. The vertical distance from the surface to the end of the burrow appears to be determined by the resistance of the under-lying material or by the water table. In the Norfolk series of sands it may be as great as the depth to the clay (usually from 4 to 12 feet) while in the Dade sands it is usually the distance to the underlying limestone (from 3 to 8 feet). Where the water table is near the surface the burrow may be quite shallow and very long (as much as 35 or 40 feet).

Although there seems to be no provision for drainage, the burrows are usually rather dry. Hallinan (1923) observes that part of the material from the burrow must be pressed radially into the sides by the tortoise, since the volume removed to the outside is not always sufficient to account for the volume of the burrow. If this is true then the tightly packed soil may assist in controlling percolation into the tunnel, and, on the other hand, would prevent rapid drying out and consequent collapse of the roof and sides. In several burrows excavated near Gainesville the sides and floor were found to be quite dry to the depth of 8 or 9 feet while the roof was damp. Under these conditions it is interesting to note that insects were found burrowing into the roof along the upper part of the tunnel.

Goff noted several cases at or near Gillette, Florida, in which part of the burrow was below the water table. In two of these instances the gopher was found at the end of the burrow completely submerged beneath the water. Normally, however, it seems that the tortoise avoids the lower grounds in which the water table is subject to fluctuation. In the northern and central region of the state it confines its burrows more or less to the Norfolk sands, while along the east and west coasts the Dade, Palm Beach, St. Lucie and other loose dune and coastal sands afford an excellent burrowing medium. Many of the coastal islands where the soil is suitable have been occupied by the tor-toise but no investigations of the commensal fauna have been made in these isolated situations.

The list of arthropods from gopher burrows given by Young and Goff includes thirty-two species.

A gopher den that probably holds the record for heterogeneity of occu-pants was one opened by Knight (1871), in which the register showed the following guests: two opossums, a raccoon, a rattlesnake more than six feet

long, two other snakes, and several of the "native black rats of the district."
Mr. E. Ross Allen of Silver Springs, Florida, found in one burrow an opossum, a rabbit, a gopher frog, and a five-foot diamond-backed rattlesnake,
besides the gopher host.

In reviewing my notes on gopher-tortoise literature I was considerably
depressed to come across my excerpts of a paper by Fletcher, entitled "The
Florida Gopher," and without doubt offering more, and more astounding,
misinformation per unit of text than any paper ever written on this particular subject before or since. The article is actually a marvel of equivoca-
tion, and I cannot resist the temptation to cite a few of the wonders it relates, selecting at random from a list of points that range from merely
moot and doubtful to the downright perverse and demoralizing and beyond doubt gained mostly through listening to the irresponsible tales of
crackers, and probably drunken ones at that. Passing over some remarks
with which one yearns to take issue, one learns the following:

The gopher cuts wire grass into segments with his jaws, swallows it, and
then later on regurgitates it to chew it at leisure like a ruminant, which
the animal further resembles in being equipped with three stomachs and
appropriate musculature for regurgitation! "It is almost comical to see him
sitting beside his doorway, his black, skinny head thrust out, chewing with
great satisfaction the morning's repast." Although an antisocial animal by
nature, the gopher is frequently seen in groups of from 10 to 20 in the
same place, grazing like cattle! One of the commensals of the gopher is a
peculiar, white, Hylalike frog with transparent limbs (an undescribed
genus, beyond question). The burrow is dug down to a point 6 inches
beyond which the water would seep in and where the sand is so moist that
the tortoise can suck particles of moisture through the sievelike serrations
of his jaws! The projecting gulars are used as a battering ram with which
one female was seen to kill another! *Transpiration* takes place through the
lower shell, and the animal dies if varnished there!

Loennberg (1894) mentioned the importance of gopher holes as a factor
in the survival of the Florida fauna when the woods are burned over each
year. There is considerable question, however, as to whether any animal
of a lower order of intelligence than the mammals has the wits to take
refuge from a fire in an available retreat, unless already inside it. I have
seen large numbers of box turtles and coachwhip snakes killed by fire in
places where gopher burrows were numerous.

Both the gopher frog (*Rana capito*) and the gopher snake (*Drymarchon
corais couperi*) owe their common names to their fondness for gopher bur-

rows. Gophers are seen with some frequency in the water, but it seems probable that in most such cases they merely fell in. The only individuals I ever saw in water more than shell deep were floating high and flapping their feet pathetically, although they seemed able to maintain a fairly regular course. On two occasions, however, I have seen them in the middle of large lakes, fully half a mile from shore.

In some thermal experiments made with the Florida gopher, Bogert and Cowles found that when placed in a desiccation chamber maintained at a temperature of 38° Centigrade and a relative humidity of 37 per cent and kept there for 79.25 hours, a young gopher (weight, 208.2 grams) lost 43 per cent of its original weight before succumbing. The same writers found a body temperature of from 34° to 35° Centigrade to be normal for active gophers.

Goin and Goff gave results of a growth study in which 131 Florida gophers were marked and released. There were 33 recoveries. The greatest growth shown was that of an individual that increased its length from 148 to 190 mm. (5.83–7.48 inches) from June, 1937, to March, 1938; during this period the weight of the turtle increased from 545 to 800 grams. The average gain in length for all 33 specimens (for an average period of a little less than a year) was 10.5 mm. (0.41 inches). The average gain in weight was 137.3 grams and the greatest gain in weight made by any turtle was 536 grams.

Breeding: Fights between both male and female gophers are sometimes seen, and it is probable that these have some bearing on courtship and mating, but copulation has not been described; among the thousands of gophers that I have seen in the wild I never saw a mated pair. I don't mean to imply that they don't copulate, but they must be very sly about it. The eggs are laid from late April to July. Most written reports state that they are always deposited at the entrance to the burrows, but I have not found this to be invariably the case. Of dozens of nests that I have run across, mostly destroyed by predators, the majority have been located at considerable distances from any burrow. The number of eggs is most frequently five or six (from four to seven) and the average greatest diameter varies between 38 and 44 mm. (1.5 and 1.73 inches) and is usually about 1 mm. greater than the least diameter. The eggs are white and brittle-shelled, and are deposited in holes about six inches deep. Whether these holes are cylindrical or flask-shaped is not known, and the laying process has not been observed, which to me is a source of humiliation and chagrin—of the former because I have been looking at gophers since I was a child, and should have seen everything they do, and of the latter because I am keen to know how the club-shaped elephant's stump of a hind foot is employed in digging a hole! The incubation period is not on record, and the young hatchlings are excessively secretive.

It has been said by two or three writers that the rutting male gopher utters impatient cries, variously described as a "short, rasping call," "a faint mew like that of a kitten," and "a low, piteous cry."

Feeding: In the stomachs of the forty or fifty gophers that I have butchered, food was always present in surprisingly large amounts and in every case consisted of grass or leaves, with occasional bits of hard fruits, bones, charcoal, and on one occasion insect chitin. Gophers spend most of their foraging time grazing, and are restricted to areas in which there is a good supply of grasses or low herbs. In captivity individuals seem very fond of watermelon and cantaloupe rinds, and if tethered on a lawn of Bermuda grass are able to keep large expanses close-cropped. On three occasions I have come across gophers hanging around the old bleached skeletons of cows or mules, and once I saw an individual biting at the edge of a scapula. My wife has watched gophers catch and chew lubber grass-hoppers, but without swallowing them.

Economic importance: Reckoned in terms of money changing hands, the economic value of the gopher is not great. I have a few times seen it for sale in small towns, usually in front of grocery stores patronized largely by Negroes. Nevertheless, the importance of this animal in the lives of the poorer rural people of Florida and south Georgia, since the days of the initial colonizations, would not be easily overestimated. One has but to look over the list of half-facetious vernacular names for the gopher—Florida chicken, Georgia bacon, and any number of others—to get a hint of its local economic role.

In a land abounding in the finest sorts of game, the Indians of early Florida revered the gopher, as do the modern Seminoles, even though it is a rarity throughout most of their wet territory. The ancient Indians had a monetary system in which gophers played a part like wampum; among both Indians and white settlers, gopher shells have served as baskets and pots and even as sun helmets.

Although gophers once in a great while cause some slight damage to crops in newly planted areas, this is altogether negligible, since the farmer can eat the gopher and more than recoup his loss. Horsemen are sometimes thrown when an old gopher hole caves in underfoot, but this too is an un-common occurrence. The gopher is on the whole a valuable and very personable animal. In the three westernmost counties of the Florida panhandle an act was passed in 1909 that made it unlawful to take or sell any gophers during the months of May, June, July, and August, and which prohibited the selling of any gopher with an under-shell length of less than nine inches. With this act as precedent, some sort of protection for the fast-diminishing remnant of the once abundant population might be feasible. It would most certainly be desirable.

Family CHELONIIDAE

The Sea Turtles

THESE large aquatic turtles are strongly adapted, and accordingly restricted, to aquatic life and seldom visit dry land except to lay their eggs. The legs are paddlelike and the carapace is covered with horny laminae, but its weight is more or less reduced by the retention, to varying degrees, of the embryonic spaces between the ribs, and the connection between upper and lower shells is not rigid. The neck is short, thick, and incompletely retractile. The temporal region of the skull is completely roofed over both dorsally and laterally.

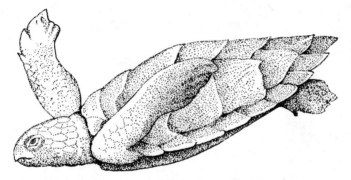

Figure 22. A hawksbill turtle, showing winglike position of fore flippers at top of upward swimming stroke.

The sea turtles are known from Upper Cretaceous fossils, and it is believed probable that they were derived from marsh-inhabiting ancestors during Triassic time. All the living genera encircle the globe in warmer seas.

KEY TO THE GENERA OF CHELONIIDAE

1 One pair of prefrontal scales; tomium of lower jaw coarsely toothed, that of the upper jaw with strong ridges on its inner surface; lateral laminae 4 . . . .*Chelonia* (p. 344)

1′ Two pairs of prefrontal scales; tomium of lower jaw smooth or only feebly toothed; that of upper jaw without strongly elevated vertical ridges on its inner surface; laterals 4–9 2

2(1) Laterals in 4 pairs; precentral not in contact with first laterals; snout elongate, narrow, not terminally toothed, the floor of the mouth deeply excavated at the mandibular symphysis; laminae of the carapace usually conspicuously imbricated ...*Eretmochelys* (p. 365)

2'   Laterals in 5 or more pairs; precentral in contact with first laterals; snout relatively short and broad, the mandibular sym-

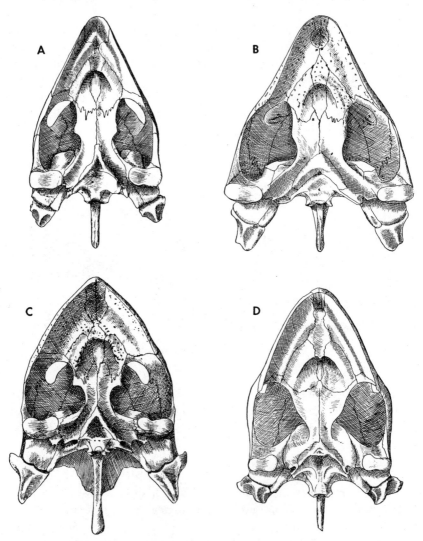

Figure 23. Skulls (ventral view) of four genera of sea turtles. *A: Eretmochelys* (*i. imbricata*). *B: Caretta* (*c. caretta*). *C: Lepidochelys* (*kempii*). *D: Chelonia* (*m. mydas*).

physis toothed terminally or blunted by wear; anterior part of roof of mouth not deeply excavated; laminae of the carapace not conspicuously imbricated except occasionally in very young . . . 3

3(2′) Bridge with 4 enlarged inframarginals; color gray to olive green; maxillary bones of skull separated by vomer . . . *Lepidochelys* (p. 396)

3′ Bridge with 3 enlarged inframarginals; color brown or reddish brown; maxillary bones of the skull in contact with each other . . . . *Caretta* (p. 381)

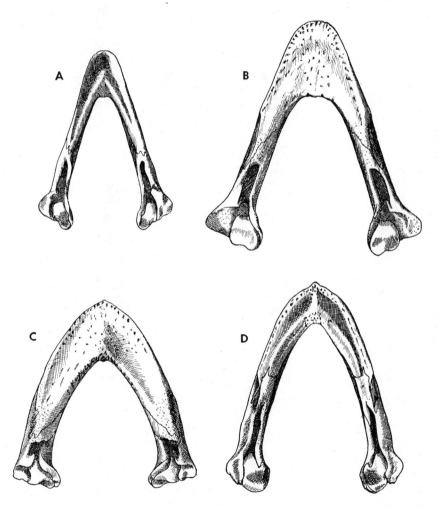

Figure 24. Lower jaws of four genera of sea turtles. *A: Eretmochelys (i. imbricata). B: Caretta (c. caretta). C: Lepidochelys (kempii). D: Chelonia (m. mydas).*

Genus *CHELONIA*

The Green Turtles

It appears at present that this genus of highly edible turtles includes but a single species, although several have been given names at various times in the past. The lateral laminae are in four pairs and the first pair is not in

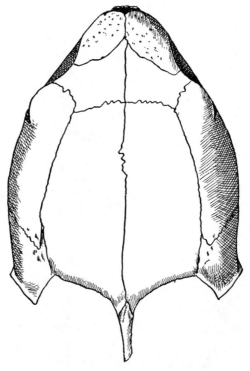

Figure 25. Dorsal view of the skull of *Chelonia m. mydas.*

contact with the precentral. There is one pair of prefrontal head scales. There are persistent spaces in the bony shell between the outer rib-ends (fontanels). The horny inner surface of the upper jaw is strongly engraved with vertical ridges; the cutting edge of the lower jaw is deeply dentate. On the alveolar surface of the lower jaw there is a median tooth that is connected with the short terminal tooth by a sharp ridge. There are four inframarginal laminae, and these lack pores.

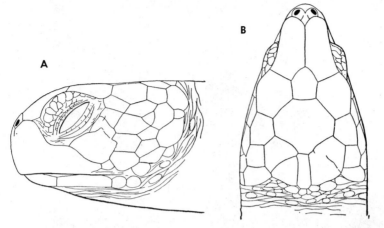

Figure 26. Young green turtle (two-thirds natural size); Florida. (From Stejneger.)

KEY TO *CHELONIA*

1 Coloration above predominantly brownish; shell margin not markedly indented above hind limb; shell less chunky and deep, especially from the middle of the laterals toward the periphery; Atlantic and Caribbean ....*mydas mydas* (p. 345)

1' Coloration above predominantly greenish or olive brown; shell often markedly indented above hind legs; shell chunky and often with the laterals completely straight from the margin to the centrals; Pacific coast of America ....*mydas agassizii* (p. 357)

Atlantic Green Turtle *Chelonia mydas mydas* (Linné)

(Plate 65. Figs. 23, 24, 25, 26.)

Range: Atlantic Ocean, Gulf of Mexico, and Mediterranean Sea, meeting the slightly different Indo-Pacific populations at the Cape of Good Hope but evidently not at Cape Horn. On American shores it has been found from the New England coast (Massachusetts) southward to about latitude 38° south, the southernmost localities apparently being on the Argentine coast at Mar del Plata and Necochea. Green turtles have been reported seen in Newfoundland waters, but apparently no specimens have been taken there. Although not a pelagic animal, straggling individuals are seen at considerable distances from land, and on such small mid-sea islands as Ascension they are often extremely abundant. It thus seems probable that a certain amount of sporadic contact between the stocks of

Plate 65. Chelonia mydas mydas. The specimen is a female. (Courtesy The American Museum of Natural History.)

the eastern and western sides of the Atlantic must take place, especially in the area between the bulge of Brazil and that of West Africa. On the European coast there are old records for English waters, but it is possible that any specimen turning up on the English coast might have been carried by the warm eastward drift of the Gulf Stream from the American side.

An interesting north-south gradient in the size of green turtles taken along the coasts of the southern United States was noted by True (1884), back in the days when these turtles were plentiful as far north as North Carolina and when one man could catch a hundred off Cape Hatteras in one day. Around Beaufort and Moorhead City, North Carolina, the turtles were small, the usual weight being about eight pounds; to the south of Charleston, South Carolina, they ranged from 5 to 15 and up to 25 pounds; at Saint Augustine, Florida, the average weight was between 20 and 25 pounds; still farther south, at the Halifax River the average was 35 pounds, while in the Indian River they came between 50 and 60 pounds, with an occasional mature individual of 200 pounds; and at Key West they ran between 40 and 100 pounds. The largest specimens came from the west coast of peninsular Florida, where at Cedar Keys a weight of from 600 to 800 pounds was not rare and where green turtles were said to reach 1,000 pounds (according to True).

In view of the fact that even in True's day green turtles only very rarely nested north of the coast of Florida and that the center of distribution of the smaller size groups thus lay in the southern part of the range rather than in the northern, this gradiant is exactly the reverse of what one might reasonably expect. Essentially the same situation is encountered with respect to the ridley, however, although there is less data available in that case; and as I have suggested in the discussion of the ridley, the explanation must lie in the fact that the young, weak turtles are more susceptible to passive migration in the northward-trending Gulf Stream than their strong-swimming elders.

Distinguishing features: The green turtles are easily recognized by their generic characters: the single pair of prefrontal scales, the four pairs of lateral laminae, the strong vertical ridges on the inner surface of the upper jaw, the single mandibular scales, and the juxtaposed laminae. The dominantly brown ground color of the carapace and skin of the legs and head and a slightly different conformation of the shell appear to differentiate this race from that of American waters in the eastern Pacific. Between these two there are thousands of miles of coastal water inhabited by green turtles, and very probably a number of recognizable races, but they will be defined only by statistical studies based upon large numbers of specimens of comparable age and sex, and to date such studies have not been feasible. The most important fact in this problem appears to be that insofar as the green turtles of the eastern and western coasts of the Americas are concerned, at least, we are not dealing with a couple of discrete entities but rather with stocks so closely allied that separating them must be largely a matter of statistics.

Description: The shell is broad, low, and more or less heart-shaped. The size is large, the heaviest recorded specimen having weighed 850 pounds, although True mentioned 1,000 pounds in a general way as the upper limit. I know nothing of such green turtles. Today a specimen with a carapace 1,200 mm. (about four feet) long and a weight of 500 pounds is a tremendous one. Measurements of a large male and female from the Gulf coast of Florida are as follows:

Sex	Shell length	Shell width	Head width
Female	842 mm.	670 mm.	116 mm.
	(33.2 inches)	(26.4 inches)	(4.57 inches)
Male	976 mm.	722 mm.	129 mm.
	(38.4 inches)	(28.4 inches)	(5.08 inches)

The carapace is smooth, keelless, and with the laminae placed side by side—not imbricated as in the hawksbills. The carapace is light to dark brown, sometimes shaded with olive, often dark fawn color, with radiating wavy or mottled markings of darker color or with large blotches of dark brown. The plastron is whitish to light yellow. The scales on the upper surface of the head are light brown in the center and the spaces between them are yellow; those on the sides of the head are also brown but with broad yellow margins, giving a yellow cast to the temporal region. The neck above is dusky; below and near the shell it is yellow. The upper surfaces of the legs and tail are colored like the shell above and are yellowish white beneath, sometimes tinged with green, and darker near the tips.

There is considerable difference between the sexes of the green turtle. The shell of the male is more elongate and more gradually tapering behind and the hind lobe of the plastron is narrower. Because of a more elongate hind quarter of the shell, the ratios of shell width, shell height, and head width to shell length are slightly lower than in the female. The tail of the male is, on a giant scale, similar in appearance to that of the male musk turtle, being very long (it extended 225 mm., or 8.85 inches, beyond the tips of the postcentral laminae in the male specimen whose measurements were given above), strongly prehensile in the vertical plane, and tipped with a very heavy, flattened nail. To augment the grappling action of the tail during courtship the single nail on each flipper is very markedly enlarged and strongly curved, quite obviously for grasping the under edge of the shell of the female. It is doubtful if females ever reach the extreme shell lengths seen in the males, although the heaviest specimens may be females.

The shell of the hatchling is about 50 mm. (1.97 inches) long and about 39 mm. (1.53 inches) wide. The egg tooth appears to be different from that of most fresh-water turtles, being merely the acutely pointed but smoothly continuous upper, anterior projection of the snout. The carapace and upper limb and head surfaces are dark brownish, the central keel and the feeble, discontinuous ridges across the laterals are light tan, as is a curious vertical keel that crosses the longitudinal keel of the first lateral. Margins of the shell and limbs are edged with white, as are the head scales in some specimens. The under surfaces are white except for the terminal areas of the flippers, which are black edged with white. A pair of keels extends down the plastron on either side of the median line.

Habitat: The optimum habitat of the green turtle is shoal water with an abundance of submerged vegetation. It is most abundant on plant beds in open water, where such water is not too deep, apparently preferring pastures under from two to four fathoms for their diurnal browsing and seeking out small, deep, rocky potholes for sleeping quarters. This habit was recently noticed by Duncan (1943), and I have seen abundant evidence that a similar diurnal migration occurs in the populations of the Gulf of Mexico and Florida Bay. Green turtles are most readily found in places in which eelgrass flats are pitted with frequent rocky holes, and such situations are available over a wide territory in Gulf and Caribbean shore waters, as well as about the oceanic islands.

Habits: Although one can find in print a fair amount of information on the biology of the green turtle of the Atlantic and Caribbean, a close inspection shows that the greater part of this is based on hearsay and local beliefs of fishermen and the like. While such data cannot be disregarded, and although much of this folk zoology will probably eventually prove to be authentic, its existence does not by any means constitute a sound basis for our understanding of the green turtle. The green turtle is the most valuable reptile in the world, and yet it would be difficult to name any animal, comparable at once in economic importance and in the depletion in numbers that it has suffered, that is so poorly known. The situation is particularly astonishing in view of the fact that a relatively small amount of basic research would almost certainly provide data on which a proper conservation program could be founded and green turtles could be restored to huge areas from which they have been virtually extirpated.

In the following summary of habits most of the points are based originally on verbal reports, in a few cases received by me directly, but for the most part handed down by long lines of general works in which each writer borrowed from his precursor until in some cases the ultimate source of the observation has been forgotten.

Like the Pacific form, the Atlantic green turtle is of generally mild temperament but is occasionally inclined to bite when first caught.

The terrestrial sunning aggregations described for the Pacific race appear not to take place in *C. m. mydas*. Such a habit may have existed formerly and have been discontinued as a result of widespread persecution by turtle hunters.

There is much fragmentary evidence that green turtles may make regular, more or less extensive migrations for one purpose or another. Such isolated facts as the former existence of a north-south size gradient, the excessive scarcity of yearling individuals in most areas, and the failure of turtles to lay on beaches adjacent to favored feeding grounds all suggest that there occur mass seasonal movements about which we are almost entirely ignorant. Such reports as the following, from Long (1774), are common and are based on beliefs that are encountered repeatedly in the countries of the Caribbean:

The instinct which directs the turtle to find these islands [the Cayman Islands], and to make this annual visitation with so much regularity, is truly wonderful. The greater part of them emigrate from the gulph of Honduras, at the distance of one hundred and fifty leagues; and, without the aid of chart or compass, perform this tedious navigation with an accuracy superior to the best efforts of human skill; insomuch that it is affirmed, that vessels, which have lost their latitude in hazy weather, have steered entirely by the noise which these creatures make in swimming, to attain the Caymana isles. . . . When the season for hatching is past, they withdraw to the shores of Cuba, and other large islands in the neighborhood; where they recruit, and in the space of about a month acquire that delicious fat for which they are so much in esteem. In these annual peregrinations across the ocean they resemble the herring shoals; which by an equally providential agency, are guided every year to the European seas. . . . The shore of the Caymanas, being very low and sandy, is perfectly well adapted to receive and hatch their eggs; and the rich submarine pastures around the larger islands afford a sufficient plenty of nourishing herbage, to repair the waste which they have necessarily undergone. Thus the inhabitants of all these islands are, by the gracious dispensation of the Almighty, benefited in their turn; so that when the fruits of the earth are deficient, an ample sustenance may still be drawn from this never failing resource of turtle, or their eggs, conducted annually as it were into their very hands.

The only systematic effort to investigate the movements of *C. m. mydas* was that of J. Schmidt (1916), who marked 65 individuals in the Danish West Indies and over a period of several months recaptured 9 of them. Only one of these individuals had covered any great distance, this being one which "wandered from the waters of St. Thomas, in a north-easterly direction to the British island Anegada, a distance of about 50 miles, but required no less than 10 months to cover the distance." Another specimen was recaught, after 11 months, in the same place in which it was liberated. It is obvious that only large-scale marking experiments will shed any important light on this intriguing problem.

Schmidt also found that young turtles of about 270 mm. (10.6 inches) were common during the winter months and concluded that these turtles were probably from one to one and a half years old. His limited marking experiments demonstrated an average gain in weight of 0.58 pounds per month by individuals weighing from 5 to 45 pounds. The smaller specimens showed somewhat higher average gains than the larger ones, but there was no discernible seasonal difference in growth rates.

Breeding: There is no authoritative account of mating and copulation in this turtle, but there is general agreement that it occurs during the laying season, as is the case with other species.

The period of nesting activity is late spring and early summer throughout much of the range. Some dates given for various localities in the Gulf of Mexico and the Caribbean are probably fairly reliable, but are all based on early data which have little bearing on the situation today, since the breeding range of the green turtle has been greatly reduced:

Bermuda	April to June
Vera Cruz	May, June, part of July
Ascension Island	Christmas to midsummer
West Indies	May to October, inclusive
Florida coast	May and June
Coast of Central America	June, July, and August

Of the above places it seems probable that only the last named and Ascension Island in the South Atlantic are at the present time of importance as sites of large-scale nesting activity. As far as I am aware, there has been

no report of a green turtle nest on the coast of the Florida mainland for
forty years.

The nest is built at night, perhaps preferably on a spring tide at half-flood
or higher. Well up the beach from high-water mark (not in wave-washed
sand, as some have claimed), the female selects a spot which may either
be the site of the nest or may be abandoned after some preliminary excavation. The diggings include a large basin, made by the front flippers and
as much as ten feet in width and two in depth, and in the bottom of this a
smaller, bottle-shaped egg hole constructed by delicate manipulation of the
hind flippers, each of which alternately scrapes out sand from the cavity
while the other presses the walls to prevent caving in. At least one writer
has mentioned the wetting of the walls of the cavity by water from the
cloacal bladder of the turtle, and this may be standard practice. Few measurements of the eggs are available. One author gives the average diameter
of the egg as 48 mm. (1.89 inches), but this may not be a valid average
size, since the egg of the closely related Pacific green turtle is known to be
considerably larger. The number of eggs laid in a given nest may vary
from about 75 to about 200. Each female is thought to lay from two to five
times per season, at intervals of about two weeks. The incubation period
is unknown.

There is no indication that this turtle ever nested north of the coast of
Florida, although it was formerly abundant at least as far north as North
Carolina. The primitive center of breeding abundance of the South Atlantic
populations appears to have been Ascension Island, while that of the
Caribbean stock was the island beaches of the Mosquito Coast of Central
America and especially the islands of the Cayman group, where apparently
no green turtles breed today. Although Garman (1884b) stated that it still
bred in the Cayman Islands as late as 1888, the phenomenal congregations
of nesting females that took place here in earlier times had ceased to occur
by the beginning of the nineteenth century. The latest mention of regular
and frequent nesting visits to Florida beaches was by writers of the 1890's.

Feeding: The adult green turtle is mainly herbivorous in its diet, subsisting largely on the marine grasses of the genera *Thalassia* and *Zostera,*
which form unbroken carpets of herbage on appropriate flats. The turtles
also eat algae and other vegetation and are not averse to stuffing themselves
occasionally (at least) on jellyfish, mollusks, or crustaceans. Captive turtles
in crawls are usually fed on fish, which they eat without hesitation. Gundlach (1880) stated that those held in Cuban pens were fed on mangrove
leaves.

Grazing turtles move about over the grass flats biting out sections of the plants near their bases, thus allowing the terminal sections to rise to the surface and accumulate in drifts, which indicate to the turtle fisherman the presence of turtles. Knight described as follows a curious belief, probably apocryphal but surprisingly widespread in former days, concerning an alleged feeding habit of the green turtle on the west coast of Florida.

The *Chelonia mydas* . . . is said by the turtle-fishers to enter the creeks which abound on that coast, and having eaten its fill of the seagrass growing there, to roll together masses of it, of the size of a man's head, which it cements with the clay on which the grass grows, and then when the turn of the tide takes it out to sea, follows it, feeding upon it. When, therefore, the fishermen find any of these balls floating down from a creek, they at once spread a strong net across the mouth and almost always secure a number of these turtles.

Various early writers mention that green turtles commonly entered "creeks" of the peninsula of Florida to feed, and if this actually occurred it probably means that they ascended the short, spring-fed runs of the area and grazed on beds of *Sagittaria* and *Vallisneria*.

Economic importance: In all parts of its range the green turtle is held in high esteem as food for man. A staple in much of the Caribbean region, it has for many years been regarded as a delicacy of the highest order in France and England and in the larger cities of the Atlantic coast of the United States. Besides regular shipments from their Caribbean possessions, in more prosperous days England and France used to import large numbers of turtles from Mexico, Cuba, and Central America. In early days the supply for the markets of the northern United States came mainly from the Indian River, Key West, and Cedar Key. Today Key West is the most active center of the industry, but most of the turtles received there are taken in Central American waters by boats from the Cayman Islands.

Although the green turtle is in no immediate danger of extinction, it will support no resurgence of the industry. It seems almost certain that with modern methods of refrigeration and food preservation to enlighten the inland public concerning the gastronomic properties of this succulent reptile, the pathetic remnant of the once-teeming hordes will be pursued with harpoon and stop net, and the centers of activity will invade ever more remote waters until the animal is backed against the wall. All the while the nests will be robbed and the laying females caught along most of the few remaining nesting beaches, where enforcement of restraining laws can come only by international agreement.

To understand something of the size of green turtle populations under primitive conditions is to wonder how the species can exist at all today—the males find the females and the new generations surmount the statistical improbability of their survival—with such markedly different levels of concentration of individuals.

The inhabitants of the Cayman Islands have always been the chief turtle fishermen of the Caribbean. They formerly fished altogether in home waters, and the abundance of turtles there may be inferred from the comments of the old writers such as that of Long quoted above. Lewis (1940) selected other early reports to illustrate the gradually lengthening forays of the turtling ships as they killed off the animals around the Cayman Islands:

Apparently the catching of turtle and collecting of their eggs was very thorough and devastating in its effect upon the turtle population in Cayman waters. A report addressed to Governor Nugent of Jamaica by Edward Corbet, dated June 21, 1802 . . . on the "population, extent, commerce, etc." of Grand Cayman, the only island of the group then inhabited, says, concerning the turtling from that island:—"Their Turtle, of which the island itself furnished but a small quantity, and which is mostly caught upon the coasts of Cuba, and brought there by small vessels and in which they have eight or nine employed of from 20 to 50 Tons burthen, they dispose of chiefly to such vessels as call there on their way to Europe or America and which they keep until opportunities offer of disposing of them in Crawls made in the sea."

This report indicates that by 1802 turtle were sought in foreign waters. These waters were Cuban. Hirst (p. 122), probably referring to the period around 1820, writes as follows:—"I cannot say how many vessels were owned in Grand Cayman, but probably not more than half a dozen and they were all of very small size. The principal industry, then, as now, was turtling. Turtle were caught in some abundance off the island itself . . . but the vessels also turtled off the uninhabited cays of Cuba. At a later date a proper arrangement was made with the Spanish Government regarding this turtling ground. The Mosquito Cays were not thought of till a much later date."

Writing about the period 1840 to 1850 Hirst (p. 200) says:—"Most of the vessels . . . were engaged in turtling off the Cays to the South of Cuba and in wrecking operations whenever the opportunity offered. At that time turtle was plentiful on the South side of Cuba, though little remained on the shores or reefs of Grand Cayman. The turtle was brought home and sold at the rate of 3d. to 4d. a pound to merchantmen on their voyage from Jamaica to England."

In Hirst's volume (pp. 271–276) there is a short review, by Edmund Parsons, of the turtle fishing along the Nicaraguan coast. Writing in 1909, the author of the article complains of the increasing scarcity of turtle threatening "a lost industry, which means ruin to Cayman."

Hirst and Parsons imply that the Cayman turtlers first went to the Nicaraguan coast in 1850. Thomas Young, however (1842, "Narrative of a Residence on the Mosquito Shore, During the years 1839, 1840 and 1841: . . ."), gives a different impression. He writes (p. 17):—"The Cape is often visited by small schooners, from the Grand Cayman's island near Jamaica to fish for turtle near the Mosquito Kays, about forty or fifty miles from the Cape, and which seldom return without a rich harvest."

A similarly melancholy story may be told of the green turtles of Bermuda, once abundant enough to excite the wonder of every writer and now of negligible commercial importance. As evidence of the early recognition of the inexorable depletion that was occurring, Garman (1884b) quoted the following act of the Bermuda Assembly for the year 1620:

AN ACT AGAYNST THE KILLINGE OF OUER YOUNG TOR-TOYSES

In regard that much waste and abuse hath been offered and yet is by sundrye lewd and impvident psons inhabitinge within these Islands who in their continuall goinges out to sea for fish doe upon all occasions, And at all tymes as they can meete with them, snatch & catch up indifferentlye all kinds of Tortoyses both yonge and old little and greate and soe kill carrye awaye and devoure them to the much decay of the breed of so excellent a fishe the daylye skarringe of them from of our shores and the danger of an utter distroyinge and losse of them.

It is therefore enacted by the Authoritie of this present Assembly That from hence forward noe manner of pson or psons of what degree or condition soeuer he be inhabitinge or remayninge at any time wthin these Islands shall pesume to kill or cause to be killed in any Bay Sound or Harbor or any other place out to Sea: being wthin five leagues round about of those Islands any young Tortoyses that are or shall not be Eighteen inches in the Breadth or Dyameter and that upon the penaltye for euerye such offence of the fforfeyture of fifteen pounds of Tobacco whereof the one half is to be bestowed in the publique uses the other upon the informer.

A sharp decline of green turtle populations in Florida waters appears to have occurred during the last two decades of the past century. It was evidently at a somewhat earlier time that the species began to abandon the shores of the mainland for nesting purposes, although it is probable that laying continued in the Tortugas until somewhat later. The fact that the numbers of adults taken did not diminish markedly for some time after laying on Florida beaches had practically stopped may be seen as evidence that the local population of mature individuals did not recruit its numbers only from new generations hatched locally, and may thus be further indication of extensive migratory movement.

Audubon (1926) gave a picture of the state of the Florida green turtle population during the early eighteen hundreds. He stated that it "approaches the shores and enters the bays, inlets and rivers early in the month of April after having spent the winter in the deep waters." He mentioned especially the Halifax and Indian rivers as favored resorts for the turtles where, he said, "great numbers are killed by turtlers and Indians, as well as by cougars, lynxes, bears and wolves." Audubon was told of one man who during the period of one year caught eight hundred turtles by pegging.

Toward the close of the century such writers as True (1884), Loennberg (1894), Brice (1897), and Munroe (1898) were all forecasting and bemoaning the early doom of the green turtle, but their prophecies and suggestions for management and legislation bore little fruit.

Today enough green turtles are still caught in Florida waters to support scattered feeble industries, but the inevitably expanding demand that quick-freezing techniques are stimulating will soon bring about the destruction of this pitiful vestige unless decisive controls are established.

As was mentioned above, and as was pointed out by J. Schmidt (1916) and later by Ingel and Smith (1949), there is little hope for the success of any pan-Caribbean negotiations for the protection and restoration of green turtles until certain fundamental aspects of the biology of the animals are better known. The specific information which in this connection seems most necessary would be available if the following questions could be answered:

(1) Where are the principal remaining breeding areas located?

(2) What is the breeding season at each of the nesting localities?

(3) How many times does the female lay during one year and how many eggs constitute her annual total? Is laying an annual occurrence, or do the turtles lay on alternate years as has been suggested?

(4) Where do the hatchling and post-hatchling stages go and what do they do prior to attaining "chicken turtle" size—five pounds or so in weight?

(5) What diurnal, seasonal, and developmental migrations, if any, take place, and to what extent might an increase in the population in one area be expected to replenish that in another? Are the Caribbean and South Atlantic stocks genetically isolated from each other?

These points could readily be clarified by marking experiments.

(6) What is the volume of the annual turtle and egg takes throughout the range of the animal. It should be emphasized that official statistics give no approximation of the actual drain, since the greatest concentrations of turtles occur in rural areas whence no data ever reach government agencies.

If adequate solutions to these problems could be obtained—and they await only a proper investigation—it seems probable that the green turtle could not only be saved from virtual extermination but might even be encouraged to regain something approaching its primitive range and abundance.

East Pacific Green Turtle *Chelonia mydas agassizii* Bocourt

(Plate 66)

Range: Pacific coast of America from southern California south to southern Chile at Chiloé Island, latitude 43° south. It is known from San Diego Bay, California, and there are old records for San Francisco, but the possibility that specimens on which these were based were brought in from farther south by fishing boats must be considered. The relationships of this race with the other Pacific forms have not been determined, and will not be until someone sets out to collect large series of specimens from a great many localities. It seems most unlikely that only one recognizable form exists in the Indo-Pacific area; if this should prove to be the case, then all the green turtles there will have to be called *japonica,* since that is an older name than *agassizii.*

In the present account it has seemed to me that the most convenient course is to follow earlier writers in regarding the green turtle of the eastern Pacific as different from that of the western Pacific. I have seen too few Asiatic specimens to have any personal convictions on the subject, and while Paul Deraniyagala has made excellent and detailed studies of the turtles of Ceylon, even good written descriptions such as his are a poor substitute for direct comparison and detailed measurements of a large number of specimens when such trivial characters are involved. I have, however, seen enough of *C. m. agassizii* on the Pacific coast of Central America to feel little hesitation in saying that it is different from *mydas* on the other side and that, moreover, these differences are merely average ones which show considerable overlapping; and for this reason it seems best to use trinomials for the two forms.

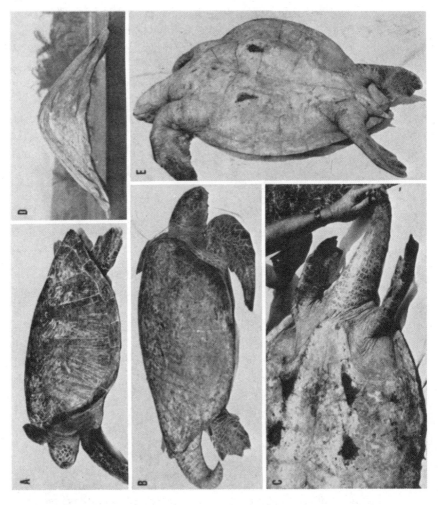

Plate 66. *Chelonia mydas agassizii. A and E:* Female. *B:* Male. *C:* Male, showing location of the vent near the nail-tipped end of the greatly elongated tail. *D:* Shell of female, showing the steeply inclined lateral laminae which may be seen in many (but not in all) female specimens of this subspecies. I have never seen a shell with these contours in the Atlantic race. (*A, B, C, and E* from photographs by Margaret Hogaboom.)

A problem of the greatest interest is that concerning the relative degrees of relationship between *mydas* and *agassizii* on one hand and between each of these and the other quasi-isolated Indo-Pacific and Atlantic stocks on the other. If we ignore geological history and merely examine existing water routes that permit communication between the green turtles of the world, then we naturally expect to find that *mydas* and *agassizii* on the two coasts of America represent the extremes of green turtle differentiation, since their near-isolation might be relieved only by tenuous contact through the Asiatic populations.

On the other hand, it is perfectly possible that such contact is negligible as far as any flow of equalizing genes between the two is concerned, and that more consequence is attached to the fact that at various times during the past *agassizii* and *mydas* have had continuous intercourse through straits across the Central American isthmus. Although such communication was cut off finally several million years ago, it nevertheless is reflected with the most astonishing clarity in the close relationship that exists between the fish faunas of the Caribbean and Pacific coasts. It is thus quite possible that a detailed study of the green turtles will show a closer affinity between American *mydas* and *agassizii* than between the latter and any of its relatives in the same ocean. The same thing may be said of all the other marine turtles.

Distinguishing features: Writers who have actually examined the few known preserved specimens of the East Pacific green turtle and compared them with *mydas* have been practically unanimous in their agreement that: the shell is higher; the shell is narrower; the marginals are more constricted over the hind legs; and the postcentral laminae are longer relative to their width. Three names have been given to East Pacific green turtles and all were based on one or more of the above features. Deraniyagala has objected that all these characters are involved in sexual dimorphism, and this is true. Their being sexual, however, does not preclude their occurring in more emphasized degree in one stock than in another, and it appears that this may be the case in *agassizii*. It is far too early to offer any positive definition of *agassizii* beyond saying that after seeing thirty-two mature specimens of it I seem to find the same differences in shell shape that others have mentioned, except that the elongate postcentrals are confined to the male and are not really longer than in fully mature *mydas,* males of which, incidentally, have been seen by very few herpetologists. Besides these features there appears to be difference in color, shades of brown predominating in Atlantic specimens (carapace and skin of the upper

surfaces of the legs and head), while *agassizii* is essentially a greenish-olive
turtle, with little or no brown in evidence in many specimens. There are
other, more subtle differences in the shape of the shell, which in the aggregate give *agassizii* a greater weight per unit of length that is not propor-
tionally correlated with the extreme height of the shell. As one of several
examples of this point of divergence that I have seen, a young female from
the Gulf of Mexico 476 mm. (18.7 inches) in shell length weighed 19¾
pounds, while another female, from the Pacific coast of Honduras, with a
shell 470 mm. (18.5 inches) long weighed 30 pounds. Even when the shells
of the Pacific specimens are not deeper along the median line, they usually
appear thicker and chunkier. It seems possible that the tails of male
agassizii may be somewhat longer than those of *mydas.* The maximum
recorded weight is 600 pounds.

Sexual differences are as in *mydas,* although they may be slightly more
exaggerated. The dimensions of a small, mature female and a good-sized
male are as follows:

Sex	Shell length	Shell width	Shell height	Head width
Female	530 mm.	420 mm.	175 mm.	78 mm.
	(20.4 inches)	(16.5 inches)	(6.90 inches)	(3.07 inches)
Male	754 mm.	554 mm.	180 mm.	100 mm.
	(29.6 inches)	(21.8 inches)	(7.10 inches)	(3.94 inches)

I have seen only one hatchling of the East Pacific green turtle, that
having been brought to San Francisco on a fishing boat that almost certainly had taken it somewhere to the southward—most probably off the
Mexican coast. The egg tooth and umbilical scar of this little specimen were
in evidence, and its measurements were as follows: shell length, 46 mm.
(1.81 inches); shell width, 35 mm. (1.38 inches); shell depth, 18 mm. (0.71
inches); head width, 15 mm. (0.592 inches). The tail of this specimen
projected 4 mm. (0.157 inches) beyond the tips of the postcentral laminae.
The three dorsal keels were very low and poorly defined, especially the
lateral ones, which were nothing more than feeble, disconnected ridges on
each of the laterals. The color above was grayish black except for yellowish
distal and posterior borders of the flippers and shell edges. Below, the color
was even yellow except for the central flipper surfaces, which were grayish
black.

Habitat: Because of the generally precipitous and rocky coastline that it
inhabits, the East Pacific green turtle is found in open water less frequently
than is *mydas.* It is most abundant in bays and about the shores of islands,

wherever protected shore water of moderate depth covers suitable beds of *Zostera,* algae, or other plants. I have seen several in narrow mangrove-bordered creeks near Amapala in the Gulf of Fonseca.

Habits: The habits of *C. m. agassizii* have not been studied. During four years' residence in Honduras I made an attempt to learn something of the natural history of this turtle in the Gulf of Fonseca, but with almost no success. Green turtles are caught there in some numbers during the months from January to June, but I was unable to get a single specimen during the summer and fall months, and although I examined the state of the reproductive organs of a considerable number of females none was found with eggs that even approached maturity. Although I kept in touch with events at the few beaches around the Gulf by occasional visits and through reports of local turtle egg hunters, I was never able to find any evidence that the green turtle nested in the area, although two or three fishermen claimed that it visits the sea beaches on the Nicaraguan coast a few miles to the southward. Many of the Fonsequeños confuse the green turtle with *Lepidochelys olivacea,* but those who do recognize it state that it never comes up to lay on the same beaches and at the same times (August through November) as the latter, and my own observations tend to support this view. Green turtles of a hundred millimeters or less in length are unknown in the Gulf. The species is generally much less abundant in the waters of Honduras, Nicaragua, and El Salvador than it is off the coasts of Mexico and Panama, and it seems possible that most specimens from middle Central America may be immigrants from the north or from the south. A study of the migratory movements is much to be desired, and although it would involve a good deal of effort, it would almost certainly yield interesting information on the biology of a valuable marine resource.

Feeding: Of ten stomachs examined by me six were empty and four contained food. Of the latter, three were packed with kelp and pieces of a sponge, while one was filled with two or three pounds of short lengths of *Zostera.*

Although the close similarity of the several green turtle stocks of the world does not necessarily imply comparable identity of habits, it is perhaps worth while to summarize the known features of the life history of the West Pacific form, since zoologists are better acquainted with it than with any of the others. The following résumé is extracted mainly from the studies of Moorhouse (1933), Hornell (1927), and Banks (1937) and from the general works of Günther (1864), Smith (1931), and Deraniyagala (1939a).

In disposition the green turtle is nearly always inoffensive but, lest the repeated unqualified statement of this fact contribute toward serious accident, I should point out that an old male individual that I secured in Honduras snapped viciously when teased with a stick. This is the only case of the sort that I have seen, but such pugnacity may be more widespread than is believed.

Several writers have referred to the fact that on remote and undisturbed beaches and rock ledges, green turtles haul out to sun themselves, sometimes in considerable numbers. This habit is, among marine turtles, apparently peculiar to *Chelonia*. In attempting to reconcile these terrestrial interludes with the widespread belief that green turtles die from asphyxiation if placed plastron-down on the deck of a ship, it has been suggested that the vibrations of the deck, transmitted to internal organs through the poorly supported plastron, may be the cause. Pope (1939), however, pointed out that a napping turtle must have far lower oxygen requirement than an agitated, struggling captive, and called attention to the several active hours that the female may spend belly-down on land when digging her nest. On the Pacific coast of Honduras the fishermen are unacquainted with the conventional method of shipping turtles plastron-up. There I have received specimens alive and in apparently good condition after at least two days in an upright position, including periods on the deck of a fishing boat and in the bed of a truck that jolted for six hours over the incredibly rocky road from San Lorenzo to Tegucigalpa. All such individuals that were sent to me remained alive when kept in the shade, either on their backs or upright. However, three of these that I placed in a pool of fresh water died for some obscure reason within a few hours, and another which I sprayed with a hose surprised me by succumbing immediately. It seems likely that there may be racial differences among green turtles with respect to the rigidity of the plastron and the amount of support that it offers to the viscera.

The migratory movements of the Pacific green turtles are little known. Some evidence, largely negative and indirect, that the population of the Mozambique Channel moves to the Aldabras for the nesting season has been seen, and such seasonal absence of green turtles as has been noticed in Ceylon and Central America almost certainly points to some sort of concerted migration.

The sparse data on growth rates in the Pacific green turtle indicate that the yearling may be from 203 to 254 mm. (eight to ten inches) in shell length, that the mature female may be about 890 mm. (35 inches) in length, and that a ten-year-old turtle may have a carapace about 1,110 mm. (44 inches) long.

Pairing and copulation occur, as apparently in other sea turtles, at the time of nesting, when the males follow the females to the nesting beaches and await them near shore. During copulation, which may occur immediately after the eggs are laid and possibly before as well, the pair float about on the surface of the water and are easily approached by boats. A greatly enlarged nail on the anterior flipper of the male aids him in grasping the fore edge of the shell of his mate.

The nesting season varies markedly with locality, chiefly, perhaps, with latitude, but probably with other less obvious factors as well. Reported seasons in several localities, together with incubation periods for the same areas, are as follows:

Locality	Latitude	Nesting season	Incubation period
Australia	23° S	late October to mid-February	65–72 days
Gulf of Siam	8°–12° N.	year around, with July–November peak	40–50 days
Ceylon	6°–9° N.	July–November	50–60 days
Malacca Straits	5° N.	December–January	——
Seychelles	5° S	year around, with March–May peak	47 days
Borneo	2° N.	May–September	——

The following account of the nesting of a female green turtle on a small island in Torres Strait is quoted from Hurley (1924: 318–319):

The sea swish laved the beach with phosphorescent fire, rustling pleasantly the shells. As we wandered our torch-light scared myriads of crabs into the foam or into holes, the excavation of which we had disturbed. A dark object came slowly from the sea, moving forward by each wave impulse. Painfully, by jerky inches, then up the steep sand bank, pausing after each effort to regain breath and energy.

Many turtles were coming ashore to deposit their eggs—a process which we studied intimately. The labor of moving their unwieldy bulk over the loose sand seemed a cruelty of nature. Above high-tide reach mother turtle stopped to roost and make her nest. Her front flippers began the excavation, throwing out showers of sand full fifteen feet. Over a ton of sand was thus shifted, and by constantly turning, a circular hole scooped out.

But it is rarely that the turtle lays in the first bunker. Probably, with the intent to camouflage and decoy she jerkily jaunts off to another site and starts all over again, where, after much sighing and heaving, she might lay, or take it into her capricious noddle to "dig in" a third time; maybe defer her intentions to another night, and move off to sea again. After watching patiently for several hours, the third time decided. Madame Shellback excavated a crater of five feet diameter and thirty inches deep, but this was only the beginning. The digging of the actual egg repository was a much more remarkable process. The hind flippers, with human hand cunning, delved vertically, extracting the sand and throwing it forward with the expertness of a past-day fireman.

In order to prevent the small shaft from collapsing and sand from sliding during the operation, the second hind flipper was lightly pressed against the sand. The adroitness, skill and exactitude of every movement amazed us more than the subsequent egg-laying. The tail gently dipped into the hole and the eggs were laid in ones and twos—there seemed to be no end to it. After depositing one hundred and sixty-nine eggs our prolific hen rested a while, and then carefully concealing her exemplary effort with sand, gently compressed it with the posterior flippers.

When completed the anterior flippers vigorously shovelled sand back and covered the nest, so that its whereabouts were completely concealed. Then she returned to the sea.[1]

The above description appears to be in general accord with the observations of other writers. The process nearly always occurs at night, and the site chosen is usually well back of high-water mark, the few reported cases of wave-washed nests probably being the result of poor judgment on the part of occasional females, whose reproductive effort in such instances almost surely comes to naught. The number of eggs laid on a given visit to the beach has been found to range from 66 to 195, with the average about 115. The average number of times eggs are deposited by a female during a single year is not known, but as many as seven laying emergences at intervals of two weeks have been reliably recorded for more than one individual. The average diameter of the soft-shelled, white eggs is probably about 45 or 46 mm. (1.77–1.81 inches), although the size undoubtedly varies with the size of the laying female. Emergence of the young hatchlings from the nest may extend over a period of 48 hours. There is little agreement among observers as to the behavior of the hatchlings on reaching the sea, some claiming that they have difficulty in diving, others maintaining that they are expert at it.

In nature the adult green turtle is apparently largely herbivorous, the

[1] Quotation used by permission of G. P. Putnam's Sons, publishers of *Pearls and Savages,* by Frank Hurley.

bulk of its diet being algae and other marine plants. Occasional stomachs contain jellyfish, sponges, mollusks, and other invertebrates, and captive specimens live well on fish and other animal food. The young are probably mainly carnivorous. Despite a certain amount of flexibility of feeding requirements in green turtles, one of the principal ecological factors that determine their distribution is the presence of extensive beds of *Zostera* and related plants, and where such marine pasturage is lacking the animals never abound.

Economic importance: Although the green turtle of the eastern Pacific is caught throughout its range for local consumption, it does not equal its relatives in the Caribbean and western Pacific in commercial importance. It is frequently brought in from Mexican waters to markets in the California ports, but this is apparently mostly incidental to other fishing endeavors. Nelson (1921) mentioned that the main export of Magdalena Bay, a village on Magdalena Island, Baja California, was at one time a monthly shipment of turtles to San Francisco, but this appears to have been discontinued. This is in my judgment a magnificent meat animal, in no way inferior to the Caribbean race, and warranting serious attention from those interested in the intelligent exploitation and conservation of a resource which under proper management might contribute materially to the welfare of poverty-stricken communities of the Pacific coast of Spanish America.

Genus *ERETMOCHELYS*

The Hawksbill Turtles

As in the case of the other marine turtles, this appears to be a monospecific genus; there is little or no evidence that further search will turn up specifically disjunct forms anywhere in the great, circumtropical territory covered. Like the green turtles the hawksbills have four pairs of lateral laminae, the first of which are separated from the precentral by the first central. There are four inframarginal lamina on either side; these are not penetrated by pores. There are two pairs of prefrontal scales on the head. The inner surface of the upper tomium is smooth and the lower jaw is not dentate. At the symphysis the lower jaw is deeply excavated and the blunt tip has no tooth. The snout is narrow and elongate.

KEY TO *ERETMOCHELYS*

1 Carapace more straight-sided and narrowly tapered behind; upper
surfaces of head and flippers with less black; Atlantic and Caribbean
.... *imbricata imbricata* (p. 366)

1' Carapace usually more heart-shaped in lateral outline; upper surface
of head and flippers almost solid black; Indian and Pacific oceans
.... *imbricata squamata* (p. 373)

Atlantic Hawksbill Turtle *Eretmochelys imbricata imbricata* (Linné)

(Figs. 22, 23, 24, 27, 28, 29.)

Range: Warmer parts of the Atlantic Ocean; shores of the Gulf of
Mexico and Caribbean Sea. On the American mainland it has been taken
from southern Brazil to Massachusetts. There are old records for the European coast, and it is very probable that they apply to *imbricata,* since hawksbills should be subject to the same forces that transport occasional specimens of the other species across the Atlantic from American waters. The
only other definite record for the eastern Atlantic that I can locate is an
early mention of a hawksbill from "Morocco," and this was merely listed
among other localities with no reference to any actual specimen. I have no
doubt that hawksbills occur on the coast of West Africa, but if so they have
escaped mention in print to an astonishing extent, and may be only very
sparsely distributed there. If this is true, the isolation between the Atlantic
and the Indo-Pacific populations is even more nearly complete than in the
case of the other sea turtles, because all the rest appear to have fairly continuous linear range from one major area to the other. Hawksbills are
known from all the Caribbean islands and from the Bermudas.

Distinguishing features: The hawksbill turtles may be recognized by
their two pairs of prefrontal scales and four pairs of laterals. The dorsal
laminae are markedly imbricate and overlap each other extensively in all
but the largest and oldest specimens, in which they may revert to a juxtaposed condition as in the green turtles. On the average it appears possible
to distinguish hawksbills from the Atlantic and Pacific coasts of America
by the shape of the carapace, which has a more evident straight-line taper
behind in *imbricata* than in East Pacific *squamata,* and perhaps by coloration, which may be deeper, more unrelieved black in the latter than in the
former.

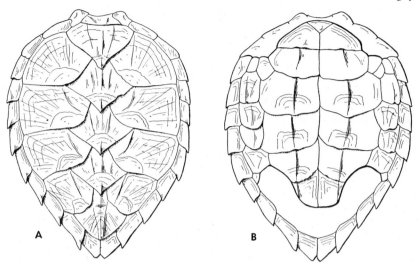

Figure 27. Shell of young hawksbill (*Eretmochelys i. imbricata*). (From Stejneger.)

Description: The shell is lance-shaped to shield-shaped, usually longer in old than in young, and shows considerable variation in lateral outline. The size is small for a sea turtle, this being the smallest of all the species except possibly the ridleys. The heaviest specimen on record was a 280-pound individual caught at Great Sound, Grand Cayman, in the West Indies, but a hundred-pound specimen is large today. The largest specimen I have seen was a male from Florida with a carapace 835 mm. (32.8 inches) long and 567 mm. (22.2 inches) wide; a good-sized female from Barbados has a shell 606 mm. (23.8 inches) long and 442 mm. (17.4 inches) wide. The carapace is decidedly heart-shaped in young and in adults is more straight-sided and tapering posteriorly; the laminae are strongly imbricate, the knifelike overlapping edges making it advisable to handle a struggling hawksbill with some care to avoid painful injury. There is a middorsal keel, usually present on the hind four centrals, at least, in all but the oldest specimens, and besides this the posterior two or three centrals may bear oblique, converging ridges. The projecting free posterior edges of the laterals are more or less pointed and receive in most young specimens a ridge that traverses the lamina from its anterior edge. As the turtle matures, the overlap of the laminae becomes progressively less until finally they may lie side by side as in *Chelonia,* and the carapaces of such specimens may be distinguished only by the shape of the precentral lamina

(of which the posterior seam is twice as long as the anterior in old specimens, and much less in young ones and in *Chelonia*). The laminae are
wrinkled in some specimens. The "tortoiseshell" coloration is not much
evident in the live animal, being seen best by transmitted light after the
laminae have been polished, but in some specimens from less protected
habitats the beautiful variegated rays of reddish, yellow, fawn color, black
or white, or even greenish may be seen. The plastron is yellow, usually with
a few black blotches on the anteriormost laminae and sometimes on a few
others. The scales of the head are chestnut brown in the center, sometimes
tinged with red, and with their margins of a lighter color; the jaws are
yellowish, with occasional streaks or bars of brown; the chin and throat
are yellow and the neck above is dusky; the legs are dusky brown above
and yellow below.

The tail of the male hawksbill is much longer than that of the female;
in the latter it projects but slightly beyond the hind margin of the carapace,
while in the male specimen whose measurements were given above, the tail
extended 95 mm. (3.74 inches) beyond the tips of the postcentral laminae.
The plastron of the male is more concave than that of the female and its
claws are longer and heavier.

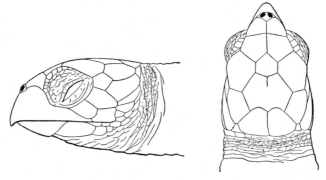

Figure 28. Young hawksbill (two-thirds natural size); Porto Rico. (From Stejneger.)

The shell dimensions of the smallest and largest hawksbill hatchlings
(with egg tooth and umbilical scar) that I have measured are as follows:

Length	*Width*
40 mm.	31 mm.
(1.57 inches)	(1.22 inches)
46 mm.	39 mm.
(1.81 inches)	(1.53 inches)

Very young specimens are usually black or brownish black above and below except for the keels, edges of the shell, upper outer areas of the neck, and an interrupted margin along the paddles, all of which are light brown.

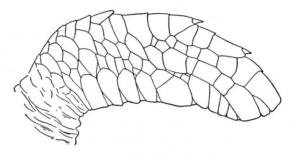

Figure 29. Right fore flipper of *Eretmochelys i. imbricata* (two-thirds natural size); Costa Rica. (From Stejneger.)

Habitat: The Atlantic hawksbill is found in shallow coastal waters. This turtle shows less specific habitat preference than the green turtle, and although sometimes found together, hawksbills are abundant in a number of situations in which green turtles cannot be found. The main difference in habitat appears to be a greater tolerance on the part of the hawksbill for shallow bays and lagoons with muddy bottom, without rock holes, and with little or no extensive beds of submarine vegetation. This is merely a difference of preference, however, for I have frequently seen *Chelonia, Eretmochelys,* and *Caretta* in the same places and at the same times.

Habits: No scientific study of the life history of the Atlantic hawksbill has been made.

There is a general belief that hawksbills do not migrate to any extent, and while this appears likely, it has yet to be authentically established.

Scattered information on growth rates indicates that this turtle may mature sexually at an age of about 3 years and on attaining a weight of about 30 pounds. According to J. Schmidt (1916), the 40- to 50-mm. (1.57- to 1.97-inch) hatchling increases its length to 70 or 80 mm. (2.76–3.14 inches) during its first winter, to about 100 mm. (3.94 inches) by the first spring, and to 200 mm. (7.88 inches) by the second fall.

By disposition the hawksbill is more irascible than the green turtle, newly caught individuals showing considerable ill humor and biting and snapping at any object thrust within reach.

Breeding: The mating process is said to take place near shore, by implication, during the nesting period and near the nesting beaches. Although this turtle still goes ashore to lay in some abundance on beaches near fairly large centers of human population, the details of its reproductive activities are known only from verbal reports of fishermen, turtle hunters, and other local inhabitants. In the Danish West Indies, Schmidt found that hawksbills bred more freely than any other sea turtle. They are still known to lay on beaches of the Cayman Islands, the island of Sacrificios near Vera Cruz, and the keys and islands of the Central American coast, and it seems likely that most of the suitable beaches of the Caribbean are visited at least occasionally. Audubon believed that in Florida hawksbills laid on shores of the outer keys only, while True stated that a gravel strand was preferred to pure sand. Published data seem to indicate that there is wide local variation in breeding season. Part of this may be the result of normal seasonal differences due to latitude, although observations are far from satisfactory for any given area. The season of oviposition in the Florida region is said to have been April to July; that in the Danish West Indies, July to October; that in the island of Grenada, July to August; and that in the vicinity of Vera Cruz, May to mid-July.

While it is generally agreed that a given female may nest more than once during the season, there is no definite information concerning the number of times eggs are deposited. One writer stated that three clutches are laid, at intervals of at least three weeks. Audubon said that two sets are deposited— one in July and one in August. Schmidt dissected a female that contained 160 shelled eggs and 300 more in yolk, and this total would seem to indicate that this particular individual might have laid three times, since the average number of eggs in a complement appears to be about 150, although here again reliable data are few. The eggs range in diameter from about 39 mm. to about 41.5 mm. (1.54–1.63 inches).

The nesting process has not been described and the incubation period is unknown.

Feeding: The hawksbill is omnivorous. While perhaps somewhat inclined to choose animal food instead of plants, where both are available, individuals in which the stomach contained only vegetable matter have frequently been taken. Algae are often eaten, and the long list of animal forms that compose the diet includes ascidians, sea thimbles, mollusks, and crustaceans. True mentioned that hawksbills, like other sea turtles, eat Portuguese men-of-war, closing their eyes to avoid the venomous tentacles, and thus rendering easy an approach by a turtle hunter. Captive specimens thrive on a diet of fish and meat.

Economic importance: While the commercial value of the hawksbill lies chiefly in the shell, the importance of the animal as a source of food for rural seaboard peoples has been too often ignored, and egg hunting is probably the most serious factor in the depletion of hawksbill populations. Although in many places the meat is not fancied, and although most writers on the subject have followed one another in referring to it as more or less inedible, it is actually perfectly good food, and in several places along the Caribbean coast of Central America I talked with experienced turtle cooks who preferred young hawksbill to all other sea turtles. There are stories which may indicate that the meat is on rare occasions unwholesome, or possibly even poisonous, but these are unsubstantiated for the Atlantic form, although it is possible that some unusual feeding aberration may make the meat of an occasional individual unfit for food.

The eggs are universally appreciated, where obtainable, and are eaten either fresh or cured. Shelled eggs are dried and sometimes strung like beads to be smoked, while those taken from butchered females are salted and smoked in sausage skins or in their own oviducts.

It is for the unusually thick, horny scales of the shell that the hawksbill is famous. These combine attractive mechanical properties—softness, flexibility, fusibility and the capacity to take a high polish—with translucence and richness of coloration (ranging from the clear yellow or light amber of the plastral scales to the various shades of brown and yellow, and even of black, white, bluish, and greenish, that dot, fleck, or splash the clear ground of the dorsals) and since earliest times have found wide use in the arts.

The scales are separated from the underlying bones either by heat, which melts the cementing layer, or by simply allowing the shell to lie exposed to the weather until separation occurs. In some localities the turtle is usually killed before removal of the *carey,* as it is known in Spanish America, but elsewhere it is taken from the living animal, either by radiant heat from low fires or by the application of hot water. This extraordinary example of conservation conscience appears to have existed in Caribbean territory since pre-Columbian times, and it, and the belief that turtles so handled survive, are prevalent among primitive peoples in various parts of the world. In the Mosquitia of Central America both the Caribs and the Mosquito Indians remove carey by this means, and of a large number of intelligent and observant turtle hunters of each group with whom I have discussed the question, all have felt certain that the scaled hawksbill survives and regenerates his epidermal armor. Some believe that the new scales are always thin and without value; others, that they coalesce to form a continuous horny layer; while still others maintain that new crops of good carey may be taken from a turtle year after year. In an attempt to appraise the extent of the injury suffered by a turtle so treated, I removed scales from living animals by several means, including dry heat, hot, wet rags applied to the shell, and a stream of boiling water directed against the scales. The last technique seemed to me the most satisfactory. It permitted removal of the scales with little blistering of the deeper layers and no bleeding, and with a minimum of evident discomfort to the turtle. I am inclined to believe that all the valuable larger scales could be thus obtained without killing the turtle; if so, I can see no reason why new carey should not gradually be deposited. On the other hand, I seriously doubt that many of the hundreds of hawksbills that annually are forced to yield their shell by being summarily half roasted before beach fires or under burning leaves actually survive for more than a short while after the operation. Even if the poorly controlled heat does not destroy the secretory cells of the epidermis, a soft-shelled sea turtle would be at the mercy of countless small nibbling fishes, and the blistered lesions of its shell surface would almost certainly attract sharks.

As to the occasionally reported hawksbills with thin, smooth laminae and those said to have the shell covered by a continuous horny layer, it seems probable that they are merely old individuals showing the senile reversion to the juxtaposed arrangement of scales described by Deraniyagala (1939).

Samuel Bard [pen name of E. G. Squier, American minister to Nicaragua] (1855) observed the process of removing the scales from live hawksbills as practiced by the Mosquito Indians of Nicaragua, and described it as follows:

The fishers do not kill the turtles; did they so, they would in a few years exterminate them. When the turtle is caught they fasten him, and cover his back with dry leaves or grass, to which they set fire. The heat causes the plates to separate at their joints. A large knife is then carefully inserted horizontally beneath them, and the laminae lifted from the back, care being taken not to injure the shell by too much heat, nor to force it off, until the heat has fully prepared it for separation. Many turtles die under this cruel operation, but instances are numerous in which they have been caught a second time, with the outer coating reproduced; but, in these cases, instead of thirteen pieces, it is a single piece. As I have already said, I could never bring myself to witness this cruelty more than once, and was glad that the process of "scaling" was carried on out of sight of the hut. Had the poor turtles the power of shrieking, they would have made that barren island a very hell, with their cries of torture.

The best markets for carey are located in the old world, where India, China, Spain, and Italy have imported it for centuries. The price varies markedly with the grade, which is determined by the size of the turtle, the habitat in which it has lived, and the method of removal, as well as by local fads and preferences. Eight pounds of shell is a very good yield for a single turtle, and the largest individuals, rarely taken nowadays, may give 15 or 16 pounds. A dollar a pound is apparently a fair price, although in the past carey has sold for as much as five dollars a pound. Hawksbills that have sold for twenty dollars and more are mentioned in reports on the industry covering at least a century.

In the trade the central and lateral scales are known collectively as the "head" of shell and individually as "blades," while the marginals are called "feet" or "noses" and may be grouped with the centrals as "hoofs and claws." The two middle laterals on either side are the most valuable. Most of the Central American *careyeros* with whom I have talked favor shell with densely arranged, very small splotches of black or very dark brown on a clear background. According to True (1884), the generally sought colors are mingled golden yellow, reddish jasper and white, or brown approaching black, except in China, where "white head" or blonde shell with a large amount of white is fancied. The famous high combs of Spanish ladies were in past times often made of the clear golden plastral carey.

While the mechanical qualities of tortoise shell are easily reproduced in plastic imitations, I have seen no plastic that even approached the brilliance and richness of the finest *carey pintado* and have never been able to understand why it has found so little favor with American jewelers.

Pacific Hawksbill Turtle *Eretmochelys imbricata squamata* Agassiz
(Plate 67)

Range: The tropical Indo-Pacific. In the western Pacific hawksbills have been taken as far north as Japan. They are known on the East African coast as far south as Madagascar, but not from the Cape of Good Hope, and it is very doubtful that they meet the stock of the South Atlantic in this region. On the American coast the range extends from Baja California to Peru; old records for "southern California" have not been corroborated by recent occurrences. It would appear that, except for the remote possibility of interbreeding between occasional transoceanic strays, the hawks-bill turtles of the Atlantic and Pacific oceans are effectively isolated.

Distinguishing features: When Agassiz described the Pacific hawksbill in 1857, he named no type locality and probably had specimens from a few widely separated areas in the western Pacific, but none apparently from American waters. His differentiating characters were, as writers on Asiatic herpetology have shown, for the most part features that are highly variable and are associated with the age and sex of the individual. While I have seen only a handful of hawksbills from the western Pacific and am un-qualified to comment on the status of those populations, I have examined a sufficient number of East Pacific specimens to justify a few tentative com-ments on the range of variation that they show, and it seems to me that these variations overlap extensively those of the Atlantic stocks with re-spect to all the characters that have been used to separate the two forms. Despite this overlap, however, I am convinced that very slight average differences could be pointed out if comparable series of specimens from, say, the northern and southern coasts of Honduras were laid side by side. The East Pacific form is darker, the top of the head and upper surface of the fore flipper being solid, deep black and that of the hind flipper nearly so. The heart-shaped shell outline, characteristic of the young of both, is, as Agassiz said, retained in more adult specimens of the Pacific form than of the Atlantic form, although whether these would constitute a majority if statistically significant numbers of specimens were examined I hesitate to say. Although no actual measurements of the angles between the hind projections of the marginal laminae and the shell edge have been made, these appear to be considerably wider in Atlantic hawksbills than in those from the East Pacific, in which the projecting points are either parallel with the margin or are actually tucked inward behind. Though these may be equally long in the two stocks, they diverge more in *imbricata,* and for this reason it seems to have the more serrate shell margin. As compared with Deraniyagala's descriptions of the coloration of the hawksbills of Ceylon, those from the Pacific coast of Honduras appear to show the same notably deep black pigmentation of the upper surfaces of the head and legs, in young and old (that distinguish them from *imbricata*), but while East Pacific hawksbills are remarkably constant in their coloration, Asiatic specimens may not be.

The coloration of fifteen young and adult female hawksbills from the Pacific coast of Honduras is as follows. The carapace is variegated in what appears to be the same manner as in Atlantic specimens, although in all

Plate 67. Eretmochelys imbricata squamata. Upper: Female. *Center:* Female with unusually ornate *carey. Lower:* Young specimens. All from Bay of Fonseca, Honduras.

the colors were hidden by an opaque layer of algae and silt, as is the case with all specimens from the mangrove creeks of the Fonseca area. The plastron is yellow, with a large black spot in the center of most or all of the anterior row of laminae. The scales of the head are deep, glossy black above, occasionally with very narrow light margins, but usually without; more usually the light rims are seen only on the sides of the head where each scale has a black central spot on an orange background. The horny sheath of the upper jaw is mostly black anteriorly and laterally, with a vertical light bar extending downward from just behind the nostril and with the upper hind edge of the black area margined with bright orange. The horny sheath of the lower jaw is yellow ventrally and anteriorly, with a large longitudinal, light-bordered black patch covering most of the side. The big mandibular scale has a large black central area rimmed with orange. The chin is dirty yellowish white with a few black-centered scales; the throat is light. The limbs below are colored like the sides of the head— each scale with a big black spot on an orange background—but above they are nearly uniform black, except that the proximal portions of the hind limbs may be colored like the under surface and that there may occasionally be very narrow orange borders to the scales.

Measurements of a fair-sized adult female from the Gulf of Fonseca, Honduras, are as follows: shell length, 615 mm. (24.2 inches); shell width, 489 mm. (19.2 inches); shell depth, 203 mm. (8.0 inches); plastron length, 480 mm. (18.9 inches); head width, 80 mm. (3.15 inches). Although the only fully mature male of the East Pacific form that I have seen was an atrociously stuffed specimen that was nailed down and that could not be turned over for inspection, sex differences appear to be the same as in *imbricata* and in the hawksbills of the western Pacific. The tail is much longer and the plastron more concave than in the female.

I have not seen a hatchling from the East Pacific. One from New Britain has a shell 41 mm. (1.61 inches) long and 32 mm. (1.26 inches) wide. Deraniyagala gave the range in shell length for Ceylon hatchlings as 39–42 mm. (1.53–1.65 inches). The largest hatchling with egg tooth and umbilical scar that I have seen is one 46 mm. (1.81 inches) in shell length. The carapace is usually reddish with a darker central area and a white margin; the plastron is black with light ridges; the flippers are black with a light margin posteriorly and terminally. The ridges on the shell are much more pronounced in young specimens than in old. Agassiz thought that he saw the following differences in the arrangement of these keels in the Atlantic and Pacific hawksbills:

imbricata	*squamata*
Median keel continuous on only last 4 centrals	Median keel continuous from first to last centrals
Only last two centrals with ridges that converge posteriorly	All centrals with ridges that converge posteriorly
Laterals without ridges running from the posterior point of each lamina downward and forward to the marginals	Laterals with ridges extending downward and forward from the hind projection of each lamina to the marginals
Keels on the two middle rows of plastrals less prominent	Keels on the two middle rows of plastrals more prominent

While these features are subject to great variation, I am not certain but that if a large series of young specimens of each population, all about 254 mm. (10 inches) in shell length, were compared carefully, they might prove to show a certain amount of constancy as regards the keel character.

Habitat: On the American coast the habitat of this hawksbill appears to be essentially shallow, mangrove-bordered bays, lagoons, and estuaries, often mud-bottomed and with or without submerged vegetation. It ranges into small, narrow creeks and passes.

Habits: Beyond a few random observations of my own, nothing is known of the habits of the hawksbill of the eastern Pacific. Although the life history of the West Pacific form is far from being completely known it has received some attention from zoologists. The following résumé is derived largely from the accounts of Hornell (1927) and Deraniyagala (1939a).

The Pacific hawksbills are believed to be strongly localized and little given to migrating. Several writers have commented on the extent to which barnacles and boring mollusks infest the shells and flippers of hawksbills, and I found similarly excessive encrustations on individuals from the Pacific coast of Central America. This susceptibility to attacks by encrusting organisms might be interpreted as further evidence that the hawksbill tends to stay in one place for long periods of time, although this does not necessarily follow. Along with this turtle's apparent disinclination to wander, Deraniyagala suggested that the hawksbill may have a well-developed homing instinct and cited the statement of an early writer on Ceylon (J. W. Bennett, *Ceylon and Its Capabilities,* 1843, p. 274) in this connection:

The homing instincts of this turtle are probably stronger than in most animals, for Bennett states that individuals stripped of their scutes were known to visit the same locality the following year when their scutes were again stripped. Conclusive evidence of these repeated visits was obtained in 1794 when the Dutch Commandant of the southeastern coastal part of Ceylon marked several turtles with brass rings. One of these rings was recovered by a renter who brought it to Bennett in 1826 and affirmed that to his certain knowledge this turtle had revisited the cove of Amaidhuva for those 32 successive years. Bennett replaced the ring upon the animal and liberated it.

Although interesting, such anecdotes as the above require corroboration by scientifically conducted marking experiments.

Hornell showed that in the Seychelle Islands a hawksbill may attain a shell length of 3,300 mm. (13 inches) in two years. Two hatchling specimens kept in captivity by Deraniyagala for 16 months increased nearly equally in size to gain about 10 pounds in weight and slightly more than 14 inches in shell length. Deraniyagala demonstrated an interesting change in the character of the horny shell armor that accompanies the increase in body proportions. In the young the scales lie side by side until a carapace length of a little more than 100 mm. (3.94 inches) has been reached, when they gradually acquire the characteristic shinglelike arrangement shown by most adult hawksbills. They remain thus imbricated until the shell attains a length of 360–400 mm. (14.2–15.7 inches), when a return to the juxtaposed condition may begin in the plastral scales, becoming progressively more marked there and, later, on the carapace until when the shell reaches a length of 600–800 mm. (23.6–31.5 inches) all may be juxtaposed.

Breeding: Mating occurs during the nesting season, at which time the males follow the females close inshore and have even been seen to come out on the beach in pursuit of them. Laying occurs in the Seychelles from September to November and on the west coast of Ceylon from November to February, with a concentration of activity during December and January. The possibility of a different season for the populations of the southeastern side of Ceylon is suggested by Deraniyagala:

Opposed to this [December–January breeding season] is the statement by Bennett (1843, p. 275) that along the southeast coast, towards Amaidhuva, this turtle came in such large numbers to nest during April, May and June that the Government farmed the right to capture them and strip them of their scutes. As the sea off this coast is very rough during the northeast monsoon, it is probable that the turtles in these waters breed during the calm southwest monsoon and that Bennett is correct. It is thus possible that in Ceylon there are either two breeding areas and two seasons for the hawksbill or that this species changes its breeding grounds according to the monsoon, frequenting the southeast coast from April to the end of May during the southwest monsoon and the west and northwest coasts from November to February during the northeast monsoon.

The only detailed account of oviposition is also from Deraniyagala, who watched a female deposit her eggs at Bentota, Ceylon, February 28, 1928, at 7 P.M. The turtle selected a site 80 yards from the edge of the water at the edge of the Pandanus brush. After flinging aside loose sand with all flippers for a while, she commenced digging methodically with the hind feet, which were used alternately in grasping and setting aside sand from the growing nest cavity. When the nest had reached a depth about equal to the length of the leg (about 500 mm. [19.4 inches]) it was goblet-shaped, with the expanded lower portion about 200 mm. (7.9 inches) in diameter. The cloaca was then lowered into the hole and the eggs were extruded in groups of 2, 3, or 4, with a few seconds' interval between each batch. After laying 115 eggs she began filling the nest, carefully lowering into it footfuls of sand, and again working the feet alternately until the hole was nearly full, when she began pulling in sand with both feet simultaneously. The digging process consumed 45 minutes, oviposition 15 minutes, and the filling and covering of the nest 45 minutes more.

There are usually between 150 and 200 eggs in a clutch, according to Hornell, and they range in diameter from 35 to 38 mm. (1.38–1.50 inches). The number of times a female lays each season is not known.

A set of eggs moved from the nest at Karaduva, Ceylon, and buried in garden soil hatched after 64 days, which probably represents an unnaturally prolonged incubation period.

In the Bay of Fonseca, where Honduras, Nicaragua, and El Salvador meet, I was told by fishermen that hawksbills nest at the same season as *Lepidochelys,* or from August to November. During my only visits to the area at this time of year I was occupied in observing the nesting activities of *Lepidochelys* and was unable to corroborate reports that on these occasions hawksbills were emerging to lay on the narrow, rocky, protected beach of a little island well inshore from the more exposed Isla de Ratones, where the ridleys appeared to be the only nesting turtles. The only adult female hawksbills from the Pacific coast of Honduras that I have dissected were eight specimens taken during the months from March through July, and in none of these were there well-developed eggs.

Deraniyagala found that a number of hatchlings kept by him were positively phototropic, tending to orient themselves toward any source of light; that they had difficulty in submerging, owing (in his opinion) to the large amounts of yolk that they carried; and that they fed actively when only two days old, biting at both chopped fish and seaweed and using the flippers to aid the jaws in pulling the food to pieces.

Feeding: These turtles are omnivorous feeders. Various kinds of algae and seaweeds are eaten, as are coelenterates, mollusks, and crustaceans, and even meat and fish are relished by captive individuals. Two specimens from the Gulf of Fonseca had the alimentary canals crammed with neatly cut, two-inch sections of the cylindrical fruits of the red mangrove. Another from the same locality had eaten some two pounds of a mixture of mangrove leaves and dead bark and wood.

Economic importance: The hawksbill of the Indo-Pacific has always been of greater importance as a source of tortoise shell than that of the Caribbean. This is due partly to the fact that shell from the Indian Ocean is rated as of superior quality but mainly to the exceptionally high regard in which carey is held by the Chinese and other peoples of the Far East. According to Deraniyagala, articles of hawksbill shell were sent as gifts by Singhalese kings to foreign courts in pre-Christian times. The emperor Nero is said to have had a bathtub made of carey.

For hundreds of years, much of the best shell has been sent to Shanghai and Singapore for exportation, and there is a flourishing trade in Nagasaki and other cities of Japan. The people of the Pacific coast of tropical America show only desultory interest in tortoise-shell work. Nearly every little community has its *careyero* [2] who makes and sells simple utilitarian articles such as combs and buttons and perhaps a few ornamental trinkets, but nowhere does the craft reach the oriental level of artistry. I know one old Honduranean shell worker whose shop is a mud-walled, floorless hut and whose only tools are a punch, a file, and a scroll saw. He lives by making things of carey for the villagers, polishing the shell with dried, crumbled leaves of a local tree, and though his overhead is negligible, his prices are so low that it is not clear how he supports himself by his profession.

Nowhere in Central America have I found shell workers welding layers of carey together to form blocks, as the Asiatic and European craftsmen do. This operation, which greatly enhances the possibilities of the shell as raw material, is apparently not easy to perform. Seale (1911) described the process as practiced by the Japanese as involving the use of heated pincers, in the jaws of which the plates of shell to be joined are placed between thin pieces of magnolia wood and subjected to moderate pressure. The Japanese expert who furnished these particulars, however, refused to divulge the proper temperature for the pincers, guarding it as a trade secret!

The eggs of the Pacific hawksbill are widely used as food. In the western Pacific the meat is not highly esteemed, perhaps mostly because of its

[2] A worker in tortoise shell. The word *carey* is widely used in Caribbean lands both for the hawksbill and for its shell.

reputed tendency to assume poisonous qualities on rare occasions. Although the explanation for the sporadic change in the usually edible flesh is not understood, it is probably due to some temporary change in the diet of the animal. There are several well-authenticated instances of deaths resulting from eating the meat. According to Deraniyagala, in Ceylon "experienced fishermen seldom eat an *Eretmochelys* without first chopping up its liver and throwing it to the crows. If the latter discard the liver, the animal is poisonous."

In the eastern Pacific, among the inhabitants of coastal Honduras, Nicaragua, and El Salvador, hawksbill meat is not suspect. In the market in Tegucigalpa it is sold for the same price as that of the green turtle. I have sampled it myself on two occasions, and can find little basis for distinction between young "chicken"-sized individuals of the hawksbill and green turtles.

Genus *CARETTA*

The Loggerheads

The range of the loggerheads is probably not markedly different from that of the ridleys except for what appears as an altogether eccentric absence of the latter in certain areas (for instance, the Caribbean). In breed-

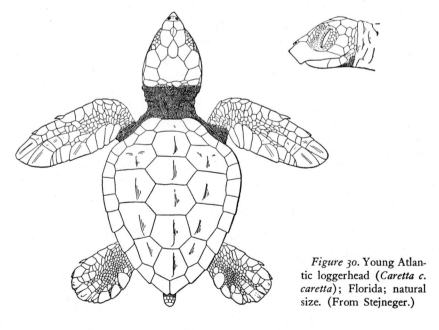

Figure 30. Young Atlantic loggerhead (*Caretta c. caretta*); Florida; natural size. (From Stejneger.)

ing range the loggerheads are evidently less confirmedly tropical than any of the other sea turtles. Like the ridleys, the genus has two pairs of prefrontal scales and the lateral laminae are in more than four pairs. There are most frequently three enlarged inframarginals on the bridge and these normally lack pores. The pterygoid bones are not markedly broadened and have no (or only very weak) lateral processes. The bony surface of the lower jaw is smooth at the hind margin of the symphysis. There are seven to twelve neural bones. The color is brown or reddish.

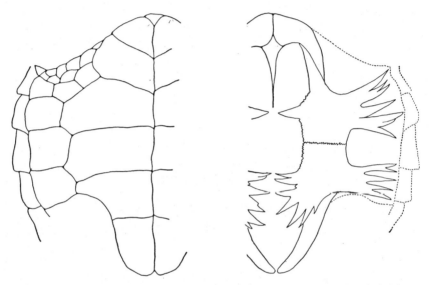

Figure 31. Plastron of *Caretta,* showing laminae (*left*) and much-reduced bony structure (*right*).

KEY TO *CARETTA*

1 Marginal laminae averaging 12 on each side; neural bones usually 7 or 8; neurals rarely interrupted by pleurals; Atlantic and Mediterranean .... *caretta caretta* (p. 382)

1' Marginal laminae usually 13 on each side; neural bones 7–12, the last 1–5 usually interrupted by pleurals; Indian and Pacific oceans .... *caretta gigas* (p. 393)

Atlantic Loggerhead Turtle *Caretta caretta caretta* (Linné)

(Plates 68, 71. Figs. 23, 24, 30, 31, 32.)

Range: Atlantic Ocean and Mediterranean Sea. Along the coasts of America it is known from Río de la Plata, Argentina, northward to Nova

Scotia. On the European coast it has been taken as far north as England and Scotland, but whether these records are of specimens derived from the western or from the eastern Atlantic is open to question, although my guess is that they arrived via the Gulf Stream. In the tropical eastern Atlantic loggerheads occur in the Canary Islands, and in Rió de Oro and the Cameroons coast in West Africa; in South Africa the Atlantic form is apparently replaced by *Caretta caretta gigas* of the Indian Ocean (and possibly of the Pacific).

Distinguishing features: The loggerheads are easily recognized by their reddish-brown color, five or more pairs of lateral laminae, three enlarged, poreless inframarginal laminae, and their elongated shell and two pairs of prefrontal scales. The Atlantic loggerhead is distinguished from the Indo-Pacific stocks by the low average number of marginal laminae (12) and neural bones (7–8), and by the infrequency of the interruption of the latter by intervening pleurals.

Description: The shell is elongate heart-shaped, slightly concave in front and over the shoulder but almost smoothly oval in lateral outline. The size is very large, this turtle probably being next in size to the giant leatherbacks, although there are indications that the green turtle may be a rival in this respect. There are records of individuals weighing between 700 and 900 pounds, and True (1884) said that they attain weights of between 1,500 and 1,600 pounds, but he had not seen any such specimens and I am inclined to think he had accepted the misidentification of someone who had caught a leatherback. As an example of the type of yarn that may have been uncritically accepted to confuse the question of the maximum size of the loggerhead, consider the following newspaper clipping, origin unknown, but datelined, "Portland, Maine, Sept. 20, (1919)."

A sea turtle of the loggerhead species, said to be the largest brought to this port in 50 years, was landed today from the fishing schooner Fannie Belle. When the back of the turtle first came into view it resembled a small U-boat.

The turtle weighed more than 1600 pounds, measured nearly 7 feet in length and had an "extreme beam" of about four feet across the back. Between the tips of the huge fins the turtle measured 12 feet. It was sold to a wholesale fish dealer for $50.

Apparently no zoologist saw this specimen, but some time later the U.S. National Museum received two excellent photographs of a large *Dermochelys* (Plate 82), and in an accompanying letter Mr. G. Flaherty wrote that the turtle had been caught off Sequin, near Portland, Maine, on September 25, 1919. Although the discrepancy in dates is puzzling, I should

Plate 68. Caretta caretta caretta. Upper left: Female. (Courtesy The American Museum of Natural History.) *Lower left:* Young specimen taken off Beaufort, North Carolina. (Photograph by W. P. Hay.) *Upper and lower right:* Female. (Courtesy Zoological Society of Philadelphia.)

be willing to bet that only one big turtle was caught off Maine that week!

But before dismissing the possibility that loggerheads of a half ton and more in weight exist, we should examine some shreds of data that might at least make it easier to maintain an open mind on the subject. In rummaging through Dr. Leonhard Stejneger's notes I came across the following startling memorandum, in Dr. Stejneger's handwriting and dated October 19, 1934:

Caretta caretta

USNM Transp. No. 131/366

received for identification from Prof. C. D. Bunker, University of Kansas, Lawrence, Kansas, who writes (Oct. 4, 1934) that

"The specimen was supposed to have been killed in the Illinois River between Wauhillau and Stilwell, Oklahoma, in 1929 [by] R. Stephenson. Specimen was given to Black by Cecil Fritsch, of Winslow, Arkansas. All efforts to check the exact circumstances of the taking have drawn a blank."

The skull is unusually large. The premaxillaries are wanting so the total length cannot be given, but the greatest width of the skull is 255 mm. [10 inches].

I should say without further delay that I can offer no data to account for the shocking locality record but shall hasten, rather, to call attention to the size of the skull, which is beyond doubt that of a loggerhead, whatever its provenance.

The largest loggerhead measured by True (1884) had a skull 203 mm. (8 inches) wide, and this specimen weighed 850 pounds. A simple (and probably fairly valid) proportion derives for the "Illinois River" specimen a weight of 1,067 pounds. And other comparable or even greater sizes can be calculated on a similar basis. DeSola (1935) mentioned a skull with what he called a "record measurement of 280 mm." (11 inches), but unfortunately failed to say whether the dimension given was length or width. In either case it belonged to an enormous turtle. The largest skull in any collection appears to be one from Australia, probably representing *C. c. gigas,* and this is 285 mm. (11.2 inches) in width. For the monster to which this skull belonged would be well within the weight class of the largest leatherbacks, since our proportion gives for it a weight of 1,192 pounds.

It should be emphasized that none of these calculated weights can stand as valid records. Whatever the upper limits of weight, a loggerhead of more than 300 pounds is a rarity today, and reliable records of larger ones are much to be desired. The dimensions of a good-sized specimen from Virginia follow: carapace length, 910 mm. (35.8 inches); carapace width,

740 mm. (29.1 inches); skull length from snout to condyle, 190 mm. (7.48 inches); greatest width of head, 193 mm. (7.61 inches).

The lateral laminae are five or more in number, the first pair in contact with the precentral; the dorsal laminae do not overlap in mature specimens, and only slightly in some young ones. The marginal laminae are variable in number, there being on each side usually 12 but occasionally as many as 15 or as few as 11. The laminae of the carapace are brown, of varying shades, occasionally slightly tinged with olive and often bordered with diffuse dirty yellow. The yellowish suffusion of the edges is often rather marked on the marginal laminae, and the whole shell may get from it a distinctly yellowish cast. There are three (usually) enlarged infra-marginal laminae on the bridge, and these are not pierced by pores at the seams. The head is extremely large, broad behind, somewhat rounded in front, and covered above with about 20 scales of various sizes, including two pairs of large prefrontals between the eyes. The scales of the head are yellowish chestnut or olive brown in their centers and yellowish toward the margins; the jaws are yellowish horn color. The upper leg and tail surfaces are dusky centrally and yellowish toward their edges; below they, like the plastron, are yellow, sometimes more or less clouded.

Large male loggerheads are but rarely captured and apparently nothing has been published regarding differences between the sexes in the Atlantic race. I have seen a number of males at close hand in the water and can say that the tail is very much longer than that of the female. Also, the shell is either narrower, or perhaps merely more gradually tapering, behind in the male.

Parker (1929) gave 48 mm. (1.89 inches) as the average shell length and 35 mm. (1.38 inches) as the shell width of hatchlings. Hildebrand and Hatsel mentioned 42 mm. (1.65 inches) as the usual shell length. A hatchling from Key West, Florida, with egg tooth and umbilical scar, has the following dimensions: shell length, 41 mm. (1.61 inches); shell width, 32 mm. (1.26 inches); shell height, 11.5 mm. (0.454 inches); head

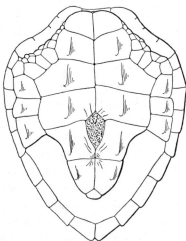

Figure 32. Ventral view of the shell of a juvenile specimen of *Caretta c. caretta.* (From Stejneger.)

width, 14.2 mm. (0.560 inches). Three dorsal keels are evident on the very young, but are usually interrupted on the first and last laterals. There is a fairly well-marked keel down the center of each of the main rows of plastrals. The coloration is uniform light brown above and dirty white beneath, with the centers of the paddle surfaces dusky below.

Habitat: The loggerhead is a confirmed wanderer. Although apparently preferring coastal bays of moderate depth, it enters streams and ascends them until the water freshens, or until they die out in the salt marsh. In the Georgia marshes I have several times come across loggerheads in little creeks no more than a half-dozen times as wide as the turtles' shells. On the other hand, they are often encountered on the high seas, as the following quotation from Catesby shows:

They range the Ocean over, an Instance of which (amongst many that I have known) happened the 20th of April, 1725, in the Latitude of 30 Degrees North, when our Boat was hoisted out, and a Loggerhead Turtle struck as it was sleeping on the surface of the water; this by our reckoning appeared to be Midway between the *Azores* and the *Bahama*-Islands, either of which Places being the nearest Land it could come from. . . .

Murphy (1914) saw large numbers between 400 and 500 miles off the coast of Uruguay during four days in November, 1912 (between Lat. 32° 54′ S. and 37° S.). Moreover, the individuals that have turned up in the British Isles have not been helpless infant waifs, as in the case of British ridleys, but large, powerful adults whose presence there must be attributed, at least in part, to their own volition. Whether loggerheads sometimes make long journeys up the Mississippi River, as the extraordinary report that I have cited above would indicate, is an altogether problematical but very interesting point.

Habits: The Atlantic loggerhead is a hardy and adaptable animal, able to make a living under all sorts of conditions. Very little is known about the wanderings of individual loggerheads, but the range of situations in which the species occurs is astonishingly great and implies an unusual ability to take things as they come. When I was a boy I used to while away whole days during vacations fishing for sheepshead on a tumbledown dock by a little salt creek in McIntosh County, Georgia. The creek was fairly deep but on very low tides no more than 15 or 20 feet across. Much of the time I spent half asleep in the hot sun waiting for bites. It is my impression today that every time I actually dozed off, a huge loggerhead that lived in the neighborhood surreptitiously sneaked up the channel, broke water under my nose with a crash, thrust out his head, gaped his cavernous mouth, and uttered a stertorous sigh that invariably scared me out of my wits. Then he went under and moved off down the bottom of the channel, presumably looking for oysters. This actually happened year after year, and I am sure it was the same turtle because of a patch of barnacles over one eye; anyway, his monstrous size alone was almost enough to identify him. This loggerhead was caught several times, usually by people fishing for drum with half a blue crab as bait. On one occasion a group of idlers hauled him to the shrimp factory scales and passed on the tradition, which I can well believe, that he weighed 500 pounds. They put him back in the creek, and several years later I caught him—on my sheepshead pole with shrimp bait. The tiny hook evidently hurt the lining of his mouth because he slowly followed the feeble tension to the surface, where he eyed me reproachfully for a moment and then made a swipe with a flipper in front of his face, breaking the line, and moved away downstream with great dignity.

In the open sea loggerheads often spend a great deal of time floating on the surface, presumably sleeping, but I never saw them do this in the sounds and estuaries, where they stay in the deeper parts of the channels and come up only for breath or for brief reconnaissance. In stop netting for diamondback terrapins in the small streams that drain the salt marshes of coastal Georgia, I often came in contact with loggerheads under exasperating circumstances. I have spent many nights at this and recall that in certain creeks our net, buoyed across the narrow channel, was frequently disrupted by passing loggerheads, sometimes with appalling damage. I also used to spend considerable time cast netting at night for shrimp or mullet in tidal streams and remember well the wreckage that resulted once when the net covered a loggerhead in the dark.

In a paper I wrote in 1942 I had the misfortune to make a remark concerning the disposition of the loggerhead that seemed to imply that I considered it good-natured. Any number of people have chivied me about this, and not without cause. I meant merely to say that the loggerhead is less outrageously aggressive than the ridley, and I still believe this, although perhaps I should never have said it. It would appear that all other herpetologists who have seen extra-alcoholic loggerheads are unanimous in their agreement that this turtle is inexcusably vicious and ornery. The following article from the New York *Herald* for September 3, 1905, appeals to me as summarizing nicely the opinion of the majority. (Headline: "Big Turtle Victor over Five Men—Loggerhead which figured in cruelty to animals case resists efforts to recapture it.")

South Norwalk, Conn., Saturday.—Five East Norwalk fishermen crawled into port last evening, using the stumps of their oars as paddles and with one of their number unconscious in the bottom of the boat as the result of an attempt to catch the famous 610-pound loggerhead turtle which escaped from Captain Charles E. Ducross, a South Norwalk marketman, and which was the largest and most vicious turtle of this species ever brought into Fulton Market.

This is the turtle whose paddles were pierced by Captain Ducross and the owner arrested and fined for cruelty to animals. The Captain was found guilty and fined $5 and costs at a time when the turtle had escaped from his pen. The mammoth chelonian tore down and crunched the planks on the side of the pen as though they were toothpicks.

Captain Ducross offered a reward of $50 for the recapture of the turtle, and Frank Petty with his two sons, Frank and George, and two other men named Swanson, set out yesterday morning in a rowboat intending to catch the chelonian. They took nets and spears and rope aplenty.

The Petty party found the turtle asleep in the harbor, not far from where it had been seen the afternoon before by the captain of the Oyster Steamer Josephine. They approached him slowly and cautiously. Swanson, who is an expert fisherman and sailor, attempted to drive an eel spear, to which a long and heavy line was attached, through the back of the turtle. The spear broke off short in the tough shell of the turtle and then there commenced a fight which lasted nearly an hour.

The chelonian seemed to have no fear of the men or of the boat. He turned upon them and with his flippers almost overturned their boat. The five men beat him over the head with the oars. These he occasionally got in his mouth and each in turn was crunched and broken off. It was in the thickest of the fight that Swanson was struck by either one of the flippers of the animal or by his beak and a long gash was torn in his arm. In spite of the wound Swanson assisted in the fight until the turtle withdrew and sank out of sight, apparently none the worse for his encounter.

It took the party nearly two hours to paddle their craft, which was nearly full of water, back to this port. They all say they do not intend to try to capture any more loggerheads, reward or no reward, and that Judge Taylor, who fined Captain Ducross for piercing the turtle's flippers, ought to be fined for not instructing the owner to cut off its head for the general welfare.

While it is barely possible that the above account may have referred to a leatherback instead of a loggerhead, the fact that the turtle had figured in a court trial and had passed through Fulton Market would seem to lend considerable authenticity to the identification. In either case it was a good scrap, and I can only conclude that loggerheads show a great deal of individuality in temperament.

Roosevelt recorded a widespread belief of fishermen that a loggerhead menaced by sharks raises its body above the surface and slaps the water with plastron and flippers to frighten the attackers away.

Little is known concerning the growth rates of loggerheads. Parker (1929) kept a hatchling the shell length of which changed from 48 to 630 mm. (1.89 to 24.8 inches) in 4½ years; during this time its weight increased from 20 grams to 37 kilograms. A yearling kept 3½ years by Hildebrand and Hatsel grew in shell length from 136 to 538 mm. (5.36 to 21.2 inches). At the age of 4½ years it weighed 45 pounds and at 6 years, 61 pounds. These writers suggested that turtles may show marked individual variation in rate of growth. It seems probable that growth in the natural environment is more rapid than in captivity.

Breeding: The loggerhead nests on the coast of the southern United States (formerly from Virginia to Florida and the Gulf states) from April to August, somewhat earlier in the southern part of the territory than in the northern, but with the peak coming nearly everywhere in June. A beach open to the sea is preferred, but narrow bay-shore beaches are sometimes used if they are of properly friable sand. It is generally stated that most laying occurs on the flood of spring tides, but I have seen several tracks cut across fifty to a hundred yards of water-packed beach at dead low tide. However, it is probable that most females do emerge during the latter half of flood tide. In both Georgia and Florida the fishermen recommend "the first full moon in June" for turtle egging, but whether this time is propitious because the hunters find the tracks more easily in the moonlight or because the turtle can see where she is going better, or (most probably) because she likes the spring tides that come with full moons, is uncertain.

The nest is usually located between high-tide mark and the dune front, but not always; the highest tides sometimes cover nests, and I have trailed many a prospecting female loggerhead among, or even behind, high dunes. The nest is not always dug the first time the turtle comes up on the beach; she may merely haul out and blink a bit, sometimes digging a few trial holes, and then return to the sea to await some subtle accordance of weather and hormones appreciable only to herself. If the night is right, however, she usually sets about digging her nest without delay. The nest is dug with the hind feet, used alternately, apparently without preliminary excavation of a shallow basin with the fore flippers as in *Chelonia* and *Lepidochelys*. When the cupped hind flipper has withdrawn sand to the greatest depth to which it can conveniently reach (18–26 inches) and to a diameter of about 10 inches, the tail is thrust into the hole, the cloacal "ovipositor" is protruded to a distance of 2 inches or so, and laying begins. The eggs fall singly or doubly, from 6 to 12 times a minute. During the laying process the turtle is usually oblivious of her surroundings and rarely is disturbed by onlookers, however obstreperous. The hole is covered by the hind flippers, which drag in sand from behind, again working alternately, and pausing at intervals to pack the sand on the eggs. When the nest has been filled the turtle packs the site by pounding it with her plastron, and flings sand about with all four flippers in an effort toward concealment. Despite the disturbance of the sand, or perhaps because of it, locating a loggerhead nest is not a simple matter for a human being, and a slim iron prospecting rod is helpful. All kinds of mammals, from skunks to bears, are able to locate the nests without difficulty, and along the Mosquito Coast of Nicaragua the jaguars actually appear to move down to the beaches for the turtle egg season.

In Florida the number of eggs laid at one time appears to average around 120 to 130; Coker (1906b) found the usual number in North Carolina to be from 120 to 150. The eggs are usually between 40 and 43 mm. (1.57 and 1.69 inches) in diameter, sometimes with one diameter slightly shorter than the other. The shell is soft and leathery and there is a dent in the side of new-laid eggs which may disappear as the shell distends under proper incubation conditions. Washing with the salt water of spring tides apparently does not kill the eggs, but desiccation is fatal and rough handling usually kills the embryos, although the eggs may be successfully transplanted if care is used. Each female probably lays at least twice a year, and possibly three times.

The incubation period is very variable, presumably being strongly affected by physical conditions of the environment. Periods of from 31 to 65 days have been noted. When hatching occurs, most of the young turtles emerge from their shells at about the same time and then remain for some hours in the nest, finally climbing out in a "solid column of turtles, 6 to 8 inches in diameter, continually moving upwards through the sand till all are out" (Hooker, letter, 1908). The cues which direct the hatchlings toward the sea have been found experimentally to include a tendency toward downhill locomotion (once the nest has been left behind) and an inclination toward the level sea horizon and away from the broken shore horizon. Once in the water young loggerheads are seldom seen, and their early activities remain a mystery.

Feeding: The loggerhead is mainly carnivorous. Although crabs and conchs of all sorts are most frequently found in stomachs, the loggerhead often makes a meal of fish if it can be found, and also eats clams, oysters, sponges, and jellyfish. There are records of its eating turtle grass (*Zostera* or *Thalassia*). Besides the fish baits that I mentioned above as having seen them take (crab and dead shrimp), I caught a small individual once on a live shrimp, hooked through the end of the abdomen, in a bayou of nearly fresh water in Mississippi. Dr. Thomas Barbour told me of watching a loggerhead close its eyes while eating a Portuguese man-of-war, to avoid contact with the vicious stinging cells, and this habit has been mentioned by others.

Economic importance: It is customary to deprecate the economic importance of this turtle. This is a mistake, however; the dietetic level of many a little rural, salt-water community in the Antillean and Caribbean areas would fall pitifully if the loggerhead and its eggs were absent. The meat of a big loggerhead does not make choice steaks, being tough and stringy, but I have yet to note the "rank" taste attributed to it by many. The "conchs" grind the meat and make of it what they call turtle balls (of late known to Key West tourists by the revolting euphuism "turtleburger"), and these are delicious if properly prepared; moreover, good soup is made of the shell, flippers, and meat. But the loggerhead is properly famous for its eggs, which raw or cooked are pleasant and worth-while eating. They are gathered annually by the thousands, and since the days of Audubon, and probably long before, people have been excitedly predicting the early extermination of the loggerhead by egg hunters. Unquestionably, the population has been reduced (Brice recorded 100 as taken in Biscayne Bay in 1894, as contrasted with 300 to 400 per year for previous years), but every season the big females lumber up the bathing beaches, sometimes in the glare of street lights, and if the weather happens to be too bad for promenading couples the tracks may be lost in the rain and a setting will be left to hatch. Even my own resistant conscience bothered me when I used to cruise the beaches of Long Island (now Sea Island), Georgia, in a model A Ford during egg season, but I have recently heard that the turtles still come up and lay there.

Loggerhead eggs make (or used to make) the best cakes in Savannah and Charleston, where people are awakened on early summer mornings by Negro hucksters passing with broad baskets piled high and balanced on their heads, and crying their wares: "Tettle-egg! Yeh crab! Eah feesh! Oistabiah! Sweempann prawn! Yeh tettle-egg!" In Cuba the eggs are dried, smoked without being removed from the oviducts, and sold by the length like sausage; in coastal Columbia they go into the popular turtle-egg *dulce,* a curious form of candy. Except for squeamish people who refuse to like them, they are everywhere sought with a reverent sort of fanaticism by man and beast. I suppose the species will eventually yield to this kind of pressure, and this impresses me as a real tragedy. There are laws protecting turtle nests in some places, but they are not adequately enforced where they are needed most.

Besides its use as food, the loggerhead has provided certain primitive people with an oxidizing oil, used as varnish and otherwise; and the shell, though thin and lackluster, passes in places as a poor substitute for hawksbill tortoise shell.

On the Honduranean coast little Black Carib boys are sometimes seen paddling about in boats made of single big loggerhead shells.

Pacific Loggerhead Turtle *Caretta caretta gigas* Deraniyagala

Range: Widely distributed in the Indian and Pacific oceans. On the American coast it is known from southern California to Chile. Geographic intergradation with *caretta* may occur along the coast of South Africa, where the loggerheads of the Indian Ocean and those of the Southeast Atlantic meet. Variation within the two stocks also brings about an overlapping of the distinguishing characters, although a majority of specimens are recognizable as belonging to one race or the other.

Distinguishing features: This race is distinguished from *C. c. caretta* by a higher number of neural bones (7-12), the last 1-5 of which are usually interrupted by pleural bones (seldom so interrupted in *caretta*), and a higher number of marginal laminae (usually 13 on a side in *gigas* and 12 in *caretta*).

Mr. Charles Bogert of the American Museum of Natural History has recently lent me a series of ten hatchling Pacific loggerheads from the Solomon Islands which appear to indicate that some of the characters of the form may be unusually stable. Dimensions and laminal counts of the ten specimens are given in Table 10. The corresponding figure for the only available hatchling of *C. c. caretta* is included in parentheses at the foot of each column, for comparison. The shell of the Atlantic hatchling is more smoothly and narrowly oval than are those of the Solomon Island specimens, with almost none of the indentation over the fore and hind limbs that is conspicuous in the latter. The first and last laterals are traversed by the lateral keel in the Pacific series but not in the Key West specimen.

Table 10. Dimensions (in millimeters) and laminal counts of ten hatchling Pacific loggerheads and one Atlantic loggerhead (in parentheses).

Shell length	Shell width	Head width	Number of marginals		Number of inframarginals		Number of laterals	
			Left	Right	Left	Right	Left	Right
43	38	15.8	12	13	3	3	5	5
44	36.5	16	13	13	4	5	5	5
46	38.3	16	13	13	3	3	5	5
46	39	15.8	13	13	3	3	5	5
45	38	16	13	13	3	3	5	5
44.5	38	15	13	13	4	4	5	5
43	38	16	13	13	3	3	5	5
46	38.2	16	13	13	3	3	5	5
45	38	16	13	13	3	3	5	5
46	38	16	13	13	4	4	5	5¼
(41.5)	(32)	(14)	(12)	(12)	(3)	(3)	(5)	(5)

It may be that the Pacific loggerhead is a larger animal than that of the Atlantic. At any rate, it holds the world's record for head width, a specimen from Australia, in the Bell Collection at Cambridge University, being 285 mm. (11.23 inches) wide (Deraniyagala, 1933).

The sexes of *gigas* evidently differ in the same ways as in *caretta,* the shell of the male being somewhat more gradually tapering posteriorly and its tail being longer, according to Deraniyagala (1939) extending past the tips of the back-stretched hind flippers, while in the female it hardly reaches the hind rim of the shell.

I believe that Professor Philippi of the University of Chile described the loggerhead of the eastern Pacific long ago (1899) as *Thalassochelys tarapacana,* but he mentioned no feature that would distinguish it from either typical *caretta* or Deraniyagala's more recently described race in the Indian Ocean. Since Philippi had only a single specimen, and that quite evidently aberrant in some respects, and since Deraniyagala's diagnosis appears to hold up throughout the whole Indo-Pacific population, it would seem best to call all Pacific loggerheads *gigas* until more specimens from the American coast have accumulated and divergence of the East Pacific stock can be demonstrated.

Shaw (1947), who published the first record of the Pacific loggerhead from waters of the United States, a specimen from 35 miles southwest of San Clemente Island, off Los Angeles County, California, suggested that another specimen taken earlier from the coast of northwestern Baja California may have been a hybrid (*Caretta* × *Lepidochelys*) because it had four enlarged inframarginals, one of which was perforated by a pore. I have seen what looked like a pore or two at the inframarginal seams of several Carettas, and the aberrant occurrence of a fourth enlarged lamina on the bridge is not rare; moreover, *Lepidochelys* may rarely lack both pores and the fourth inframarginal. But these two genera are very divergent in their habits and are probably not nearly so closely related as their external similarities imply; an authentic case of hybridization between them would be most surprising, to me at least.

It is of the greatest interest to note Shaw's mention of the use, by the Mayo Indians of Sinaloa, of the old Carib name *caguama* for the loggerhead. In pre-Columbian times this term was used by the sea Caribs to indicate not only the loggerhead turtles, the eggs of which they preferred above all others, but also the separated dorsal laminae, which they employed as a substitute for hawksbill tortoise shell. Descendants of the Siboney Indians of Cuba knew the loggerhead as *caguama* until comparatively recent times, and Dunn (1945) recorded the expression from coastal Columbia. I found it still in wide use among the Black Caribs of Central America. In latinized form it was proposed as one of the early specific names of the loggerhead and once as a generic name. There is no question but that it is etymologically a much more appropriate designation for the loggerheads than *caretta,* derived from French *caret* and Spanish *carey,* and since time immemorial applied to the hawksbill and to its shell, and only by error to the loggerhead.

Habitat: This turtle, like the Atlantic form, roams widely. It is found in bays, estuaries, and lagoons, but apparently shows no marked predilection for them, being at home anywhere on the continental shelf and often venturing into the open sea.

Habits: Due to the almost universal confusion of this turtle with the Pacific ridley until quite recently, there are practically no reliable published data on the habits or economic importance of *C. c. gigas.* Its life history in American waters is wholly unknown. I have collected considerable information from Central American fishermen, but what applies to *Caretta* and what to *Lepidochelys* it is impossible to say with certainty, although the latter is much the more abundant in Honduras and in Mexico. Some sort of sea turtle is rumored to lay as far south as the coast of Chile, and if the rumor is correct I believe that it most probably refers to *Caretta.*

Genus *LEPIDOCHELYS*

The Ridleys

Although in general distributed through the warmer parts of the major seas, this group shows certain peculiarities in range that are discussed under the specific headings. In this genus the laterals range from five to nine, and there are four enlarged inframarginals, some of which, at least, usually are perforated by a pore near the hind margin. The precentral and lateral laminae are usually in contact. The pterygoid bones of the skull are markedly broadened anteriorly, with strong ectopterygoid processes. In the lower jaw there is a strong median elevation at the posterior edge of the bony alveolar surface. The neural bones are eleven to fifteen in number. The color is gray or greenish.

KEY TO *LEPIDOCHELYS*

1 Color olive; laterals usually in more than five pairs; bony alveolar surface of upper jaw with or without a gentle elevation extending parallel to the cutting edge, but lacking a conspicuous ridge; Indian and Pacific oceans .... *olivacea* (p. 403)
1′ Color gray; laterals usually in five pairs; bony alveolar surface of upper jaw with a conspicuous ridge parallel to cutting edge; Gulf of Mexico and the Atlantic coast from Florida occasionally to Massachusetts, England, Ireland, and the Azores .... *kempii* (p. 396)

Atlantic Ridley *Lepidochelys kempii* (Garman)
(Plates 69, 70, 71. Figs. 23, 24.)

Range: Of all the sea turtles this species has the most extraordinary distribution. It is known from the Gulf of Mexico, mostly in the northern portion, northward along the Atlantic coast to Massachusetts, whence it is carried with some frequency to England, Ireland, the Scilly Isles, and the Azores. In 1944 Dodge listed seven records of ridleys from the coast of Massachusetts. Dunn (1918b) said that a specimen, the skull of which was then in the collection of Smith College, was brought in alive to Port Antonio, Jamaica, in 1894. This is the only record for the Caribbean. Lewis (1940) expressed the belief that it does not now occur in Jamaica. Major Chapman Grant, also, who knows *kempii* well and who is quite

Plate 69. Lepidochelys kempii. Upper: Young specimen taken off Beaufort, North Carolina. (Photograph by W. P. Hay.) *Lower:* Female. (Courtesy Zoological Society of Philadelphia.)

familiar with the region about Jamaica, is convinced that there are none there. However, I should like to know more about the allegedly hybrid turtle that does occur there and that goes by the singular name of Mc-Queggie, apparently in all the British islands of the vicinity. The ridley is so universally spoken of as a "mulatto" or "bastard" that it would be most interesting to see how the McQueggie differs from the ridley. C. J. Maynard had the following to say about the Little Cayman McQueggie (field notes quoted by Garman [1888]):

I was told of a hybrid between the hawkbill and the loggerhead, on which the shell was often good, but not always, and the head resembled that of the loggerhead. I asked the fishermen why the shell was not always good and was informed that when the offspring "took after the mother" (always the logger-head) the shell was poor, but that when they "took after the father," the hawk-bill, the shell could be used.

From our present knowledge of its distribution the ridley would appear to be really at home only in the Gulf of Mexico, whence some individuals are each year lured, or swept, northward by the reassuringly warm current of the Gulf Stream. Existence in the teeming Gulf Stream for indefinite periods of time would probably be perfectly feasible for a ridley, and the oceanographers have shown what might happen to anything that is sur-rendered to that current. In 1921 and 1922 Dr. Henry Bigelow released a large number of drift bottles in the coastal waters of New England. Of these, seven were carried to the European side of the Atlantic and were picked up in the following localities: Azores, 1; Canary Islands, 1; Ireland, 3; English Channel, 1; Scotland, 1; France, 1. It is perhaps significant that all the New England and European records of the ridley were based on juvenile specimens, which, once in the sweep of the Gulf Stream might have little more control over their ultimate destination than a drift bottle. One of the more extraordinary anomalies in the distribution of the ridley is its apparent absence just across the Gulf Stream from Florida in the Bahamas, where no one seems to know it.

Distinguishing features: The following combination of characters will serve to distinguish the ridley from any other sea turtle in its range: the predominantly (and very constantly) gray color of shell and dorsal skin surfaces; the four pairs of enlarged inframarginals, each usually pierced by a pore near the posterior seam; the large head and convex outer surfaces of the jaws; the two pairs of prefrontals; the short, chunky, broadly heart-shaped or nearly circular shell (as seen from above); the five pairs of

Plate 70. Lepidochelys kempii. Young specimen from coast of Massachusetts. (Photographs by George Nelson.)

laterals (very constant); the single large scale on the mandible; and any number of osteological characters that are properly generic in nature. Grant (1946) recommended a simple rap with the knuckles on the shell of a suspected ridley to clinch the determination, the true ridley being said to sound hollow, like a dead log, when so tapped, while the other species sound like "living things." Certainly no diagnosis could be simpler, if it really holds up. From the Pacific ridley, *kempii* is distinguished by its predominantly gray color (apparently a juvenile holdover), by the consistent presence of five (instead of 6–8) laterals on each side of the shell, and by certain skull characters that appear to include the presence in *kempii* of a ridge on the bony alveolar surface parallel to the cutting edge, this being merely a gentle elevation in *olivacea*. Garman (1880) laid great stress on the "humps" or shoulders on the anterior part of the ridley's shell as a character for rapid recognition, but these develop only in old individuals, and fully mature specimens are only rarely seen, for some incomprehensible reason.

This may be the smallest of the Atlantic sea turtles, and is probably even smaller than its close relative, the Pacific ridley. It reaches sexual maturity at a shell length of about 600 mm. (slightly less than two feet). The record shell length is 703 mm. (27.6 inches). It seems probable that both the Atlantic ridley and that of the eastern Pacific are on the average somewhat smaller than the West Pacific population.

Plate 71. Young specimens of *Caretta c. caretta* (West Indies; *left*) and *Lepidochelys kempii* (Tortugas, Florida; *right*), showing marked difference in shell outline. (Photograph by Frank Carpenter.)

The only known hatchling was collected by Prince Maximilian zu Wied heaven knows how many years ago and is now in the collection of the American Museum of Natural History, where the following carapace measurements were taken by Schmidt and Dunn in 1917: length, 47 mm.

(1.85 inches); width, 41 mm. (1.61 inches). Shell measurements of another small specimen are as follows: length, 185 mm. (7.29 inches); width, 172 mm. (6.77 inches); greatest height, 90 mm. (3.54 inches).

In the second specimen the head, shell, and legs are dark gray above, the laminae and flipper edges with some light margining; some of the scales on the sides of the head are light with dark centers, others are wholly yellowish; toward the orbit the horny beak is dark, but elsewhere it is light. The ventral surface is generally yellow with some dark color on the terminal parts of the flippers and on the posterior lower marginals. Dorsally, keels are strong on all the centrals, especially high and sharp on the second and third, and fairly well marked on the second, third, and fourth laterals. Below, the keels are strong on all the plastrals except the gulars, where they are lacking. The hind margin of the shell is deeply serrate. Pores are conspicuous in all four inframarginals on each side.

Habitat: The optimum habitat of the Atlantic ridley appears to be represented by the section of the Bay of Florida around the little island known as Sand Key. To quote my account of 1942:

Here, alone, it is said to be more abundant than the loggerhead. [Stewart] Springer and I accompanied shark fishermen on a bait-catching trip to Sand Key in June, 1941, and while the results—one ridley, one loggerhead and three green turtles—were inconclusive, we had nevertheless gone in search of ridleys and had seen one. While it is difficult to understand what attraction not afforded by a hundred others this small island might hold, the word of the people who know the animal best cannot be disregarded. It seems to prefer shallow water, and is intimately associated with the subtropical shoreline of red mangrove. Away from the tepid, marly waters of Florida Bay it is most frequently seen on the Gulf Coast of the Florida Peninsula, being apparently less abundant on the Atlantic side. In 1906 Coker described it as common about Beaufort, North Carolina, but since this area is far away from the breeding range these North Carolina individuals must be regarded as expatriates, probably destined never to reproduce their kind in such cold water, but otherwise leading a more or less normal existence in a strange habitat. Specimens with full stomachs have been taken as far north as New England.

Habits: With respect to the behavior of the ridley when caught, I commented as follows in a former paper (1942):

All who are acquainted with *kempii* agree that temperamentally it is unstable and irascible. While captive loggerheads and green turtles may be handled with comparative (the latter with complete) impunity, the ridley exhibits almost hysterical violence and obstinacy when caught. It is frequently pointed out by the people on the Keys that a ridley will die, apparently of rage and frustration, if placed on its carapace out of water. A large, uninjured and apparently healthy male specimen which we left in a shaded spot on the deck of a boat at Lower Matacumbe died within five hours. Throughout this period it thrashed its flippers frantically, snapped at every object within reach, and never for a moment relaxed frenzied efforts to right itself. A small individual observed in a tank at the turtle dock at Key West behaved in a similarly inconsolable fashion.

My selection of the loggerhead to contrast with the ridley in the above comment was unfortunate (see section on *Caretta*), since even they are by no means always placid. There is, however, a desperate, hysterical character to the reactions of a captive ridley that I never saw in the loggerhead.

Breeding: Mr. Richard M. Kemp, who sent Garman the original specimens of the Atlantic ridley, and whose name the form bears, said in a letter, "We know that they come out on the beach to lay [in the Florida Keys] in the months of December, January and February, but cannot tell how often or how many eggs." Strange to relate, next to nothing can be added to this feeble information, either by way of corroboration or of extension. While stationed at Saint Marks, on the Gulf coast of Florida, a locality where ridleys abound, my friend J. C. Dickinson made extensive inquiry among local fishermen concerning the breeding range of this turtle but found no one with any opinion on the subject. At Cedar Key, also, the ridley is common and is frequently caught, and yet no one knows where it nests. DeSola and Abrams dismissed the subject by saying that nesting occurs in the same places and at the same times as in the loggerhead; this implication that Georgia beaches are visited may be correct, but was not supported by definite evidence. DeSola and Abrams also mentioned finding eggs in two-foot-long Georgia specimens, but their statement that these individuals weighed only "about eight pounds" is confusing, to say the least. It seems likely that breeding occurs all along the Florida coast, where suitable beaches exist, and that the lack of data means merely that the ridley is confused with the loggerhead by most people who find turtles nesting. On the mid-peninsular Gulf coast, where the form is well known to all the fishermen, there is a dearth of good beaches.

This is a first-class problem, rivaling in interest the one posed by the amazing apparent scarcity of ridleys anywhere south of the Florida Keys.

Feeding: The only food that has been found in stomachs examined has been crabs, in one case the spotted lady crab (*Ovalipes*) and in the other the Dolly Varden crab (*Hepatus*).

Economic importance: Opinion regarding the edibility of the ridley is varied. Most published statements rank it as inferior to all the other turtles except the leatherback, but these comparisons are based mostly on hearsay. The fishermen and turtle hunters on the Gulf coast of Florida regard it as next to the green turtle in flavor. I have eaten several and would rate the young "chicken" ridley along with the young of any of the others, although the adult seems to me somewhat inferior to the green turtle both for soup and for steaks.

Pacific Ridley *Lepidochelys olivacea* (Eschscholtz)

(Plates 1, 2, 72. Fig. 33.)

Range: Warmer parts of the Indian and Pacific oceans. It ranges northward in the western Pacific to southern Japan, while in the eastern Pacific it is known from Baja California to Chile. I can find no records of *olivacea* from the western Indian Ocean, although it certainly should occur there.

Plate 72. Lepidochelys olivacea. Female on nesting beach at Isla de Ratones, Bay of Fonseca, Honduras. (Flashlight picture by Margaret Hogaboom.)

There is a record of some sort of a ridley from "South Africa," and Deraniyagala and others record it from the coast of West Africa, but whether or not they based their identifications on comparison with the characters of *kempii* is not known. Thus, since the southernmost locality in the eastern Atlantic from which *kempii* has been definitely recorded is the Azores, the geographic continuity of the ranges of *kempii* and *olivacea* has by no means been demonstrated. While the two are certainly very closely related, and it is quite possible that their distinguishing characters may overlap completely or even intergrade geographically, it seems best for the present to retain specific rank for these two very widely separated populations. I have been unable to corroborate the opinion of Hay (1908b) and of Philippi (1899) that the ridley of the East Pacific is different from that of the West Pacific. The former of these writers described a ridley from the Pacific coast of Mexico as *Caretta remivaga* and the latter gave one from the Chilean coast the name *Thalassochelys contraversa;* it appears likely that one of these names will someday come into use.

Distinguishing features: This turtle is easily recognized by its broad, rather flat-topped shell, large head with two pairs of prefrontals, four pairs of enlarged inframarginals, each perforated by a pore, its high lateral count (usually 6 or more on each side), and the uniform olive color of the shell of the adult.

Description: The shell is short and wide, often nearly circular in lateral outline, and fairly high, but flattened or even dished in along the centrals. The size is moderate. The largest East Pacific female that I have seen had a shell 690 mm. (27.2 inches) long and 610 mm. (24.0 inches) wide, and that of the smallest sexually mature female was 645 mm. (25.4 inches) long by 580 mm. (22.8 inches) wide; each of these individuals had seven laterals on the left side and nine on the right side. The only male of the eastern population that I have seen at close hand had a shell 620 mm. (24.4 inches) long and was obviously an old individual. Data on four adult females of medium size from the Pacific coast of Honduras are given in Table 11. The largest male and female measured by Deraniyagala in the Indian Ocean had the following shell dimensions:

Sex	Length	Width
Male	730 mm.	620 mm.
	(28.7 inches)	(24.4 inches)
Female	790 mm.	680 mm.
	(31.1 inches)	(26.8 inches)

The carapace is smoothly domed, the highest point anterior to the middle, and very broadly heart-shaped as seen from above. The color is uniform olive with none of the mottling or variegation seen in the other species. The laterals range in number from 6 to 8, occasionally being 5 or 9; they are often higher on one side than on the other; in East Pacific specimens (at least) the higher number is usually found on the left side. The marginals

Table 11. Data on four adult females of the Pacific ridley
(measurements in millimeters).

Shell length	Shell width	Shell height	Greatest head width	Number of laterals	
				Right	Left
685	575	246	110	6	6
648	615	264	125	6	6
687	615	267	130	6	6
665	628	274	133	6	6

are from 12 to 14. The bridge is covered by four enlarged inframarginal laminae, each of which bears a perforation near the posterior seam; a similar pore may occur on the axial lamina. The plastron is light greenish yellow or greenish white. The head is wide and with markedly concave sides, especially the outer surfaces of the horny upper beak. The alveolar surface of the upper jaw (bone) is smooth. There is one enlarged scale on the mandible. The legs and neck are olive above, lighter below.

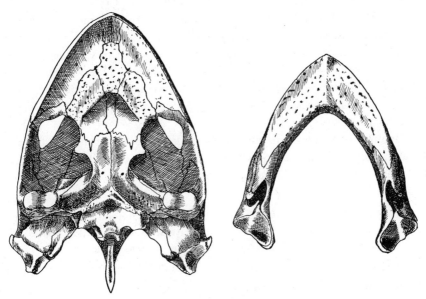

Figure 33. Ventral view of the skull and dorsal view of the lower jaw of *Lepidochelys olivacea.* Note the expanded ectopterygoid processes and the close similarity to the skull of *L. kempii* (Figure 23).

The tail of the mature male is much longer than that of the female, extending at least as far back as the tips of the back-stretched flippers in the former and not passing the hind shell rim in the latter; the single claw on each foot is also much stronger and more curved. The plastron of the male is more concave, and the lateral profile shows a more gradual and a straighter slope from the anterior highest point back and down to the hind shell margin. Males may easily be recognized as such in the water from some distance away. Deraniyagala says the male is the more heavily pigmented.

The smallest and largest of 17 Indian Ocean hatchlings measured by Deraniyagala had the following shell dimensions:

Length	Width
40 mm.	31 mm.
(1.57 inches)	(1.22 inches)
46 mm.	36 mm.
(1.81 inches)	(1.42 inches)

Two specimens from San Blas, Sinaloa, Mexico, with yolk scar and egg tooth are nearly uniform grayish black in color, except for a lighter shade on the ventral keels, which are strong and sharp from humerals to anals. The dorsal keels are strong on all the centrals (but absent on the precentral) and on all of the six laterals of both sides of both specimens. Both have two yellow toes on each fore and hind flipper. The two specimens have shells 40 mm. (1.57 inches) long and 41 mm. (1.61 inches) long, respectively. The keels on their backs appear stronger than those of West Pacific specimens.

Habitat: I have seen this turtle only in protected and relatively shallow water. Oliver (1946) observed great numbers of ridleys in the open sea off the Mexican coast. Deraniyagala gave the following appraisal of the habitat preference:

Lepidochelys is probably more of a bottom dweller and less given to floating at the surface [than the other sea turtles]. It inhabits the shallow water between the reef and shore and is by no means uncommon in the larger bays and lagoons, where it appears to frequent certain restricted areas.

Habits: Oliver reported what may have been some sort of migratory aggregation of ridleys in deep water off Guerrero, Mexico, on November 28, 1945. The turtles were adults and were scattered over a tremendous area, since they remained in view of the moving ship from 9:30 A.M. until at least midafternoon. They floated at the surface, and birds, probably masked boobies, stood on the shells of about half of them. Oliver's account of collecting some of these ridleys is quoted below:

At 1 P.M., a small motor launch was put in the water in an endeavor to collect some of the turtles. The method used was to approach to within 10 or 15 yards of the turtle, to shoot it in the anterior part of the carapace with a forty-five calibre gun, and then have a swimmer grab the turtle by the carapace to bring it to the boat. The turtles were easily approached, making little or no effort to escape until the boat was nearly upon them. When we approached one turtle that had a bird on its back, the bird flew away when the boat was within 15 yards range. The turtle seemed completely oblivious of the bird's departure, but when the boat closed to six yards, the turtle raised its head from the water, sighted the boat and hurriedly started to swim away. The shooting served to stun the turtle so that the swimmer could reach it before it escaped. Once the swimmer had secured the turtle by the carapace and turned it upside down, it was easy to handle. Even a large turtle seemed relatively helpless when on its back. The turtles made little effort to bite, although in their efforts to escape, the claw of the front flipper inflicts a dangerous wound. One made a two-inch long cut in the wrist of the swimmer, while another tore an inch-thick piece of wood out of the gunwale of the motorboat.

Three turtles were caught in a little over one hour's time and a fourth escaped after being grabbed by the swimmer. The smallest specimen was a male with a carapace length of 24 inches. The other two were females, the larger having a carapace length of 28 inches and an estimated weight of 80 pounds. At the time the turtles were caught the ship's position was 15° 57′ North latitude, 99° 46′ West longitude. The depth of the water, obtained from the U.S. Hydrographic chart, was 2738 fathoms. The air temperature was 88°F.; water temperature 86°F.

Although the mating season is not definitely known, it seems probable that it coincides with the time of nesting, as in other sea turtles, since I saw several adult males loafing in a narrow channel off a Honduranean beach on which females were laying every night. Boatmen told me that these males were not to be found in the channel at other seasons.

Eggs are laid from mid-August through November in the eastern Pacific; from September to December on the Indian coast; from September to January on the coast of Ceylon; and according to Theobald (1868) during March and April on the coast of Burma.

There appears to be no definite information concerning the number of times a female may emerge during a single year to lay, although it is generally thought to be more than once. In Ceylon Deraniyagala found the eggs to number from 90 to 135 per complement and to range in diameter from 38 to 43 mm. (1.50–1.69 inches). The same writer gave the incubation period as 50 and rarely 60 days.

Accompanied by Mrs. Henry Hogaboom and Messrs. Al Chable and

Allen Arnold, my colleagues at the time on the staff of Escuela Agrícola Panamericana, Honduras, my wife and I had the opportunity of observing the nesting behavior of a number of Pacific ridleys on the beach of Isla de Ratones in Honduranean territory in the Gulf of Fonseca. An account of our trips to this island has been published elsewhere (Fauna, 1948), and the following paragraphs are taken from that article. The date was October 11, 1947.

We had gone perhaps five hundred yards when Chable, . . . who was lower on the beach than the rest, stopped before a dark bulk that loomed dimly before him. His headlight went on and revealed at his feet a big ridley blinking in the light as the backwash sucked sand from beneath her plastron. We stayed with this turtle and made detailed notes on her activities.

At 10:04 P.M. the turtle left the water at right angles and headed directly for the grass line. She crawled like a fresh-water turtle, moving alternate fore and hind flippers together. She rested three times during the fifty-two-yard traversal of the beach and stopped at 10:09 P.M. in dry sand among high tide litter sixteen feet from the grass. Without . . . reconnaissance she began throwing sand with fore and hind flippers, changing the orientation of her body slightly until a shallow basin had been excavated around her. Since she dug more effectively with her hind [feet] than with her fore feet, this pit was deeper behind than before, and the long axis of her body inclined downward toward the rear. Eventually the posterior margin of her shell had sunk about five inches below the level of the surrounding sand.

After a short pause, during which the turtle applied her tail to several spots in apparent appraisal of the quality of the sand, she began excavating the nest cavity. The left hind flipper was lifted, brought in beneath the hind margin of the shell, and its edge pushed into the soft sand. It was then curled to enclose perhaps half a teacup of sand which was carried out laterally and dropped with a little flip. As this fell, the other hind flipper kicked sand straight back. The process was then repeated in reverse. This was the digging procedure, and it was continued throughout with little variation.

As the hole deepened it was made wider, and asymmetrically so, the increased dimension being that toward the turtle's head. We peered down into the cavity from time to time and saw that [this asymmetrical growth was due to the fact that] the fore wall received the most active scraping, not only from the curved end and edge of the flipper but also from the strong toenail which projected from the margin several inches back from the end. We determined definitely that this was frequently brought to bear . . . against the front wall of the cavity and that it was very effective, both in removing sand and in breaking or dislodging grass roots and other obstructions. The alternate use of the two flippers with no change in the position of the body produced a hole that was noticeably square in cross section, with four definite, though rounded, corners.

As the flippers had to reach [progressively] farther into the hole, digging proceeded more slowly. [While the nest cavity] deepened the hind part of the turtle's shell sank lower, owing to continued gradual excavation of one side of the upper pit by the off flipper as each load of sand was removed from the lower hole [by the other flipper]. When the nest was as deep as the flippers could reach, digging ceased.

The turtle slapped at her tail and cloaca two or three times with each flipper, knocking away adhering sand. The tail was dropped vertically into the hole. The cloaca, temporarily prolonged for its function as an ovipositor, extended considerably beyond the tip of the tail, its opening being about four inches below the plastral surface. The first egg fell into the nest at 10:25 P.M. Thereafter, eggs came every four to ten seconds, either singly or in groups of two, three, or four, most frequently two or three.

The curious "crying" by nesting females [apparently a device to keep the eyes washed free of sand] has been noted in other species. The present turtle began secreting copious tears shortly after she left the water, and these continued to flow as the nest was dug. By the time she had begun to lay, her eyes were closed and plastered over with tear-soaked sand and the effect was doleful in the extreme. Her behavior as the eggs were deposited heightened the melancholy atmosphere. Each time, just before eggs were laid, her head was elevated at an unnatural angle and her mouth was slightly opened, frequently allowing a loud gasp or sigh to escape. With the contraction which pushed out the eggs, came a rapid lowering of her head until her chin pressed into the sand, after which she lay for a while heaving slightly. It was difficult to believe that she was not suffering acutely, but impossible to explain why.

The last eggs were laid at 10:37 P.M. after a few somewhat increased intervals between extrusions. The turtle immediately began to fill the hole, raking in sand from the surrounding ramp with her hind flippers. There appeared to be some selection with respect to the quality of sand used, [since] the cloacal opening was pressed against each section of the ramp just before it was dragged into the nest, and sometimes the lot was rejected. We poured a small amount of perfectly dry sand over a part of the ramp about to be pushed into the hole. The slightly everted cloaca was touched to this and the flipper immediately shifted to another area. As sand fell into the hole it was packed by the back of first one curled flipper and then the other. When the hole was full, the turtle began to pound the sand with her plastron, at first tilting her shell [to one side and the other] and then, lifting herself with her fore and hind flippers, she dragged in more sand from behind and let herself fall sharply upon it. As the surface hardened, the thumping sound which these falls produced became audible from a distance of thirty feet or more. The packing operation lasted four and a half minutes and was performed with what seemed to us excessive thoroughness.

At 10:45 the turtle stopped pounding and began flipping sand backward with her fore feet, meanwhile rotating her body laterally to bring in sand evenly from all sides. She then crawled across the nest site twice and started back to the water at 10:52.

When she was half way down the beach we turned several lights on her. This seemed to confuse her completely, and she doubled back and made for the grass. We extinguished the lights and she immediately turned [back] toward the sea, entering the water at 10:56 P.M.

Five nests that we measured on this beach ranged in over-all depth from 482 to 550 mm. (18.9–21.6 inches), while the lateral diameter of the egg cavity varied between 178 and 205 mm. (7.0–8.07 inches) and the front-to-back diameter between 250 and 304 mm. (9.85–12.0 inches). The number of eggs ranged from 72 to 132. Measurements of a random sample of 50 eggs from the five nests showed a maximum diameter of 45.4 mm. (1.79 inches) and a minimum diameter of 32.1 mm. (1.26 inches), with the average 37.5 mm. (1.48 inches). One diameter of an egg was usually 1–2 mm. greater than the other.

Feeding: Although perhaps mainly vegetarian, this turtle is known to eat sea urchins and mollusks. Deraniyagala found the young to be hardy in captivity, feeding on bread, fish, and meat and on occasion biting pieces from the flippers of their brood mates.

Economic importance: In Central America and Mexico this turtle's eggs are widely appreciated. The most active egg hunters of the Fonseca area are Salvadorans who paddle their *cayucos* great distances to camp on the beach in thatched *ranchos* for the three months of the egg season. In El Salvador turtle eggs sometimes sell for higher prices than hen's eggs. Although the egg hunters appear never to kill nesting turtles, ridleys caught between seasons are usually butchered and sold. Around Acapulco, where this species is called *caguama*—the old Carib name for the loggerhead—it is often killed on the laying beaches.

In most places the meat of *Lepidochelys* is regarded as better than that of the hawksbill but inferior to that of the green turtle. Günther (1864), however, said that "its flesh though relished by Chinese, is unpalatable to Europeans." The eggs are everywhere acceptable, and Smith (1931) mentioned an annual harvest of 1,500,000 eggs in the Irrawaddy Division of Burma.

The shell is sometimes used in veneering and inlaying; according to Seale (1911), in former days the laminae sold in the Philippines for from two to four pesos per kilogram.

Family TRIONYCHIDAE
The Soft-shelled Turtles

THE relationships of this peculiar family of turtles have been the subject of almost as much discussion and debate as those of *Dermochelys,* and for somewhat similar reasons. The family or, rather, the more inclusive soft-shell group—superfamily or suborder, as one may choose— has existed in much its present shape since the middle Cretaceous, and probably split off from the main turtle line even earlier. Its more striking features involve a secondary reduction of the shell as an adaptation to life in the water, and more specifically to burrowing into soft bottom. Whereas the more highly locomotor aquatic turtles have lightened the shell by leaving spaces between the ends of the pleurals and the peripherals, the soft-shells have markedly reduced the number of peripherals or even dispensed with them entirely. The shell also lacks laminae, being covered instead by an undivided leathery skin. The plastron is incompletely ossified, with a broad median space covered only by thin cartilage and skin and a flexible union with the carapace. The shell is low and usually nearly circular in lateral outline. The legs are broadly webbed and three-clawed (hence the name, Trionychidae) and the nostrils are placed at the end of a long, tubular extension of the snout. The neck is withdrawn into the shell as in the Cryptodiroidea, where it bends in a vertical series of curves.

Although for a time the contrary was believed, it now seems evident that the characters of the soft-shells, instead of reflecting a primitive con-

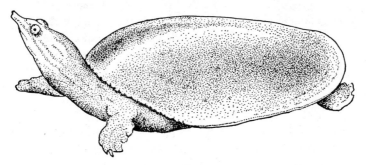

Figure 34. A soft-shell turtle.

411

dition, are those of a highly specialized animal. The group is at present known from the warmer parts of Asia, Africa, and North America. It is not represented in Australia and apparently has never existed in South America. Although now extinct in Europe, it lived there in great abundance during Tertiary time. There is one genus in the United States.

Genus *AMYDA*

To this genus all the living American soft-shelled turtles belong. It includes, besides, numerous African and Asiatic species and a great number of fossil forms, some of which probably date back as far as the Eocene.

Soft-shells are able to remain submerged for long periods of time—several hours, if not indefinitely—owing to their exceptional capacity for pharyngeal (and probably for anal) respiration. By means of pumping movements of the hyoid mechanism, fresh water is constantly supplied to the pharyngeal chamber, where oxygen is taken up by the highly vascular folds and papillae.

KEY TO *AMYDA*

1　　Nostrils rounded, no longitudinal ridge projecting from the septum; carapace without tubercles on its anterior leathery surface; Mississippi Valley from South Dakota, western Wisconsin, northern Illinois, and extreme western Pennsylvania southward to central Mississippi, Louisiana, and south-central Texas ....
....*mutica* (p. 436)

1'　　Nostrils crescentic, with a longitudinal ridge projecting from the septum; carapace usually with tubercles, at least on the anterior leathery surface 2

2(1')　Bones of carapace coarsely granular above, with more or less distinct longitudinal ridges which run together; young with the leathery surface of the carapace with longitudinal rows of tubercles, with a light margin but no marginal dark lines, and with the dorsal surface covered with large dark spots separated by narrow light intervening areas; southern Georgia southward through peninsular Florida ....*ferox ferox* (p. 413)

2'　　Bones of carapace finely granular above, without longitudinal ridges; young usually with one or more dark lines around edge of carapace, or with dark ocelli or solid light spots but with no such pattern as that described above 3

3(2') Anterior edge of carapace with well-developed, more or less pointed and triangular tubercles; anterior and posterior flaps of carapace covered with distinct tubercles 4

3'   Anterior edge of carapace with poorly developed, short, and bluntly rounded tubercles; tubercles on anterior and posterior flaps of carapace very small except in very old specimens; Texas and southern Oklahoma westward to western Arizona and northeastern Baja California; in Mexico southward in Tamaulipas to the Gulf ....*f. emoryi* (p. 423)

4(3) Carapace with a single more or less continuous line around its margin 6

4'   Carapace with at least two more or less continuous dark lines around the rear and lateral margins 5

5(4') Yellow line from snout through eye to neck conspicuous, dark-bordered above and below; other light head markings usually lacking except for a light labial stripe; southern North Carolina southward to east-central Georgia ....*f. agassizii* (p. 420)

5'   Yellow line from snout to neck usually inconspicuous, the part on the neck bordered only above with black; a conspicuous sub-ocular stripe often present; Florida panhandle and southwestern Georgia westward along the Gulf to Louisiana .....*f. aspera* (p. 431)

6(4) Dark spots or ocelli of carapace almost or quite the same size in the center of the carapace as near its border; Nebraska, Kansas, and Oklahoma eastward toward the Mississippi River in southern Minnesota, Iowa, Missouri, and Arkansas and northwestward to Wyoming and Montana ....*f. hartwegi* (p. 433)

6'   Dark ocelli of the carapace distinctly larger in the central area than near the borders; northeastern part of the Mississippi Basin from Wisconsin eastward across the southern half of Michigan and Ontario to the St. Lawrence drainage and southward through the eastern edges of West Virginia, Kentucky, and Tennessee ....*f. spinifera* (p. 426)

Southern Soft-shelled Turtle *Amyda ferox ferox* (Schneider)

(Plate 73. Map 22.)

Range: Florida, westward to the Apalachicola River and northward into southern Georgia, where it intergrades with *agassizii* in the Altamaha drainage. Although the type locality is usually stated to be the Savannah River, Georgia, Dr. Coleman Goin has shown me convincing evidence that this restriction may be unjustifiable, and Neill (1951) expressed a similar opinion.

Map 22. Distribution of the subspecies of *Amyda ferox.*

Distinguishing features: Fresh specimens of this soft-shell are usually identifiable by the curiously sculptured and longitudinally ridged or wrinkled carapace and by the striking coloration of the young, as described in some detail below.

Description: This is the largest of the North American soft-shelled turtles, a shell length of close to 450 mm. (17.6 inches) or even slightly more perhaps representing the upper limit for the species. Agassiz (1857: 1: 401) said the largest he had ever seen was 18 inches long. Stejneger (1944) gave the length of the largest specimen in the United States National Museum as "17 inches [432 mm.] along the curvature," with a skull 68 mm. (2.28 inches) wide; another specimen, represented by a plaster cast, measured 435 mm. in carapace length (460 mm. along the curvature) and had a skull 67 mm. wide. But the same author pointed out that

larger specimens may exist or have been living in Florida not long ago. The National Museum has a series of sixteen weathered skulls picked up by Dr. A. E. Mearns near Kissimmee about the beginning of this century, fourteen

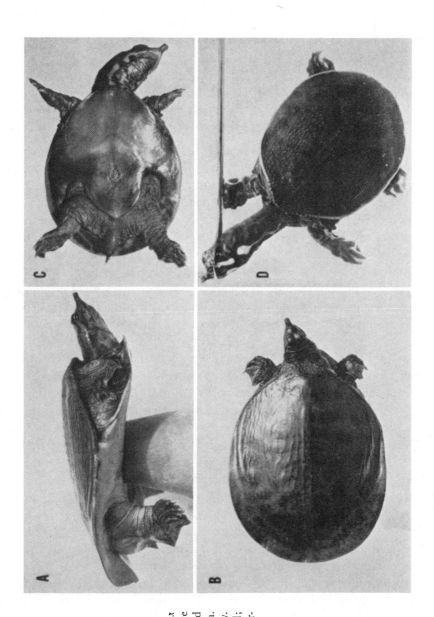

Plate 73. Amyda ferox ferox. A and B: Mature specimen of undetermined sex; Everglades, Florida. C: Young; same locality. D: Young individual; Okefinokee Swamp, Georgia.

415

of them larger than that of the 18-inch specimen mentioned above. The basicranial length of this [latter] one is 92 mm. (3.62 inches) [the basicranial length as used by Dr. Stejneger was the distance from the posterior edge of the occipital condyle to the tip of the snout]; the corresponding dimension of the fourteen Kissimmee skulls ranges from 95 to 114 mm. (average 103).

These skulls varied in width between 65 and 80 mm., with an average of 73 mm. What may represent the ultimate in size records for the southern soft-shell is the 90-mm. (3.54-inch) width of a skull measured by Wright and Funkhouser in the Okefinokee Swamp. These writers gave the dimensions of a large specimen as follows: length, 381 mm. (15.0 inches); width of carapace, 279 mm. (11.0 inches); length of plastron, 292 mm. (11.5 inches); height of shell, 127 mm. (5.00 inches); width of head, 63.5 mm. (2.50 inches).

The carapace of the adult is somewhat elongate, with blunt tubercles along the anterior edge and clustered on the fore flap; its surface is strikingly sculptured with a complex dispersal of indentations and elevations, some of which often tend to arrange themselves in linear fashion and to form on the shell a series of longitudinal ridges. The ground color of the larger adults is grayish to brown, with or without darker blotches. The plastron, and under surface generally, are uniform white. Head and limbs are brownish, sometimes marked with some vestiges of the juvenile pattern but often plain in older specimens.

The females of this species are larger than the males and have shorter tails. Besides these differences, which are common in the genus, some, at least, of the old males develop the most astonishingly expanded crushing surfaces in the upper jaw. These may broaden to the extent that they almost meet in the median line and may, as well, push out the outer maxillary surface until it becomes sharply convex instead of straight or even concave as in females. It has been suggested that this extraordinary condition may indicate a preference among old males for a certain kind of jaw-developing diet. It is interesting to note that an almost identical condition exists in *Graptemys barbouri* and in the Gulf coast diamondback terrapins, but here it is found only in females!

The coloration of young southern soft-shelled turtles is easily the most distinctive in the genus. The carapace is yellowish olive with dusky spots and a narrow, well-defined, yellowish outer edge. These spots are so large that the narrow light lines separating them, in reality the background, appear to be a reticulated light pattern. The under side of carapace and plastron is slate gray, with lighter spots on the former and on the anterior edge of the latter; the light outer border is present also on the under side of the carapace. The upper surface of the head, neck, and legs is olive mottled with lighter color, and a Y-shaped figure extends from the anterior edge of each orbit to the middle of the proboscis. On the side of the head, on the olive ground color there is a well-defined band of yellowish from the posterior corner of the eye to the base of the lower jaw; another band, more orange in color, originates behind the former and descends downward beyond the middle of the neck; a third band curves around the mouth to extend backward on the neck; and the fourth band is usually discernible between the last two. There is great individual variation as regards retention of the juvenile coloration. Specimens of 100 mm. in carapace length usually show it, and occasionally it is present to a greater or lesser degree in individuals as long as 300 mm. Four very young specimens from Polk County, Florida, had the following dimensions (in millimeters):

Length of soft carapace	38.0	40.5	40.0	40.9
Width of soft carapace	33.5	35.5	34.0	37.0
Height of shell	12.5	12.0	12.5	14.0

These measurements indicate that the lateral outline of the hatchling may be much less circular than in other species.

Habitat: This is possibly the least confirmedly fluviatile of all the North American soft-shell turtles, perhaps due chiefly to the fact that the lowland streams characteristic of its range are mostly slow and silt-laden and not unlike lakes or sloughs as environments. In Florida it is found in streams, lakes, big springs, and canals alike. To judge from the numbers of their soft-shell inhabitants some of the larger canals in the Everglades must represent something like an optimum habitat. The species is not uncommon near the mouths of streams in brackish water, where the tide must occasionally take it to sea. Roger Conant (letter) tells me that a specimen was recently found at sea in Bahamian waters, but it seems unlikely that this individual actually weathered a trip across the Gulf Stream under its own power.

Habits: The following interesting account of the underwater habits and personality of the southern soft-shelled turtle is quoted from Marchand (1942), who has observed these animals by means of water goggles on numerous occasions and for long periods of time, and is probably better acquainted with their more obscure traits than anyone else:

This, in my opinion, is one of the most interesting turtles of Florida. It is to be found most frequently buried down in the bottom, with only the head protruding. Often individuals may be seen swimming to the surface for air, their long necks stretched to an astonishing degree. To capture a soft-shell bare-handed underwater is somewhat difficult. The best procedure when one is found buried in the mud, is to thrust the hand into the mud until the carapace is felt, then to slide the hand forward until the anterior margin of the shell is reached. The fingers are then gently slipped around the animal's neck, and when a secure hold has been obtained, all possible pressure is applied and the turtle is pulled up through the water and thrown into the boat. The largest individuals are extremely difficult to handle for when the fingers tighten around the neck, the instantaneous reaction of the turtle is to withdraw its head, and in the case of a large specimen the head is frequently pulled through the hand and the turtle escapes. When buried in the mud the soft-shell appears to regard itself as perfectly concealed and protected, for the carapace may be handled or even pulled about, with the turtle remaining apparently uncon-cerned. After the animal has been grasped by the neck the head has a certain amount of freedom of movement because of the looseness of the skin, and a vigorous choking grip may be necessary to discourage attempts at biting.

On March 24, 1940, while marking turtles at Crystal Springs, Pasco County, Florida, I noticed several *Pseudemys* hidden in a maze of bushes growing in the water, and attempted to squeeze in and reach the turtles. I was forced to lie flat and crawl into the bushes, and while doing so I chanced to look down into the mud directly in front of me, where I saw the proboscis of a good-sized soft-shell. I enlarged a place in the overhead vegetation and then proceeded to observe the creature. The soft-shell extended its head until its jaws and eyes were exposed and stared at the glass of my mask which was about a foot away. Several times I reached down and rubbed the slender proboscis, with no effect other than to make it blink its eyes. It was only after I had punched the pos-terior part of its shell several times that the animal shook the mud from its back and swam away.

On another occasion at Crystal Springs I located several soft-shells in a group of three bubbling mud-sand springs. I assumed a position on the rim of the spring basin and pushed my legs calf-deep into the mud so that I could sub-merge with little disturbance and watch the turtles. One of the turtles in front of me emerged from the mud of the bottom, headed up toward shore, circled, and when about three feet above the bottom dived suddenly and completely disappeared. After some minutes the head of this individual appeared some three feet from where it had disappeared in the mud. This animal must have disturbed another whose presence I had not suspected for I saw a path of animation appear in the mud and presently a head came out about a foot from where I was standing. Only the proboscis protruded, its dark color blend-ing perfectly with the silt and mud bottom, and the exposed white nostrils causing it to resemble the broken end of a small twig. I dropped a small amount of silt directly in front of the nostrils, and the currents of water set up by the taking-in and expelling of water for pharyngeal respiration were easy to detect. Often the turtles would be seen gulping water in and out of the mouth.

Some areas on the bottom of Crystal Springs have been so thoroughly rooted up by the burrowing of soft-shells that the entire bottom is bare and soft. An area thus frequented by the turtles assumes a characteristic appearance and can be easily recognized.

At Rainbow Run [Marion County, Florida] an interesting situation has been noted. There are some areas in the main stream in which the bottom is bare white sand, and the soft-shells are fond of burrowing in these places. Whenever they do so a group of fish, of several species, invariably surrounds them. Thus, to locate the soft-shells in one of these areas it is only necessary to stand in the boat and look for a school of fish [similar associations between fish and other species of soft-shelled turtles have been observed].

The soft-shell underwater seems to exhibit a certain amount of curiosity. Once at Rainbow Run I located an individual beneath a pile of old lumber on the bottom. Investigating more closely, I was surprised to find a male *P. f. suwanniensis* resting on top of the soft-shell, the two apparently having no distrust of each other. It was difficult to extricate the *P. f. suwanniensis* because of the interest displayed by the soft-shell, which showed great concern and stretched its head and neck every time the other turtle moved.

Mr. O. Lloyd Meehean, at one time engaged in government fisheries investigations at Welaka, Florida, made an unusually interesting observation in connection with his use of rotenone in fish population studies in Florida lakes. He found that, along with the paralyzed fish, numbers of soft-shelled turtles usually turned up incapacitated by the poison, while no other species of turtle was affected. This surprising turn of events is difficult to explain. Though it would appear probably to have something to do with the pharyngeal and anal respiration practiced by soft-shells, it is by no means clear why rotenone, which is said to act through suffocation, should affect an animal in which the surfaces for aquatic respiration are of limited function and merely auxiliary to a pair of efficient lungs.

Soft-shell turtles used to be phenomenally numerous in the canal along the Tamiami Trail through the Everglades, and elsewhere I have mentioned the free-for-all fights one might see there when numbers of soft-shells and garfish formed semicircular ranks around the delta of each little brook draining the 'glades and fought furiously among themselves over anything edible that floated down.

Breeding: The eggs are laid from about the middle of March to July in nests usually excavated in the sand and on open shores. The egg complement is usually from seventeen to twenty-two in number, and the average diameter of the eggs is about 26 mm. (1.02 inches). The only data on the incubation period is that of Goff and Goff, who hatched twenty eggs (removed from a Lake County soft-shell) in sixty-four days. The nesting process was recently observed by Hamilton (1947), who watched two individuals lay in a sandy road bed near Fort Myers, Florida. A nest examined by him was at the mouth only slightly larger than the diameter of the egg and about four inches in diameter at a depth of five inches. Cloacal water appeared to have been used to dampen the soil. Only seven eggs were present in this nest. Hamilton found a great number of nests destroyed by mammals, mostly spotted skunks, raccoons, and foxes, and the shells about all the nests robbed by predators numbered from four to seventeen, which may indicate that the soft-shells of southern Florida either lay fewer eggs than those farther north or possibly that they more frequently split the season's quota into two layings.

Feeding: No systematic study of the feeding habits of this turtle has been made. Enough has been observed to indicate that it is omnivorous, though not overly fond of plant material; that it is quite predaceous, relying for the bulk of its captured diet on invertebrates; and that it does some scavenging. At one time or another it can be found eating nearly any kind of small animal that lives in or ventures near water.

Economic importance: This turtle has always been one of the favorite foods of the Indians, and justly so, for it is delicious when properly cooked. I have seen it in markets in several of the larger cities. Its relation to fish management has not been adequately determined. It has been known to cause great damage in duck ponds.

Agassiz' Soft-shelled Turtle *Amyda ferox agassizii* (Baur)

(Plate 74. Map 22.)

Range: Southern North Carolina southward through most of South Carolina to the Altamaha River drainage of Georgia, where intergradation with *ferox* begins, and westward to central Georgia, where intergradation with *aspera* occurs.

Distinguishing features: Stejneger (1944) placed *agassizii* in a separate group, which he regarded as more closely related to Asiatic than to other American soft-shelled turtles. Neill (1951), however, suggested that the skull characters upon which this opinion was based are associated with hypertrophied crushing surfaces of the jaws, and these are known to be

Plate 74. Amyda ferox agassizii. Females from near Columbia, South Carolina. (Courtesy Zoological Society of Philadelphia.)

much influenced by feeding habits. All things considered, there seems little doubt that this turtle is one of the complex of weakly differentiated races of *A. ferox.* The crescentic nostrils, double-lined, dark shell border (in young specimens), and spotted carapace will distinguish it from all other soft-shells except *aspera,* from which the occurrence, before and behind the eye, of a conspicuous yellow stripe, with black border above and below, and the lack of a subocular stripe are said to differentiate it. To identify an old, secondarily blotched or faded specimen will usually be impossible in our present poor state of comprehension of the group. This is a weak race.

Description: The maximum size of Agassiz' soft-shell appears to be in the neighborhood of 380 mm. (14.9 inches). Neill (1951) recorded female specimens weighing 30, 32, and 35 pounds from the Salkehatchie River, South Carolina. A good-sized female from South Carolina has the leathery carapace 362 mm. (14.2 inches) long and 284 mm. (11.2 inches) wide. The carapace is very low, and oval in lateral outline, slightly elongate, and with the widest part behind the middle; there are patches of tubercles on the anterior and posterior sections. The ground color is light brown, either with two more or less definite dark lines around the posterior and latero-posterior borders and a scattering of dark spots on the carapace, some of which are ocellated, or with the dorsal surface blotched with irregular areas of black pigment. The plastron is plain white or yellowish; it and the under edge of the carapace are unmarked in adults. The soft parts are olive to tan, mottled and streaked with dark above. On each side of the head there may be a postorbital and postmandibular light stripe, these black-bordered and often meeting on the neck, but tending to become obscured with age. There is a ridge on the septum dividing the nostrils.

Two characters that Stejneger used for distinguishing this turtle have to do with proportions of the skull which are probably subject to wide variation with age and sex and perhaps with diet; these are: (1) The longitudinal diameter of the internal nasal openings is equal to their distance from the intermaxillary foramen. In other species this diameter is said to be much greater than the same distance. (2) The crushing surfaces of the jaws are unusually wide. The median width of this surface at the tip of the lower jaw is greater than the long diameter of the internal nasal openings. This dimension was stated to be usually about twice as great as in other species in the United States. The upper alveolar surface is also very broad, but may be exceeded by that of old males of the Southern soft-shell.

Nothing has been published concerning sex and age differences in Agassiz' soft-shell, beyond the observation that the tail of the male is bigger, with more nearly terminal vent, and that, while the females show greater tendency to lose the juvenile markings, both sexes are practically indistinguishable from *ferox* in old age. The young have a more nearly circular shell and more accentuated markings than the adult. The hatchling is not known.

Habitat: As far as is known, this turtle is confined to rivers.

Habits: Apparently there is no published information on the habits of this soft-shell.

Emory's Soft-shelled Turtle *Amyda ferox emoryi* (Agassiz)
(Plate 75. Map 22.)

Range: Texas and southern Oklahoma eastward to southwestern Arkansas and western Louisiana, where intergradation with the eastern races occurs; from the Texas panhandle southward to the Gulf of Mexico in Tamaulipas, and westward through northern Mexico and southern New Mexico to the Gila-Colorado drainage of western Arizona, southern Nevada, southeastern California, and northeastern Baja California. The occurrence of this turtle in the tributaries of the Colorado River is thought to be due to recent introduction, and Miller (1946) suggested that the species reached that area via the upper Gila River, a tributary of which it invaded when the dam of an artificially stocked pond on a ranch in Grant County, New Mexico, gave way.

Distinguishing features: Although very closely related to the other races of *A. ferox*, Emory's soft-shell may be identified in a majority of cases by its crescentic nostrils (not rounded) and by the presence on the carapace of a few widely spaced, poorly developed tubercles on its fore edge and of only very minute tubercles on the soft flaps. If the juvenile pattern is evident it usually includes a single dark line within the light shell border.

Description: This is a good-sized soft-shell, the record length apparently being that of a specimen from the Colorado River with a shell 370 mm. (14.6 inches) long. Measurements of a large adult male from Houston, Texas, are as follows: length of leathery carapace, 354 mm. (13.9 inches); width of leathery carapace, 288 mm. (11.3 inches); height of shell, 84 mm. (3.30 inches); tail length beyond hind margin of carapace, 32 mm. (1.26 inches). The carapace is moderately elongate and wide behind, with numerous minute, white tubercles on its surface in many specimens and with the larger tubercles at the anterior edge obtusely triangular and widely spaced. The ground color of the carapace is brown, and it may either be marked with the youthful pattern, comprising a light border behind and at the sides, with a dark line inside this, or it may be blotched with irregular dark spots that often run together. The plastron, under surface of the carapace, and lower soft parts are white, except for the fingers and webs, which are mottled greenish. The head and neck are dark bluish above and yellow on the sides, fading into whitish below. There may be spots of various sizes, shapes, and degree and manner of dispersal on the head, neck, and legs.

Little or nothing has been written about sex differences in Emory's soft

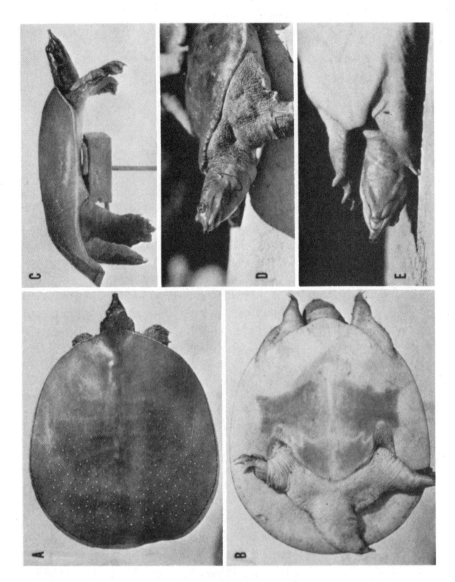

Plate 75. Amyda ferox emoryi. A, B, and C: Male; Orange, Texas. D and E: Specimen of undetermined sex; El Paso, Texas.

shell, and the only data available on the very young are those of Agassiz (1857, pl. 4), who figured a little specimen with a shell 36.5 mm. (1.44 inches) long, 31.5 mm. (1.24 inches) wide, and 11 mm. (0.434 inches) high.

Habitat: Like most other soft-shells, Emory's turtle appears to prefer rivers, but it is found also in ponds and reservoirs and seems to thrive in impounded water. In northwestern Texas, Cope (1892) found it "abundant in all permanent water."

Habits: This subspecies of soft-shell appears to have the same vigor and redoubtable, aggressive disposition that characterize its relatives in the East. I remember well almost losing a finger while trying to remove a hook from the mouth of one I caught in the Brazos River, Texas, when I was very young.

Breeding: The presence of an undescribed soft-shell in the western states was called to the attention of Agassiz (1857, 1: 407) by the unusually large size of some eggs that he received, as he relates below:

The first intimation I had of the existence of another species of *Aspidonectes* within the boundaries of the United States was from the sight of two eggs collected in Texas by Dr. Heerman and presented by him to Dr. Holbrook, who gave them to me. These eggs were so much larger than those of either of the three other species of the family which I then knew that I did not hesitate to consider them as derived from an unknown species. My supposition was very soon changed into certainty, after I had received from the Smithsonian Institution all the specimens of turtles collected in Texas during the operations of the Boundary Survey, under the command of Colonel Emory, among which were young and adult specimens of this species, collected in the Lower Rio Grande of Texas, near Brownsville.

These eggs, figured by Agassiz in his Plate 20, are 29 mm. (1.14 inches) in diameter, and apparently are the only ones ever seen by a zoologist. At any rate, no one has added any data on any aspect of the breeding habits of this species.

Feeding: I have caught these turtles on worms. Strecker (1927c) said that captive specimens preferred fish but could exist on raw beef; he also mentioned a specimen's killing and eating a young lined snake.

Economic importance: These soft-shells are eaten locally in Texas, and Pope (1939) mentioned finding them in a market at Ciudad Juarez, Mexico.

Eastern Spiny Soft-shelled Turtle *Amyda ferox spinifera* (Le Sueur)

(Plate 76. Map 22.)

Range: Northeastern part of the Mississippi Valley from Wisconsin eastward across the southern half of Michigan and southern Ontario to the St. Lawrence River and tributaries, Vermont, western New York, and Pennsylvania, whence the eastern limit of distribution extends southward, through the eastern edges of West Virginia, Kentucky, and Tennessee. Along the upper Mississippi from Minnesota southward to eastern Arkansas and western Tennessee, there is a broad area in which *spinifera* intergrades with the more western *hartwegi*. In the extreme southern sections of the Mississippi drainage intergradation between *spinifera* and *aspera* occurs. In Mississippi, Louisiana, and Arkansas "there is an extremely complex population of soft shells that exhibits characteristics of the three subspecies, *hartwegi, spinifera,* and *aspera*" (Conant and Goin). Neill (1951) stated that in this latter area the characters of the southwestern subspecies, *A. f. emoryi,* also appear.

Distinguishing features: The leathery shell, the ridge on the nasal septum, the ocelli, and the presence of a marginal dark line on the carapace and of tubercles on its anterior margin serve to distinguish *spinifera* from any other North American turtle except the subspecies *aspera* and *hartwegi*. From these it is set off by the combination of a single marginal line on the carapace, the centrally increased size of the dorsal spots, and the two separate light lines on each side of the head.

Description: The low, flattened carapace of this soft-shell is a wide oval in lateral outline and nearly as broad as long, the greatest width being over the hind legs. The anterior margin of the carapace is studded with prominent conical tubercles and, besides these, either the whole surface of the carapace or at least its anterior and posterior sections are roughened by small, sharp projections that impart a sandpaperlike surface. The carapace is olive to grayish in ground color with numerous dark, often ocellated, spots that increase in size toward the center of the shell, and with a single dark line concentric with the margin and bordering a wider light outer band. This shell pattern is usually lacking in older females, in which the carapace may become irregularly blotched and mottled with dark areas. Maximum size is probably close to 400 mm. (15.7 inches). The largest Ohio specimen measured by Conant (1938a) was a female with a carapace 367 mm. (14.4 inches) long. Lagler (1943b) mentioned "more than 16 inches" as the maximum size, and the same writer measured ten Michigan

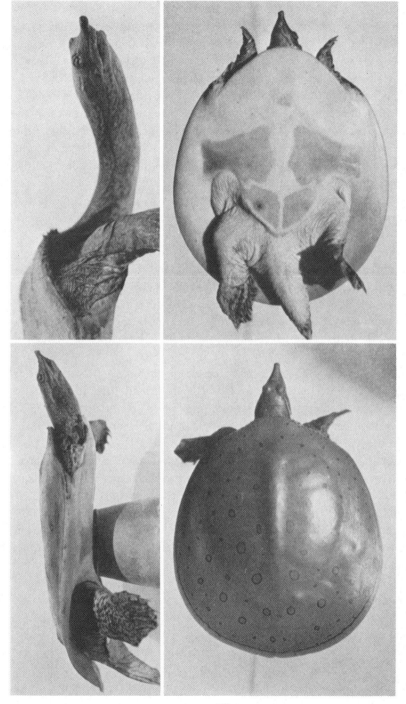

*Plate 76. Amyda ferox spinifera. Upper: Male specimen; unknown locality. (General Biological Supply Co.)
Lower: Young; Green Castle, Indiana.*

427

specimens that ranged in length of carapace from 162 mm. to 407 mm., with an average of 236 mm. A large specimen from Indiana, also a female, had the following measurements: length of carapace, 330 mm. (13.0 inches); width of carapace, 267 mm. (10.5 inches); length of head and neck, 228 mm. (8.98 inches); length of tail, 89 mm. (3.50 inches). The plastron is uniform white or yellow in color; the under edges of the carapace are occasionally dark-streaked or spotted. The soft parts are olive gray in ground color; the head is spotted or streaked with black and with two dark-edged light stripes on the side, one extending backward from the eye and one from the angle of the jaw, with the two usually not meeting on the neck. On the septum that separates the nostrils there is a narrow ridge that projects into each nostril and extends posteriorly into the nasal chamber. The lips are yellow and dark-spotted. The feet are mottled or streaked with yellow.

There is marked disparity in the sizes of the sexes. Of the seventy-one Ohio specimens that he examined, Conant found the largest female to be 367 mm. (14.4 inches) in shell length, while the biggest male was only 165 mm. (6.50 inches) long. The coloration of the sexes is often strikingly different; the males tend to keep the juvenile markings of spots and lines, whereas the females lose these and acquire with advancing age a blotched or mottled carapace.

Hatchlings are probably usually between 38 and 40 mm. (1.50–1.57 inches) in shell length. Young individuals are usually paler in color than adults and with the characteristic markings more accentuated. They may have a slight central ridge, and the snout is usually more angular than in the adult.

Habitat: Most writers agree that this very aquatic species is partial to larger, slow-moving rivers. It is, however, occasionally seen in ponds, lakes, or smaller streams. It rarely leaves the immediate vicinity of water.

Habits: In spite of its drastic adaptations to aquatic existence, the soft-shell generally, and this species in particular, is fond of emerging from the water in protected places and basking in the sun for hours at a time, sometimes congregating in large numbers in a favored spot. The normal period of activity in Indiana, Illinois, and Ohio appears to be from April to October, although individuals have been seen in the open water as late as December on several occasions. Toward the southern edge of its range it may pass entire winters without going into deep hibernation. Actually it might be somewhat difficult to recognize hibernation in a soft-shell, since the winter dormant period is apparently often passed under a shallow covering of sand in much the same manner in which the animals spend a large part of their "active" lives, except that in the latter case they usually burrow where the water is not too deep to allow the long neck to extend the snout tip to the surface for an occasional breath of air. Although the soft-margined pancake shell of *Amyda* is often mentioned as an adaptation to aquatic locomotion, it seems more likely that its principal advantage to the turtles is the ease with which it may be shuffled into the bottom soil by sidewise movements and the inconspicuously low mound that it makes when so covered.

Breeding: Apparently copulation has not been observed, although Conant twice saw pairs in which the feet of the male were resting on the carapace of the female; the dates of these observations were not given. Laying occurs from June until the last of July. The eggs vary in number from ten or a dozen to about twenty-five, although there is one record of "about thirty." The average dimensions of twenty-one eggs of Ohio specimens were found by Conant to be 27.0 by 25.6 mm. (1.06 by 1.01 inches), indicating that they are not always perfectly spherical; the shell is described as rather thick but not brittle. The nest is flask-shaped, ranging in depth from four to ten inches, in greatest diameter from three to five inches, and with a narrowed neck around one and one-half inches in diameter. The only available account of the nesting process that may be regarded as certainly applying to this subspecies is the following much-quoted excerpt from Newman (1906a) whose observation was made at Lake Maxinkuckee in northern Indiana, at 11:10 A.M., June 22, 1903:

A warm sunny day. Place: the "old road" about ten feet from the water's edge and concealed from view on one side by tall grass. A large female *Aspidonectes* [*Amyda*] has just emerged from the grass and is commencing to make a nest. No time is lost in selecting a spot. She scratches out foot-holds for the fore feet and begins to excavate with the hind feet, using right and left feet alternately with a circular, gouging movement. At intervals she pushes aside the accumulated earth with the hind feet. As the hole becomes deeper it is necessary for her to raise the anterior part of the body to its full height in order to give a more nearly perpendicular thrust with the hind feet. In less than forty minutes the nest is completed and she commences to lay the eggs, letting the tail down into the narrow hole as far as possible. After depositing several eggs she arranges them with the hind feet and then rakes in some earth previously wet up with water from the accessory bladders. The earth is gently packed in before any more eggs are laid. The remainder of the eggs are deposited, and the hole is filled up with earth and tramped down quite firmly with the knuckles of the hind feet, right and left feet being used alternately. This treading movement continues for some minutes and seems to be quite thorough. Although not in any way disturbed, the tortoise left without attempting to cover up the traces of scratching feet, and anyone who is familiar with the appearance of a tortoise nest would have no difficulty in detecting this one. At 12:25 she turned and started for the water, but was captured with a landing net. The nest was examined and found to be flask-shaped with a narrow neck only an inch and a half in diameter. The depth of the nest was a trifle over six inches and the diameter at the bottom about three inches.

The nest contained eighteen rather large spherical eggs of a delicate pink color and with a very thin brittle shell.

The incubation period is not known. It is believed that in some nests the young do not emerge until the spring following deposition of the eggs.

Feeding: This soft-shell appears to be omnivorous in its feeding, but with a marked predilection for crayfish and other invertebrate animals. Despite its pugnacity and strength of jaw, and its sinister habit of lying concealed in the sand as if to ambush passing fish, it has never been proved definitely to be important as a fish predator. Vegetation is eaten with some frequency and that it is not merely ingested incidental to the taking of other food is shown by the fact that Surface (1908) found in the stomach of an Ohio specimen "yellow and red field corn, to the extent of almost as much corn as would be produced upon two average ears of that plant." Lagler (1943a) mentioned watermelon rind as good bait for soft-shell traps, and the same author (1943b) gave the statistics shown in Table 12, based on an examination of the stomachs and colons of fifteen Michigan specimens, and probably affording a fair representation of the diet of the turtle, in that area, at least.

Table 12. Contents of stomachs and colons of specimens of the eastern spiny soft-shelled turtle.

Food item	Stomachs		Colons	
	Composition by volume (%)	Frequency of occurrence (%)	Composition by volume (%)	Frequency of occurrence (%)
Fish remains	trace	1.9	——	——
Crayfish	47.4	45.5	46.7	66.7
Insects	52.4	90.9	53.3	83.3
Snails	0.2	9.1	——	——
Lower plants	unmeasured trace	9.1	unmeasured trace	33.3
Vegetable debris	0.2	27.3	trace	33.3

Economic importance: Evermann and Clark (1916) called this the "most highly esteemed for food of all Indiana turtles." In most parts of its range the demand for this turtle nearly always far exceeds the supply. As was mentioned above, the importance of this soft-shell as an enemy of valuable fishes has not been demonstrated and is probably not nearly so great as is popularly supposed. It quite possibly is of more significance as a competitor with game fish than as an enemy, since the preponderance of crayfish in its diet implies a dependence on a partly similar food supply.

Southern Spiny Soft-shelled Turtle *Amyda ferox aspera* (Agassiz)

(Plate 77. Map 22.)

Range: This East-Gulf subspecies is known from tributaries of the lower Mississippi River in Louisiana, from Mississippi, Alabama, and southwestern Georgia, and from certain if not most of the streams of the Florida panhandle, where it is known as far to the eastward as the Chipola River.

Distinguishing features: For various reasons, mostly owing to a lack of sufficient material for study and because of a confusing series of individual, age, and sex variations, the characters by which *A. f. aspera* differs from the other subspecies of *A. ferox* have not been satisfactorily defined. The name *aspera,* which means "rough," was given to the form by Agassiz (1857: 1: 406) because of "the very coarse and large tubercles of the front

Plate 77. Amyda ferox aspera. Young specimen (preserved); Milton, Florida. (Photograph by Coleman J. Goin.)

and hind part of the carapace, which extend, behind, even over the bony shield and are there supported by prominent warts on the bony plates." The same writer called attention to the two or three dark lines around the light border of the carapace, which in males and young specimens contrast with the single line of *spinifera* (and also of *hartwegi*). As Conant and Goin pointed out, the lateral head stripes, if present, do not meet on the neck in *aspera,* while they do meet in the other forms. The objection to these characters is that they leave unidentifiable a certain percentage of specimens in which the diagnostic structures or pattern may have been lost through age or which because of extreme youth may not have developed.

Sexual differences were noted by Stejneger (1944), who found that in the male the calluslike areas of the plastron are more extensive than in the female, that the tail of the male is longer (with more distally located vent), and that the "sandpaper" effect of the back may be more pronounced in certain young males. The same writer believed that "the greater development of the tubercles or 'spines' may be correlated with sex or age, or with both, remembering that the females are larger than the males."

Dimensions of a large male specimen from Natchez, Mississippi, are as follows: length of leathery carapace, 450 mm. (17.7 inches); width of leathery carapace, 370 mm. (14.6 inches); length of bony central disk of carapace, 240 mm. (9.45 inches); width of bony central disk, 240 mm. (9.45 inches); height of shell, 88 mm. (3.42 inches).

Table 13, based on data taken by Stejneger from young adult specimens,

Table 13. Variation in certain characters of young adult specimens of the southern spiny soft-shelled turtle (measurements in millimeters).

	Mississippi locality			
	Pearl River	Black River	Pearl River	Black River
Sex	female	female	male	male
Length of leathery carapace	286	210	180	158
Width of leathery carapace	242	182	152	140
Length of bony carapace	117	119	111	99
Width of bony carapace	131	85	110	98
Extent of plastral calluses	small	small	large	large
Spines on edge of carapace	large	medium	small	small
Tubercles on median line of fore flap of carapace	yes	yes	no	no
Skin of carapace "sandpapered"	no	no	yes	yes
Lateral outline of carapace	ovate	oval	ovate	oval

shows the range of variation in certain characters to be expected in this subspecies.

Habitat: Like *A. f. spinifera,* this turtle is a haunter of deeper, silt-bearing rivers. Nearly all records are from rivers and larger creeks and their associated flood-plain impoundages.

Habits: Nothing is known of the habits of these turtles. The fishermen who catch them on trotlines regard them as very good eating, but apparently they rarely are caught in numbers large enough to build up a commercial demand.

Western Spiny Soft-shelled Turtle *Amyda ferox hartwegi* Conant and Goin

(Plate 78. Map 22.)

Range: "Nebraska, Kansas and Oklahoma eastward toward the Mississippi River in southern Minnesota, Iowa, Missouri and Arkansas; northwest to Wyoming and Montana" (Conant and Goin). The describers of this subspecies mentioned examining intergrades between *hartwegi* and *spinifera* in the following localities: Minneapolis, Minnesota; Allamakee County, Iowa; Jackson County, Illinois; Reelfoot Lake, Tennessee; and Lawrence County, Arkansas. Intergradation with *emoryi* almost certainly occurs but has not yet been demonstrated.

Distinguishing features: The western spiny soft-shell is similar in general appearance to the eastern and southern subspecies, from both of which Conant and Goin differentiated it as follows:

A soft-shelled turtle of the genus *Amyda* in which the carapace is marked with numerous dark spots or small ocelli (or both), such markings being the same size (or only slightly larger) in the center of the carapace as they are near its perimeter.

In the subspecies *spinifera*, ocelli are well developed, and those toward the center of the carapace are distinctly larger than the ones near the perimeter of the shell. In the subspecies *aspera*, there are also ocelli on the upper shell, but turtles of that race are characterised by the possession of two or more dark, usually interrupted lines around the rear and lateral edges of the carapace. In both *spinifera* and *hartwegi* there is a single, more or less continuous dark line parallel to the edge of the carapace. Also in *aspera*, the postocular and postlabial light lines usually meet on the side of the head; in the other two species they do not normally meet.

As in other members of the genus *Amyda,* the pattern in all these forms may be seen best in juvenile females and in males; in large females, which undergo

Plate 78. Amyda ferox hartwegi. Young adult male specimen; Wichita, Kansas. (Photographs by Isabelle Hunt Conant.)

a remarkable change in coloration and pattern as they grow older, the markings may be made out only with difficulty, if at all.

Dimensions of the type specimen of *hartwegi,* a male, are as follows: length of carapace, 168 mm. (6.63 inches); width of carapace, 136 mm. (5.35 inches); depth of shell, 41 mm. (1.61 inches); length of tail, from posterior edge of plastron to tip, 77 mm. (3.03 inches).

Sex differences are much the same as described for *spinifera,* involving a smaller size for the male, together with longer retention of the juvenile color pattern and a relatively longer, thicker tail with the vent nearer the tip. As was mentioned above, the old female often loses the original color pattern; this is replaced by dark, irregular mottling or blotching of the carapace.

Young individuals are said to have heavier ventral pigmentation of the carapace and in many cases to lack the dorsal tubercles.

Habitat: This is evidently a fluviatile turtle, like most of the other soft-shells.

Habits: Life history data applying to undoubted specimens of *hartwegi* are very scarce. Although Cahn (1937) recorded considerable original information on the habits of this species in Illinois, that state is a part of the territory in which *spinifera* and *hartwegi* meet, and until Illinois populations have been more thoroughly studied it is difficult to assign Cahn's data to either form with any certainty. It seems most likely that the Illinois River stock may be typical *hartwegi,* in which case the good account of nesting by a Meredosia female provided by Cahn would be apropos here. As far as comparable points are stressed, however, Cahn's account and that of Newman (1906a), who described the same process as carried out by an Indiana specimen of *spinifera,* bring out no marked differences, and it is probable that the two races have essentially the same habits. Cahn's record of nesting was for July 11, 1931. Force (1930) found 32 ripe eggs in the oviduct of a female on May 20, 1929; the eggs had an average diameter of 30.8 mm. (1.21 inches) and would probably have been laid within the near future. Blanchard (1923) found an individual digging a nest in Dickinson County, Iowa, on June 3; this may or may not have been *hartwegi.*

Nothing is known of the feeding habits of this race.

Economic importance: Throughout its range this turtle, like most other soft-shells, is enthusiastically sought for food and occasionally is found on the market, although most are eaten by those who catch them. The bearing of the habits of this race on fish culture and management has not been determined.

Spineless Soft-shelled Turtle *Amyda mutica* (Le Sueur)

(Plate 79. Map 23.)

Range: Mississippi Valley, from western Wisconsin, southern Minnesota, and southern South Dakota southward to central Mississippi and Louisiana, and toward the west, southward through the eastern half of Kansas and across Oklahoma into south-central Texas. Toward the east the limits of its range pass through northern Illinois, central Indiana, and southeastern Ohio, extreme western Pennsylvania and West Virginia, eastern Ken-

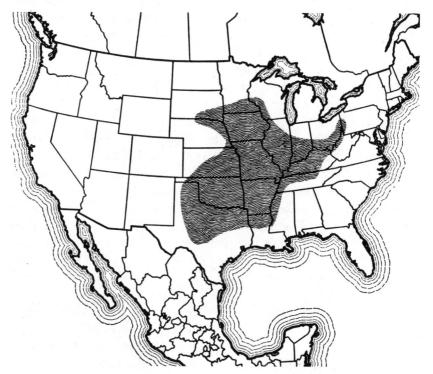

Map 23. Distribution of *Amyda mutica.*

tucky and northwestern Tennessee. Concerning the extraordinarily rare occurrence of this turtle in Pennsylvania, Netting (1944) commented as follows:

It appears safe to conclude that *Amyda mutica* once occurred in the un-glaciated portions of western Pennsylvania in the Ohio and Allegheny rivers, that adequate collecting may bring to light relict populations of this turtle in suitable habitats, and that unpolluted stretches of the larger streams, especially below dams, merit particular attention.

Apparently there is in existence at present only one authentic Pennsylvania specimen.

Distinguishing features: This very distinct species of soft-shell is easily identified by the lack of spines on the smooth, leather-edged carapace and by the circular outline of the nostrils, in which no ridge projects from the septum.

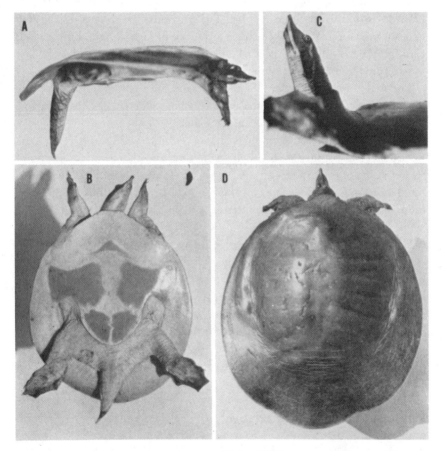

Plate 79. Amyda mutica. A, B, and C: Males; Black River, Powhatan, Arkansas. *D:* Female; same locality.

Description: This is said to be the smallest of the North American soft-shelled turtles, which it probably is, although the maximum recorded shell length for a specimen of *mutica* [356 mm. (14.0 inches)], approximates the upper limits for some of the others. The usual length for an adult female is perhaps around 180 mm. (7.10 inches) and for a male, 140 mm. (5.52 inches). The carapace is very low and smooth, with no dermal tubercles, and is more nearly circular in lateral outline than that of any other soft-shell; that of a specimen from Missouri is 175 mm. (6.90 inches) long and 156 mm. (6.15 inches) wide. The ground color of the carapace is olive to brown, with irregular dots or dashes of darker color and sometimes with a marginal light band bordered within by a dark line; this is often replaced in large females by a mottled or blotched pattern. The plastron is uniformly light in color. The soft parts are gray to olive, with dots or mottling of a darker shade. A black-bordered light line extends through the eye and onto the neck. The throat and lips are yellowish. The nostrils are circular; there is no ridge on the septum.

The males are smaller than the females, and with longer tails. According to Cahn (1937), the claws of the hind feet are longer in the female while those of the fore feet are longer in the male.

There is less difference in shell shape between young and mature individuals of the spineless soft-shell turtle than in other species because of the retention of the juvenile circular outline by the adult *mutica*. The hatchling has a carapace between 34 mm. (1.34 inches) and 35 mm. (1.38 inches) long. The markings are more intense and the upper lateral edges of the snout more angular.

Habitat: The habitat preference of *A. mutica* is apparently similar to that of the spiny soft-shell. It is essentially a fluviatile animal, being found both in larger, more slowly flowing rivers and in clear, sandy creeks; it occurs but rarely in lakes and flood-plain sloughs. P. Smith (1947) suggested that the marked preference shown by this species for clean, sandy river bottoms may account for its absence from the prairie counties of Illinois, where such situations are scarce. This same writer found that in the Embarrass River of Illinois *mutica* and *spinifera* were equally abundant, but that *mutica*, unlike *spinifera*, was not to be found there in lakes or in smaller streams.

Habits: Remarkably few observations on the life history of this turtle have been recorded. As compared with *Amyda spinifera*, the present species, besides being smaller, appears to be even more aquatic, and even more restricted to fluvial environments, where it may occasionally bask in the sunshine on the bank, but never on protruding logs or rocks and usually within a few inches of the security of the river. Although extremely active, vigorous and rapid in all its movements, and demanding respect for its jaws from anyone handling it, this species is perhaps a little less pugnacious than *spinifera*. Cahn (1937) pointed out that on level ground it can outrun a man, and if this is true *mutica* must be regarded as among the speediest of all turtles. Most of its time, when not foraging, it spends buried in the mud, or preferably sand, with only the head exposed, and in water sufficiently shallow to allow the snout to be thrust above the surface from time to time to augment the oxygen that it takes from the water.

Breeding: Muller (1921) gave the time of nesting as the latter half of June and the early part of July. His observations were made on an island in the Mississippi River near Fairport, Iowa. Hurter (1893) said that they lay near St. Louis "about the third week in June." Cahn believed the season in the Illinois River to be early June. The eggs are usually 18 or 20 in number, although as many as 33 have been found; as in most species the size of the egg complement varies with the size of the individual. The average diameter of 116 eggs measured by Cahn was 22.6 mm. (0.89 inches), which is somewhat smaller than the average size of the eggs of *spinifera*. The nests are usually located on sandy bars or banks where vegetation does not impede free view of the surrounding territory. In sandy soil the covered nests appear as small craters, while in pebbly areas they may be detected as circular patchs of clear sand. Muller found the temperature in nests to be quite constantly 90° Fahrenheit. The incubation period varies with the hygric condition of the medium but is about 70 days in Iowa.

Feeding: This is another of the suspected fish enemies whose real depredations should be demonstrated by a careful study before the animal is condemned as inimical to conservation programs. Certainly it eats fish— fish have been found in any number of stomachs, and Cahn saw one individual chase down and catch a brook trout. This took place in a tank, however, and anyway, the real question is: how many fish does it eat? Only when the answer has been found can the biologist presume to sit in judgment. It is probable, from what is known, that crayfish and aquatic insect larvae bear the main burden of feeding *mutica,* and that among the fish eaten a considerable number may be taken when incapacitated in some way, or even after death. Cahn remarked on the fact that this turtle spends considerable time foraging around sunken trees and brush piles. A small percentage of its food is vegetable material of various sorts; it has been known to eat fruits and one observer even reported that "hard nuts" are taken.

Economic importance: Although it is the unanimous agreement of all who have eaten these turtles that they leave little to be desired in the way of palatability, they are apparently of little commercial importance, and are not often seen in city markets. Clark and Southall commented that in Chicago one of these soft-shells "could not be given away, much less sold." I doubt that this apathy of the market is any great source of grief to the fisherman lucky enough to catch a *mutica,* however, and from my own experience in such cases would expect the prospective buyer of such a prize to find that it could not be bought, much less given away.

Under the above section on feeding habits, mention is made of the possible role of this turtle as an enemy of desirable fishes.

Suborder ATHECAE

The Leatherback Marine Turtles

THE name Athecoidea, proposed by Cope for the leatherback turtles in 1871, means "without a shell" and is in a way an unfortunate choice of terms, since the leatherbacks do have a shell of a sort. Unlike all other modern turtles, however, the group is characterized by a shell of osteoderms, which is to say the carapace and plastron are formed by a single layer of irregularly polygonal bones that form a mosaic imbedded in the skin and entirely free from the internal skeleton. There is a free proneural bone at the anterior end of the carapace. The internal skeleton is relatively unmodified.

These extraordinary animals have received a large amount of attention from students of reptilian evolution, and two directly opposed views as to their origin have been debated for many years. These are as follows:

(1) The leatherbacks represent a distinct ancestral line that diverged from the main chelonian stem at least as early as Triassic time.

(2) The group was derived from the principal thecophoran line by way of the giant marine *Archelon* or some similar Cretaceous sea turtle.

Whichever of these views we accept, there can be little doubt that the leatherback is the most remarkable, as well as the largest, living turtle; and whatever the history of the characters involved, they add up to what must be regarded as the taxonomic equivalent of all the rest of the turtles combined. Moreover, the fact that the turtle student finds constant need for a term under which to group all the rest of the turtles when contrasting them with the leatherback to my mind justifies retention of this subordinal scheme on a practical basis.

Family DERMOCHELIDAE

There is a single genus in the family, *Dermochelys*.

Genus *DERMOCHELYS*

This genus, which appears to include only one species, is circumtropical in distribution, but being quite pelagic it frequently straggles far to the north, and probably far to the south, of its breeding range. The characters of the family and genus are as follows:

Certain enlarged bones of the dermal mosaic of both carapace and plastron are arranged in longitudinal rows which form strong keels extending the length of the shell; there are seven of these keels above and five below. The limbs are very strong and highly modified for swimming; the anterior pair are much larger than the posterior, which in the adult are broadly connected with the tail by a web. There are no claws.

Whether we regard these turtles as primitive or as modern it must be conceded that they are the most completely specialized for aquatic existence of all the members of the order.

There are probably at least two races of living leatherbacks, but as yet they have not been adequately defined. The statistical treatment that will probably be needed to show the minor differences involved must await the accumulation of far more material than is available at present. For reasons given on page 453, I have followed Stejneger and Barbour in recognizing different forms in the Atlantic and Pacific oceans, although I have concluded that these can be little more than weakly differentiated races, and have accordingly used trinomials for them.

Atlantic Leatherback Turtle *Dermochelys coriacea coriacea* (Linné)
(Plates 80, 81, 82. Figs. 36, 37.)

Range: Atlantic Ocean, Mediterranean Sea, Gulf of Mexico, and the Caribbean. It ranges southward at least as far as Mar del Plata, Argentina, and it has been recorded from as far north as the coast of Nova Scotia. In

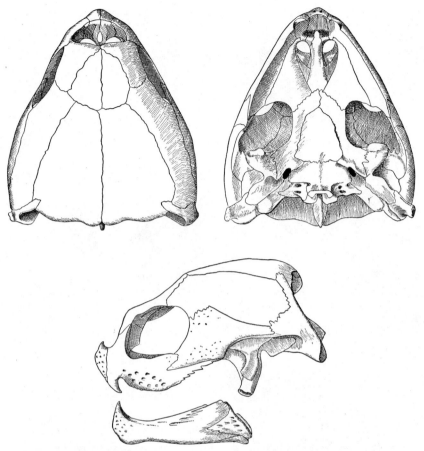

Figure 35. Dorsal, ventral, and lateral views of the skull of the leatherback turtle (*Dermochelys*).

the eastern Atlantic it is known from the British Isles to the Cape of Good Hope.

Distinguishing features: This extraordinary turtle, one of the most remarkable of all living reptiles, could be confused with nothing else. It is distinguished from all other sea turtles within its range by the smooth, scaleless, black skin of its back and the seven narrow ridges that extend down it, by the five longitudinal ridges on the plastron, and by the cusp on either side of the upper jaw. Differentiating between it and its relatives in the Indo-Pacific, however, is another matter, and for a discussion of this question the reader is referred to the section on *D. c. schlegelii.*

The size attained by the leatherback turtles cannot be determined with

Plate 80. An amazingly lucky turtle hunt on Flagler Beach, Florida. *Upper:* Posed picture of *Dermochelys coriacea coriacea* taken near the nesting site. *Lower:* The egg-laying process completed. The nest cavity can barely be discerned beneath the hind end of the shell. Note the closed eyes and compare the figures of laying *Lepidochelys olivacea.* (Photographs by J. Carver Harris, apparently the first ever made of the egg laying of *Dermochelys c. coriacea.*)

Plate 81. Additional photographs of the turtle hunt. *Left:* Removing the eggs from the nest. *Right:* Polishing up the sand-covered turtle for a picture. (Photographs by J. Carver Harris.)

accuracy from existing records because of the difficulty of distinguishing between estimated and actually measured dimensions and weights. The greatest weight ever mentioned in print, as far as I know, is that in Agassiz' statement (1857, 1: 324): "I have seen specimens weighing over a ton." There are several apparently reliable records of weights of between 1,000 and 1,200 pounds, and a newspaper account of a Maine specimen said to have weighed 1,500 pounds is quoted here in the section on *Caretta c. caretta*. Between 1,200 pounds and Agassiz' "over a ton" there is a conspicuous scarcity of trustworthy weight reports. A large specimen today weighs around 800 pounds, not much bigger (or possibly even somewhat smaller) than top-sized loggerheads or green turtles, whose own maximum sizes are also largely a matter of conjecture. Most published measurements of length, unless specifically explained, appear to refer to total length—snout to tail tip—and not to shell length. After looking over available sets of dimensions for leatherbacks taken on the Atlantic coast of the United States (including newspaper reports) I can offer the data shown in Table 14, which at least may give some general idea of the average proportions

Table 14. Reported dimensions of leatherback turtles.

Total length	Carapace length	Carapace width	Width flipper to flipper	Weight
5½ ft.	——	——	6½ ft.	——
7⅓ ft.	——	——	——	——
7⅓ ft.	5 ft. 7 in.	——	8 ft. 10 in.	1,130 lb.
7 ft. 1 in.	——	——	8¾ ft.	——
8¼ ft.	——	——	10 ft.	1,000 lb.
7½ ft.	——	——	9 ft.	1,200 lb.
——	8 ft.	——	——	"nearly a ton"
10¼ ft.	——	6 ft. 7 in.	——	1,087 lb.
6½ ft.	——	3½ ft.	——	800 lb.
——	——	——	——	1,280 lb.
7½ ft.	——	——	——	"between 1,200 and 1,500 lb."
7 ft.	——	4 ft.	12 ft.	1,600 lb.

of the animal. Although it is so stated in only a few of the above cases, I believe that all the measurements were made with flexible tapes over the curvatures.

There is considerable variation in the coloration of mature specimens of

Figure 36. Young *Dermochelys c. coriacea* (natural size) showing juvenile scales on the shell. (From Stejneger.)

the leatherback. The usual condition seems to be a ground color of dark brown or black, mostly unmarked on the dorsal surfaces of head, legs, and carapace, but sometimes with desultory spots of yellow or whitish. These are more pronounced in small adults than in large ones, and are most crowded on the neck and lower jaw, where there is usually a mottled coloration. A very young specimen (total length, 97 mm.; from West Africa) has the following coloration: ground color, slaty black with narrow yellow margins on the flippers and shell. The dorsal keels are also yellow, as is the top of the snout, and there are two or three yellow spots above each eye. Beneath, only the central areas of the flippers and bands between the broad ventral keels are black; all the rest, including the lower surfaces of upper and lower jaws, is yellow. The tail projects 3 mm. beyond the posterior edge of the carapace. Each maxillary cusp terminates in a long and very acuminate spine.

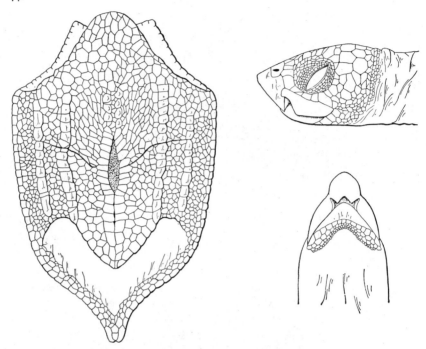

Figure 37. Young *Dermochelys c. coriacea. A:* Ventral view of shell. *B:* Lateral view of head. *C:* Ventral view of head. (From Stejneger.)

The male is more slender posteriorly than the female and has a much longer tail.

Young individuals are much more conspicuously marked than adults. They are covered with small scales, which later are shed, and the maxillary cusps terminate below in sharp spines.

Habitat: Although confirmedly pelagic and built to cruise warm seas unhindered, the leatherback apparently finds some attraction in the shallow water near land from time to time. The first specimen ever recorded from the United States was caught in Chesapeake Bay (in 1811), and there are numerous records for this and other bays of our northern coast. It may be that the increasing chill of the water in these extralimital parts of its range in some way influences it to turn shoreward, since the same thing happens in the extreme south, in Argentina, where it sometimes ascends the Plata river to La Plata.

Habits: Almost the only information on the way of life of the Atlantic leatherback is that which must be gleaned from reports of nonzoological observers. Although much of this must be discounted until confirmed by

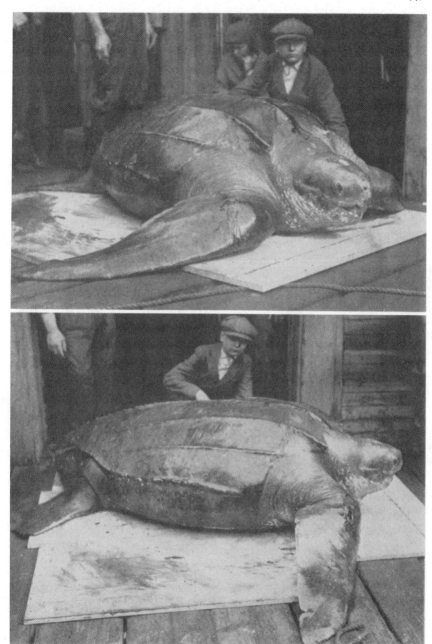

Plate 82. Dermochelys coriacea coriacea, taken off Portland, Maine, in 1919. (Photographs by M. G. Flaherty.)

scientific investigation, it is possible to piece together a fragmentary picture of the life history of the animal.

Nearly all who have taken part in the capture of one of these huge creatures have agreed on two points: first, that it is a strong and at times dangerous adversary that uses both flippers and jaws with telling effect when attacked from a small boat; and, second, that it emits cries of pain or rage or both when hurt. The cries have been variously described as pathetic wails, groans, roars, and bellows, and it would be of the greatest interest to learn whether they are merely an accentuation of the snorelike exhalation that can be heard when a *Caretta* breaks water, or are some more specialized vocalization.

The extraordinary swimming ability of the leatherback and the frequency with which it is seen in the open sea suggest that individuals may travel great distances, but whether as aimless wanderers or as scheduled migrants is unfortunately not known. Although at least one instance of the capture of a leatherback is reported nearly every year, it can be readily found nowhere in the Atlantic. The first record for the coast of the United States appears to be one for the year 1811, and from data given by DeKay (1842) and Ford (1879) it is possible to list the instances in which the capture of leatherbacks was made public during the succeeding sixty years:

Year	Place
1816	Sandy Hook, New Jersey
1824	Massachusetts Bay
1826	Long Island Sound
1840	Chesapeake Bay
1872	Delaware Bay
1878	New Jersey (four specimens)
1879	Wilmington, North Carolina

The fact that leatherbacks have been more often encountered in recent than in earlier times probably is merely a reflection of the increased traffic in United States waters and the greater frequency with which accounts of such encounters appear in print.

Breeding: The mating and laying seasons are alleged to be the same, but no details are available, and there are conflicting statements concerning the time when laying occurs. Garman (1883) was told that nesting took place in the vicinity of Florida from December through January, while most other accounts describe the season as spring and early summer months. Several Mosquito and Black Carib fishermen whom I questioned agreed that in Honduras and Nicaragua the season extends from May to August. Although the leatherback appears to nest nowhere in numbers, it still makes occasional visits to widely separated points in the West Indies and on Caribbean shores. In the days of Audubon it apparently nested regularly on the Florida Keys, where it reportedly was the latest arrival among the sea turtles, laying in the late summer two sets of eggs that might total 350. Audubon stated that it was incautious in selecting a nesting site, like the loggerhead, in contrast to the hawksbill and green turtle, which chose the wildest and most remote spots. J. Schmidt's inquiries led him to believe that in the Danish West Indies nesting occurs from March to June, and mentioned especially the islands of Tortola and St. Croix as localities where leatherbacks laid. He stated that the eggs are 60–61 mm. (2.36–2.40 inches) in diameter, apparently taking his data from Reinhardt and Lütken. The Dry Tortugas and the Bahamas were formerly favored as breeding sites, and several old writers mentioned the coast of Brazil in this regard.

As far as I am aware, the only definite record of the laying of a leatherback on the coast of the mainland of the United States during the past hundred years was one quite recently established, and as is often the case when an arresting herpetological discovery is made in Florida, Mr. E. Ross Allen of Silver Springs was on the scene.

My friend Ross Allen is an exceptionally gifted zoological collector, and besides his unsurpassed skill in the field his operations there are attended by good luck with a consistency little short of uncanny.

Not long ago Ross handed me a set of excellent flashlight photographs of a nesting leatherback, suggesting that I might like to use them in the turtle book. We were in a meeting that was just convening at the time, and I made quite a spectacle of myself. In fact, I broke up the meeting. When I had got a grip on myself, Ross remarked that he had never been on a sea turtle hunt before; my reason began to totter again, and it still troubles me occasionally.

In answer to a letter from me requesting details of the above observation, Wilfred Neill, Ross Allen's associate, wrote as follows:

Four of the eggs were preserved. The accompanying label reads: "Eggs of *Dermochelys coriacea;* laid near Flager Beach, Flagler County, Florida, June 6, 1947, at 10:45 P.M.; 80 eggs laid at a depth of about 3 feet." Ross tells me that the site was about 150 feet back from the water.

I have measured the eggs for you. They are nearly, but not quite spherical. I tried to locate the maximum and minimum diameters. The measurements are as follows (in millimeters): 57.7 by 54.8; 58.0 by 54.9; 54.7 by 55.0. The egg shells, in formalin, seem to be somewhat brittle. They are so thin, however, that the egg is a bit flexible, not rigid.

The party that discovered the eggs was headed by John Maloney, writer. The photos were made by Carver Harris. The party was guided by Johnny Pappy. The turtle was captured and given to Marineland.

The only additional details on nesting procedure in the Atlantic leatherback that I can find are in the following account, quoted by Gosse (1851) from a Jamaica newspaper (the *Morning Journal,* for April 30, 1846):

> The anxiety of the fishermen in this little village [Negril Bay, at the western end of Jamaica] was aroused on the 30th of last month by the track of a huge Sea-monster, called a Trunk-turtle, which came on the seabeach for the purpose of laying her eggs. A search was made, when a hole in the sand was discovered, about 4 feet in depth, and as wide as the mouth of a half-barrel, whence five or six dozen white eggs were taken out. The eggs were of different sizes [cf. Deraniyagala's reference to the occurrence of eggs of "abnormal" sizes and shapes in egg complements of the Pacific leatherback], the largest the size of a duck's egg. On the morning of the 10th of this month, at half-past five o'clock, she was discovered by Mr. Crow on the beach, near the spot where she first came up. He gave the alarm, when all the neighbors assembled and got her turned on her back. She took twelve men to haul her about 200 yards.

Feeding: The scanty information available on the feeding habits of the leatherback indicates that it is probably omnivorous, since traces of both animal and plant food have been found in the few stomachs that have been examined.

Economic importance: Owing to its rare and sporadic occurrence, the Atlantic leatherback is of little value to man, although in the past the skin and shell covering have been rendered for the extraction of an oil that they contain in abundance. The eggs are eaten wherever found, but the meat is usually not esteemed.

Pacific Leatherback Turtle *Dermochelys coriacea schlegelii* (Garman)

Range: Pacific and Indian oceans. In the Eastern Hemisphere it ranges from Japan to southern Australia and the Cape of Good Hope. In the Western Hemisphere it has visited coastal waters from British Columbia to New Zealand and to Chile, where it is known from as far south as Chiloé Island.

Distinguishing features: The retention here of Garman's old and not especially reputable name [practically a *nomen nudum,* since it was supported by no description, but was based merely on a reference to figures in Temminck's and Schlegel's *Fauna Japonica*] does not imply that I (alone among herpetologists) know how to tell Pacific leatherbacks from Atlantic ones or that I even vouch for the distinction. I separate them, following the Stejneger and Barbour check lists, merely because (1) it is fairly common practice to do so; (2) it helps to avoid the confusion and meaningless generalizations of lumped life-history data; (3) it is my own conviction that there are slight differences that have eluded demonstration because of lack of series of comparable specimens; (4) people who feel strongly that there is only one leatherback may become indignant and jump on the problem tooth and claw to prove that I am wrong, and we might wind up knowing a lot more about the zoogeography of the leatherbacks.

It seems clear that the two (or more) stocks are not concealing any obscure but trenchant points of divergence, and that whatever characters may separate them are racial ones of trivial importance; for this reason it seems best to follow Garman in using subspecific nomenclature for them.

With respect to characters by which the two stocks may differ, I have collected the following more or less pertinent bits of information.

Van Denburgh (1905) said that the coloration looked different, basing his opinion on a comparison of California *schlegelii* with photographs of *coriacea.*

Philippi (1899) thought that he had from Chilean waters both typical *coriacea* and a new species (*Sphargis angusta*), and for the latter gave the following diagnostic characters: fore flippers shorter, their length equal to 59 per cent of the shell length in the new species and, according to him, 67 per cent in *coriacea;* shell narrower; tail longer; neck longer; color darker. It is probable that the main reason *angusta* looked queer to Philippi was that it was a male, although the different flipper length may be significant, as, indeed, may the narrower shell, despite its involvement in sexual dimorphism.

Babcock (1931) pointed out that the only divergence from *coriacea* shown by the figures of a Japanese leatherback in Temminck and Schlegel, on which Garman based his "variety *schlegelii,*" is a somewhat greater skull length, which may be individual. The same writer compared young specimens of *coriacea* with material from Ceylon and found the latter to show a less symmetrical arrangement of head scales and to have shorter fore flippers proportional to total length.

My own meager observations probably have little value, since I have seen mature *schlegelii* only in photographs, but a comparison of all available data seems to me to indicate that the Atlantic form is darker, with less light mottling of the back (usually solid black in adults) and especially of the lower jaw and throat.

Some time ago, before I had worked up any particular interest in the problem, I compared a small leatherback from Ceylon with one of similar size from the Atlantic Ocean off West Africa and entered in my notebook the following comments, which may add some slight weight to the suspected color difference:

The little Ceylon leatherback is 80 mm. in total length when unbent and agrees with the African specimen (total length 97 mm.) except that its dark ground color is more nearly black and its yellow is lighter and more generally and irregularly distributed over the ventral surface; its maxillary spines are not quite so long.

Inasmuch as the taxonomic subdivision of the leatherbacks will probably prove to be largely a matter of coloration, it is perhaps worth while to quote here the detailed account of the color pattern of the Indian Ocean form as given by Deraniyagala (1939a), who has seen more leatherbacks than any other herpetologist:

The adult is colored as follows:—Dorsally a slaty black with four or three longitudinal rows of small white spots not larger than the iris and extending between each pair of carapace ridges. These spots are more numerous at the bases of flippers. Head black with a few white blotches; jaws white clouded with black; neck with five longitudinal rows of white spots. Ventrally pinkish white or white, usually with dark reticulation representing the scale marks. A black lateral band usually extends from inguinal bay to cloaca. Sometimes in females the black disappears more or less completely from the plastron. Top of caudal crest white. The newly hatched young are an intense blue black marked with white to which the encroaching black imparts a bluish appearance. Dorsally the upper eyelid, outer halves of the supraocular scales, and jaws, are bluish white. The carapace ridges are bluish white and the neural, the narrowest of these white bands, has the scale behind the nuchoscapular hump, namely the seventh or eighth in the series, dark or nearly so. The inner three bands are continued on to the neck as uniserial rows of spots with two other bands between them. The flippers have white margins interrupted by black spots at the terminations of the first and second digits. The bases of the flippers and the anterior part of the lateral supramarginal area of the carapace are also feebly spotted with pale blue. A thin, bluish white line runs along the crest of the tail.

Ventrally the throat is white but a dark band runs behind the postmandibular scales and there is a dark median gular band. The bases of the flippers are blotched with white, the rest of each limb being black. The epiplastral prominence is white and from it radiate the white plastral ridges. Ecdysis [shedding] occurs after the first three or four weeks; the animal then becomes lighter in color, and within five months changes from a deep purplish indigo blue to a slaty grey. During this period the white carapace bands are broken up into large spots. A white spot appears on each carapace scale and on the scales at the base of each limb, while the white margins of the flippers widen; the plastron has the black interspaces considerably reduced in females, less so in males.

The sexual characters as outlined by Deraniyagala are as follows: The tail of the male is much longer, the back-stretched legs barely reaching the cloaca, while in the female the tail tip reaches barely halfway down the length of the legs. The shell of the male is lower, narrower across the hips, and with the plastron concave. The last osteoderm of each ventral keel forms a strong prominence which is probably useful in some way during copulation. The dark interspaces of the plastron are more persistent in the male than in the female.

Deraniyagala found the shells of hatchling Pacific leatherbacks to range in length from 58–60 mm. (2.28–2.36 inches).

It appears that leatherbacks of the Pacific may reach a larger size than those of the Atlantic, for there have been a disproportionately large number of specimens weighing over a thousand pounds among the relatively few records for the Pacific coast of the United States. Moreover, the Pacific form holds the record for total length—a reported nine feet for an Australian specimen—and for definitely recorded weight; there are a good record of a 1,500-pound specimen and a somewhat questionable report of a 1,900-pound individual, details of which are given below. One or the other of these may be the world's record if we exclude Agassiz' unexplained statement that he had seen leatherbacks weighing "more than a ton."

Table 15 summarizes dimensions of Pacific loggerheads caught on the Pacific coast of the United States; the figures are taken partially from newspaper accounts. I believe that all of the above measurements were made along the curve of the shell, but I cannot be entirely sure of this. Dimensions of the two largest specimens measured by Deraniyagala in Ceylon were as follows:

Total length	Curved carapace length	Curved carapace width	Arm spread	Weight
2100 mm. (82.7 inches)	1650 mm. (65 inches)	820 mm. (32.3 inches)	—	448 kilos (988 pounds)
2290 mm. (90.3 inches)	1560 mm. (61.5 inches)	870 mm. (34.2 inches)	2470 mm. (97.4 inches)	—

Table 15. Reported dimensions of Pacific loggerhead turtles.

Total length	Carapace length	Carapace width	Width flipper to flipper	Weight (lbs.)
—	4 ft. 5 in.	3 ft. 5 in.	—	—
6 ft. 7 in.	—	—	—	800
7 ft. 5 in.	5¼ ft.	2 ft. 10½ in.	4¼ ft.	—
8 ft. 7 in.	5 ft. 2 in.	3½ ft.	—	1,286
6¼ ft.	—	—	—	1,000
—	—	—	10 ft.	1,000
7⅓ ft.	—	3 ft. 2 in.	—	1,100
—	6 ft.	5 ft.	—	1,200
7⅔ ft.	—	—	9 ft.	1,575
6⅔ ft.	—	—	9 ft.	—
—	5 ft. 2 in.	—	—	1,902½

Habitat: The Pacific leatherback is pelagic. It is a swimmer of great speed and endurance, and its individual wanderings may actually contribute toward a breaking down of distance barriers and may thus explain the relative homogeneity of the Atlantic and Pacific stocks.

Habits: Beyond a few word-of-mouth accounts of the behavior of the occasional individuals that have been caught on the Pacific coast of the United States, nothing is known of the habits of this turtle in the eastern Pacific. The tremendous strength and speed of the animal lead one to imagine that it must roam the seas at will, but just how extensive the wanderings of a given individual might be, and whether any seasonal or concerted migrations occur, are matters of conjecture. The following clipping from the Oakland (California) *Tribune* for June 21, 1907, has the tantalizing ring of a perfectly circumstantial story, but is of course impossible to authenticate:

San Diego, June 21.—One of the puzzles of the sea came to the surface yesterday when a giant sea turtle was captured by B. Solissa, a fisherman, in whose net it had become entangled. The monster weighs 1,902½ pounds, almost a ton, and its shell is five feet two inches [1] from tip to tip. On its back somewhat plainly is burned or carved this inscription:

"British ship, Sea Bride, August 12, 1881, 3 south 86 west. If found please notify Thomas Fletcher, Brawley Road, Riverton, England." It would appear from this that the turtle was captured twenty-six years ago in the South Pacific and that it was released after the lettering had been burned in its shell.

As in the case of the Atlantic form, the majority of fishermen who have had encounters with the Pacific leatherback have referred to its voice and its pugnacity. Van Denburgh (1905) quoted from a letter sent by one of the participants giving the following exciting account of the capture of a large female off the California coast near Santa Barbara:

The turtle was first seen swimming on the surface about two miles off shore and to the southwestward of Santa Barbara whistling buoy. I went after it (accompanied by a boy) in an 18 foot sailboat. I had a gaff with a hook on the end of it and bent about 200 feet of rope onto the handle. I had also prepared a number of other ropes with nooses on them to be ready for quick work.

On approaching the turtle it did not hear the wash of the boat until we were within about 25 feet of it, when it made a rush to windward and started to dive, but the momentum of the boat when I luffed into the wind carried her right along side of him and I dropped the tiller and got forward with the gaff-hook and swung over the side in the weather rigging and got the hook fast in the leathery part of his neck. He immediately sounded and run [sic] out the full length of the line—about 200 feet—and towed the boat about half a mile further out to sea. He then came to the surface and we overhanded the line and pulled up close to him again. When he caught sight of the boat he turned and came toward us and threw one of his flippers over the gunwale of the boat, nearly capsizing her. I climbed up on the upper side and shoved him off with an oar. He grabbed the end of the oar and bit the end of it off like a piece of cheese. His movements in the water were very swift; using his fore flippers he could turn almost instantly from one side to the other and his head would project almost 18 inches from the body. I succeeded in throwing a noose over his head and later by attracting his attention in the opposite direction got ropes around both flippers—finally having five lines on him—and started to tow him toward the shore. He repeatedly slipped the ropes off from his neck and flippers—several times getting almost entirely free. We were from 11:30 A.M. till nearly 4 P.M. in finally landing him. When about half way to shore he suddenly turned and made a break out to sea, towing the boat stern first with all sail drawing full for several hundred yards with little effort. He emitted at intervals a noise resembling the grunt of a wild boar. There were (when we

[1] The credibility of this story is weakened by the evident incongruity in the moderate shell length and the tremendous weight attributed to the turtle.

first tackled him) about a dozen remoras attached to different parts of the body. Most of them stayed with him all through the struggle and only deserted him when I hoisted him to the deck of the dock. The turtle lived for four days after taking out of the water.

Breeding: The breeding habits of the leatherback in the eastern Pacific are not known. The only herpetologist who has had the enviable experience of observing the entire nesting and laying procedure of *Dermochelys* is Paul Deraniyagala of Ceylon. On the southern coasts of Ceylon nesting is frequent, and during the peak of the season in May and June bands of half a dozen or more leatherbacks may emerge to lay within a few miles. On the night of May 29, 1934, on the beach at Tangalla, in the Southern Province of Ceylon, Deraniyagala witnessed the whole nesting performance, and although his description of it has already appeared in print three times (Deraniyagala, 1936 and 1939a; Pope, 1939), it is such a unique contribution to natural history that I cannot resist the temptation to include it here in full:

Glistening silvery in the moonlight, the turtle ascended the beach in a straight line to the sandy embankment created by the scouring action of the waves. Through this obstacle she cut her path with simultaneous jerks of her powerful fore flippers and gained the dry sand. Here she commenced what Sinhalese fishermen term a "sand bath," flinging up a shower of sand over her back to a distance of about three meters by strong, simultaneous jerks of her fore limbs. The upward direction of these movements differed from her usual shuffle. The turtle probably tested the density of the sand while "sand bathing" for each jerk of her fore flippers excavated hollows which were 16 to 26 cm. deep. Her course was zig-zag and she even doubled back upon her track searching for a suitable place in which to nest. Meanwhile she was completely coated with sand, except for her eyes, which were washed by a copious flow of tears. After a satisfactory place was found, she dispelled the loose sand with a few preliminary sweeps of the flippers; a shallow cavity was next hollowed out posteriorly with a side to side movement of her carapace, facilitated by the outstretched hind limbs and cruro-caudal fold of skin. During this operation her fore flippers sank into the sand, and apparently acted as anchors, while a cushion-like mound of sand lay heaped behind each. After these preliminaries she excavated a smaller nest hole within the crater by working the hind limbs alternately, flinging the sand to a distance of 30 cm. or more as it was brought up. When the combined depth of the crater and the nest hole was about 100 cm. and she could no longer reach the bottom of the pit, she began to lay.

Anchored by the fore limbs, the turtle sloped her body into the pit at an angle of about 35 degrees, protruded her cloaca, and then deposited the eggs in batches of two or three at a time, moving her head up and down as she strained. Her breathing was somewhat stertorous and a peculiar fishy odor was noticeable. Her eggs laid, she began to fill the nest hole working the hind limbs alternately, each taking up a flipperful of sand which was gently placed upon the eggs. This was continued until the eggs were well covered, after which the sand was pushed in rapidly. Eventually, with her fore limbs still buried, the animal demolished the brink of the nest pit by swinging her outstretched hind quarters and tail rapidly from side to side, although the carapace was stationary. During this procedure, every time a hind flipper touched the ground, it flung the sand crosswise toward its fellow with a rapid, scooping movement.

Although the turtle is said to be cautious in her approach to land, once oviposition commenced the animal was completely indifferent to the presence of man, noise or lights, and this indifference persisted even after she had covered up the eggs, and had begun to churn up the beach all around the nest, in spite of blows.

When the nest was nearly covered she moved her front limbs for the first time since oviposition commenced. Both were jerked back simultaneously, showering sand over her back and into the pit, but without visibly altering her position. She worked the front and hind limbs in turn over five minute intervals, the former always with a simultaneous jerk, the latter generally alternately. Eventually she gyrated upon her plastron upon the nest area and flung up great scoopfuls of sand with her fore flippers, occasionally employing her head to push down any ridges of sand created during this process. Throughout this phase the turtle did not appear to move from the nest and it was only by comparing the animal's position with a haversack I had laid down when she first commenced to dig that it became apparent that she had moved quite two meters during ten minutes. At this stage I struck her a sharp blow on the head with a stick and sat upon her, but undeterred she continued to churn up the sand and worked shoreward instead of towards the sea.

After a time she doubled back on her tracks, and slowly and laboriously repeated the process. Finally she decided that her duty was done, and it was certainly very thorough, for after she had gone, three of us dug for an hour with our hands but were unable to locate the eggs. The departing turtle no longered showered sand with her fore limbs, but wearily made for the sea, stopping after every two or three shuffles, blowing most of the time. Gradually she recovered her energy and rested only after every ten or fifteen shuffles. She approached the wave line and there paused. When the surf reached her she allowed herself to be washed away into the waves without exerting herself. At her first attempt she failed to get past the breakers, but as the next wave rose she hugged the ground, escaped under the wave and was gone.

The eggs of the Pacific leatherback are, according to measurements given by both Günther (1864) and Deraniyagala, smaller than those of the

Atlantic form. In Ceylon normal eggs vary in diameter between 50 and 54 mm. (1.94–2.12 inches) and are usually spherical, soft-shelled, and white, although some may be sprinkled with small green spots. The complements range from 90 to 130, and of these 10 or 15 are usually abnormal, being small or ellipsoid or dumbbell-shaped and without yolk. Each turtle is thought to lay three or four times a season, although conclusive data on this point are lacking. The incubation period is from 55 to 65 days.

Contrasting with the early summer nesting season of Ceylonese leatherbacks is the record given by Günther of a nesting emergence on February 1 on the coast of Tenasserim.

Feeding: The Pacific leatherback is apparently omnivorous; dissections of stomachs have revealed jellyfish, crustacea, and algae. Deraniyagala successfully raised young specimens on a diet of fish, eggs, bread, young octopus, and fleshy green algae.

Economic importance: The oil yielded by the skin and shell is widely valued as varnish, used often on boats. The eggs are eaten wherever available, and the meat likewise is used as food, although estimates of its palatability vary surprisingly. A specimen was bought by a California hotel and fed to patrons who pronounced it "very fine eating." There are indications that it may sporadically be slightly poisonous.

BIBLIOGRAPHY

Abbey, C. D. 1882. Longevity of the turtle. Amer. Nat., 16: 243.

Abbott, Charles C. 1868. Catalogue of vertebrate animals of New Jersey. Geology of New Jersey, pp. 751–830.

——. 1882. Notes on the habits of the "savannah cricket frog." Amer. Nat., 16: 707–711.

——. 1884. Hibernation of the lower vertebrates. Science, 4: 36–39.

Abbott, W. L. 1894. Notes on the natural history of Aldabra, Assumption and Glorioso Islands, Indian Ocean. Proc. U.S. Nat. Mus., 16: 759–764.

Abel, Othenio. 1919. Die Stämme der Wirbeltiere. Berlin and Leipzig. Pp. i–xviii, 1–914, 669 figs.

Agassiz, Louis. 1857. Contribution to the natural history of the United States. Boston, Little, Brown and Co. Vols. 1–4.

Albert, I., Prince of Monaco. 1898. Sur le développement des tortues (*T. caretta*). Soc. Biologie Paris, 5: 10–11.

Alexander, A. B. 1902. Statistics of the fisheries of the South-Atlantic states. Rept. U.S. Comm. Fish and Fisheries, 29: 343–410.

Allard, H. A. 1909. Notes on two common turtles of the eastern United States. Science, 30: 453–454.

——. 1935. The natural history of the box turtle. Sci. Monthly, 41: 325–338.

——. 1939. Mating of the box turtle ending in death to the male. Copeia, no. 2, p. 109.

——. 1948. The eastern box turtle and its behavior. Jour. Tenn. Acad. Sci., 23: 307–321.

——. 1949. The eastern box turtle and its behavior. *Ibid.,* 24: 146–152.

Allen, D. L. 1938. Ecological studies on the vertebrate fauna of a 500-acre farm in Kalamazoo County, Michigan. Ecol. Monogr., 8: 347–436, 28 figs.

Allen, E. Ross. 1938. Notes on feeding and egg laying habits of the Pseudemys. Proc. Fla. Acad. Sci., 3: 105–108, 9 figs.

——. [n.d.]. The gopher tortoise. All Pets Magazine.

——, and Ralph Slatten. 1945. A herpetological collection from the vicinity of Key West, Florida. Herpetologica, 3: 25–26.

Allen, Glover M. 1899. Notes on the reptiles and amphibians of Intervale, New Hampshire. Proc. Boston Soc. Nat. Hist., 29: 63–75.

Allen, J. A. 1868. Catalogue of the reptiles and batrachians found in the vicinity of Springfield, Massachusetts, with notices of all the other species known to inhabit the state. Proc. Boston Soc. Nat. Hist., 12: 171–203.

——. 1869. Appendix to the catalogue of reptiles and batrachians found in the vicinity of Springfield, Massachusetts. *Ibid.,* 248–250.

——. 1870. Notes on Massachusetts reptiles and batrachians. *Ibid.,* 13: 200–263.

——. 1874. Notes on the natural history of portions of Dakota and Montana territories. *Ibid.,* 17: 33–70.

Allen, Morrow J. 1932. A survey of the amphibians and reptiles of Harrison County, Mississippi. Amer. Mus. Novitates, no. 542, pp. 1–20.

——. 1933. Report on a collection of amphibians and reptiles from Sonora, Mexico, with the description of a new lizard. Occ. Papers Mus. Zool. Univ. Mich., 259: 1–15.

Allin, A. E. 1940. The vertebrate fauna of Darlington Township, Durham County, Ontario. Trans. Roy. Can. Inst., 23: 83–118.

"Amphibius." 1938. Land and water tortoises. Their care in captivity. London, Marshall Press. Pp. 1–31, 9 figs.

Anderson, Oscar I., and James R. Slater. 1941. Life zone distribution of the Oregon reptiles. Occ. Papers Dept. Biol. College Puget Sound, 15: 109–119, tables 5–6.

Anderson, Paul. 1942. Amphibians and reptiles of Jackson County, Missouri. Bull. Chicago Acad. Sci., 6: 203–222.

Angel, Fernand. 1946. Reptiles and amphibians. Faune de France. Paris, Feder. Fran. Soc. Sci. Nat. Pp. 1–204, 83 figs.

Annandale, N., and M. H. Shastri. 1914. Relics of the worship of mud turtles (Trionychidae) in India and Burma. Jour. Proc. Asiatic Soc. Bengal, n.s., 10: 131–138.

Anonymous. 1917. Animal and plant life of Oklahoma. Okla. Geol. Survey, circ. no. 6, pp. 1–68 (reptiles, pp. 34–35).

——. 1918. The reptiles of Rhode Island. Part I. Turtles. Park Mus. Bull. Providence, Rhode Island, 10: 81–84.

——. 1932. A remarkable capture of leatherneck turtles off Bajo Reef, near Nootka Sound, west coast of Vancouver Island, B.C. Rept. Prov. Mus. Nat. Hist., Victoria, B.C., p. 6, pl.

——. 1945. Nevada: Studies of habits of tortoise. Desert Mag., 8: 30.

——. 1946a. Mammoth turtle captured in Blue River, Oklahoma. Game and Fish News, 2: 20, 1 fig.

——. 1946b. Fifty gallons of soup on the hoof. *Ibid.,* 22, 1 fig.

——. 1946c. Deep fork fishermen capture 100-pound turtle. *Ibid.,* 16, 1 fig.

——. 1947a. Giant turtles captured in eastern Oklahoma. *Ibid.,* 3: 11, 2 figs.

——. 1947b. Conservation goes to the fair. *Ibid.,* 12–13, 6 figs.

Ashley, Harold R. 1948. Muhlenberg's turtle in southern New York. Copeia, no. 3, p. 220.

Atkinson, D. A. 1901. The reptiles of Allegheny County, Pennsylvania. Ann. Carnegie Mus., 1: 145–157.

Audubon, John James. 1926. Delineations of American scenery and character. New York, G. A. Baker and Co. Pp. 194–202.

Babbitt, Lewis H. 1932. Some remarks on Connecticut herpetology. Bull. Boston Soc. Nat. Hist., 63: 23–28.

——. 1936. Soft shelled turtles in Vermont. *Ibid.*, 78: 10.

Babcock, Harold L. 1916a. Notes on an unusual feeding habit of the snapping turtle, *Chelydra serpentina* (Linné). Copeia, no. 37, pp. 89–90.

——. 1916b. An addition to the Chelonian fauna of Massachusetts. *Ibid.*, no. 38, pp. 95–98.

——. 1917a. An extension of the range of *Clemmys muhlenbergii*. *Ibid.*, no. 42, p. 32.

——. 1917b. Further notes on *Pseudemys* at Plymouth, Massachusetts. *Ibid.*, no. 44, p. 52.

——. 1917c. Habits of a box turtle. *Ibid.*, no. 49, pp. 91–92.

——. 1918. Notes on some New England turtles. *Ibid.*, no. 53, pp. 15–16.

——. 1919. The turtles of New England. Mem. Boston Soc. Nat. Hist., 8: 323–431, pls. 17–32.

——. 1920. Some reptile records from New England. Copeia, no. 85, pp. 73–74.

——. 1926a. New England turtles. Bull. Boston Soc. Nat. Hist., 39: 5–9.

——. 1926b. The diamond-back terrapin in Massachusetts. Copeia, no. 150, pp. 101–104.

——. 1928. The long life of turtles. Bull. Boston Soc. Nat. Hist., 46: 9–10.

——. 1930a. *Caretta kempii* from Massachusetts. Copeia, no. 1, p. 21.

——. 1930b. Variation in the number of costal shields in *Caretta*. Amer. Nat., 64: 1–2.

——. 1931. Notes on *Dermochelys*. Copeia, no. 3, p. 142.

——. 1932. The American snapping turtles of the genus *Chelydra* in the collection of the Museum of Comparative Zoology, Cambridge, Massachusetts, U.S.A. Proc. Zool. Soc. London, 4: 873–874.

——. 1933. The eastern limit of range for *Chrysemys picta marginata*. Copeia, no. 2, p. 101.

——. 1937a. A new subspecies of the red-bellied terrapin, *Pseudemys rubriventris* (Le Conte). Occ. Papers Boston Soc. Nat. Hist., 8: 293–294.

——. 1937b. The sea-turtles of the Bermuda Islands, with a survey of the present state of the turtle-fishing industry. Proc. Zool. Soc. London, ser. A, 107: 595–601.

——. 1938. Field guide to New England turtles. New England Mus. Nat. Hist., Nat. Hist. Guide, 2: 1–56, 9 pls.

——. 1939a. How to make turtle soup. New England Field Nat., 2: 16–17, 2 figs.

——. 1939b. Growth of an individual box turtle. Copeia, no. 3, p. 175.

——. 1940. Tortoise shell and the arts. New England Nat., 8: 21–23, illustrated.

——. 1941. *Chrysemys* from Nantucket Island. Copeia, no. 4, p. 262.

Bailey, Reeve M. 1941. The occurrence of the wood turtle in Iowa. *Ibid.,* p. 265.

Bailey, Vernon. 1913. Life zones and the crop zones of New Mexico. North Amer. Fauna, 35: 1–100, 6 figs., 16 pls.

——. 1923. Sources of water supply for desert animals. Sci. Monthly, 17: 66–86.

——. 1928. The desert tortoise. Nat. Mag., 12: 372–374, 4 figs.

——. 1933. Cave life of Kentucky. Amer. Midland Nat., 14: 385–635, 90 figs.

Baird, Spencer F. 1857. Report of the reptiles collected on the survey (No. 3). Expl. Surv. R.R. Route Pacific Ocean, Zool., Reptiles, vol. 10, no. 4 (1853–1856), pp. 16–20, pls. 17, 18, 23, 24.

——. 1859. Reptiles of the boundary. *In* W. H. Emory, Report of the United States and Mexican survey made under the direction of the Secretary of the Interior. Vol. 2, pt. 2, pp. 1–35, 41 pls.

——, and Charles Girard. 1852. Descriptions of new species of reptiles collected by the U.S. exploring expedition under the command of Capt. Charles Wilkes, U.S.N. Proc. Acad. Nat. Sci. Philadelphia, pp. 174–177.

Baldwin, F. M. 1925. The relation of body to environmental temperatures in turtles, *Chrysemys marginata* and *Chelydra serpentina*. Biol. Bull., 48: 432–445.

Bangs, Outram. 1896. An important addition to the fauna of Massachusetts. Proc. Boston Soc. Nat. Hist., 27: 159–161.

Banks, E. 1937. The breeding of the edible turtle (*Chelonia mydas*). Sarawak Mus. Jour., 4: 523–532.

Barbier, H. 1905. Les chéloniens du Musée d'Histoire Naturelle d'Elbeuf. Bull. Soc. Etude Sci. Nat. Mus. Hist. Nat. Elbeuf, 23: 69–102, 2 pls.

Barbour, Thomas. 1914. On some Australasian reptiles. Proc. Biol. Soc. Washington, 27: 201–206.

——. 1934. Reptiles and amphibians: their habits and adaptations. Rev. ed. Boston, Houghton, Mifflin Co. Pp. i–xx, 1–129, 142 figs.

——. 1942. New records for ridleys. Copeia, no. 4, pp. 257–258.

——, and A. F. Carr, Jr. 1940. Antillean terrapins. Mem. Mus. Comp. Zool., 54: 381–413, 5 figs., 9 pls.

——, and H. C. Stetson. 1931. A revision of the Pleistocene species of *Terrapene* of Florida. Bull. Mus. Comp. Zool., 72: 295–299, 3 pls.

Bard, Samuel A. 1855. Waikna; or adventures on the Mosquito Shore. New York, Harper and Bros. Pp. i–ix, 13–366, 56 figs.

Barkalow, Frederick S. 1948. Notes on the breeding habits of the turtle *Kinosternon s. subrubrum*. Copeia, no. 2, p. 130.

Barney, R. L. 1922. Further notes on the natural history and the artificial propagation of the diamond-back terrapin. Bull. U.S. Bureau Fish., 38: 91–111, figs. 76–84, tables 1–4.

Barrett, C. 1924. Reptile life in Australia. Nat. Hist., 24: 43–59.

Bartram, William. 1794. Travels through North and South Carolina, Georgia, East and West Florida. 2d ed. London. Pp. i–xxiv, 1–520, 7 pls.

—— (annotated by Francis Harper). 1943. Travels in Georgia and Florida, 1773–74, a report to Dr. John Fothergill. Trans. Amer. Phil. Soc., n.s., 33: 121–242, 26 pls.

Baur, G. 1887a. Ueber die Stellung der Trionychidae zu den übrigen Testudinata. Zool. Anz., 10: 96.

——. 1887b. On the morphology of the carapace in Testudinata. Amer. Nat., 21: 89.

——. 1888a. Unusual dermal ossifications. Science, 11: 144.

——. 1888b. Notes on the American Trionychidae. Amer. Nat., 22: 1121–1122.

——. 1888c. Osteologische Notizen über Reptilien. Zool. Anz., 258: 1–7.

——. 1889. The relationship of the genus *Dirochelys*. Amer. Nat., 23: 1099–1100.

——. 1890a. The genera of the Cheloniidae. *Ibid.*, 24: 486–487.

——. 1890b. On the classification of the Testudinata. *Ibid.*, 530–536.

——. 1890c. Two new species of tortoises from the South. Science, n.s., 16, 262.

——. 1891. American box-tortoises. *Ibid.*, 17: 190–191.

——. 1893a. Notes on the classification and taxonomy of the Testudinata. Proc. Amer. Phil. Soc., 31: 210–225.

——. 1893b. Notes on the classification of the Cryptodira. Amer. Nat., 27: 672–675.

——. 1893c. Two new species of North American Testudinata. *Ibid.*, 675–676.

——. 1893d. Further notes on American box turtles. *Ibid.*, 677–678.

——. 1893e. Notes on the American Trionychidae. *Ibid.*, 1121.

Beck, William M. 1938. Notes on the reptiles of Paynes Prairie, Alachua County, Florida. Fla. Nat., 11: 85–87.

Beebe, William. 1924. Galápagos: world's end. New York, G. P. Putnam's Sons. Pp. i–xxi, 1–433, 8 pls., 82 figs.

——. 1925. One quarter of a square mile of jungle at Kartabo, British Guiana. Contrib. N.Y. Zool. Soc., 6: 5–193, 16 figs.

——. 1937. Turtle sanctuary. Harpers, Nov., pp. 653–660.

Belding, L. 1887. Collecting in the Cape Region of Lower California. West. Amer. Sci., 3: 93–99.

Bell, Thomas. 1831. A monograph of the Testudinata. London. Pp. i–xxiv, 1–81, 40 pls.

Bello y Espinosa, ——. 1871. Zoologische Notizen aus Puerto Rico. Zool. Garten, 12: 348–351.

Benedetti, Delia. 1925. Richerche sull' accrescimento della *Testudo graeca* (L.). Boll. Inst. Zool., R. Univ. Roma, 3: 108–125.

Benedict, Francis G. 1932. The physiology of large reptiles. Carnegie Inst. Washington Publ. no. 425, pp. i–x, 3–539, 106 figs.

Berridge, W. S. 1926. Marvels of reptile life. London, Thornton Butterworth. Pp. 1–250.

Beyer, George S. 1900. Louisiana herpetology. Proc. Louisiana Soc. Nat., 1897–1899: 24–46.

Beyer, H. G. 1885. The influence of variations of temperature upon the rate and the work of the heart of the slider terrapin (*Pseudemys rugosa*). Proc. U.S. Nat. Mus., 8: 225–230.

Bishop, Sherman C. 1921. The map turtle, *Graptemys geographica* (Le Sueur) in New York. Copeia, no. 100, pp. 80–81.

——. 1923. Notes on the herpetology of Albany County, New York. III. The snakes and turtles. *Ibid.*, no. 125, pp. 117–120.

——, and W. P. Alexander. 1927. The amphibians and reptiles of Allegany State Park. N.Y. State Mus. Handbook, 3: 31–141, 58 figs., map.

——, and F. J. W. Schmidt. 1931. The painted turtles of the genus *Chrysemys*. Field Mus. Nat. Hist., Zool. Ser. Publ. no. 293, 18: 123–139, 27 figs.

——, and W. J. Schoonmacher. 1921. Turtle hunting in midwinter. Copeia, no. 96, pp. 37–38.

Blake, S. F. 1921. Sexual differences in coloration in the spotted turtle, *Clemmys guttata*. Proc. U.S. Nat. Mus., 59: 463–469, 1 pl.

Blanchard, Frank N. 1921. A collection of amphibians and reptiles from northeastern Washington. Copeia, no. 90, pp. 5–6.

——. 1922. The amphibians and reptiles of western Tennessee. Occ. Papers Mus. Zool. Univ. Mich., 117: 1–18.

——. 1923. The amphibians and reptiles of Dickinson County, Iowa. Univ. Iowa Studies Nat. Hist., 10: 19–26.

——. 1924. A collection of amphibians and reptiles from southeastern Missouri and southern Illinois. Papers Mich. Acad. Sci., Arts, Lett., 4: 533–541.

——. 1925. A collection of amphibians and reptiles from southern Indiana and adjacent Kentucky. Papers Mich. Acad. Sci., 5: 367–388, 2 pls.

——. 1928. Amphibians and reptiles of the Douglas Lake region in northern Michigan. Copeia, no. 167, pp. 42–51.

Blatchley, W. S. 1891. Notes on the batrachians and reptiles of Vigo County, Indiana. Jour. Cincinnati Soc. Nat. Hist., 14: 22–35.

——. 1899. Notes on the batrachians and reptiles of Vigo County, Indiana. II. Ann. Rept. Dept. Geog. Nat. Res. Indiana, 24: 537–552.

——. 1900. The lakes of northern Indiana. 25th Ann. Rept. Geol. Nat. Res. Indiana.

——. 1906. On some reptilian freaks from Indiana. Proc. Acad. Nat. Sci. Philadelphia 58: 419–422.

Bleeker, P. 1858. Opsomming der tot dus verre van het eiland Sumatra bekend gewordene Reptilien. Nat. Tijds. Nederland. Indië, 15: 260–263.

Bocourt, Marie Firmin. 1868. Description de quelques chéloniens nouveaux appartenant à la faune mexicaine. Ann. Sci. Nat., ser. 5, Zool., 10: 121–122.

——. 1870–1909. Etudes sur les reptiles. *In* Duméril, Bocourt, and Mocquard, Recherches zoologiques pour servir à l'histoire de la faune de l'Amérique Centrale et du Mexique. Paris, Mission Scientifique au Mexique et dans l'Amérique Centrale. Pt. 3, sect. 1. Pp. i–xiv, 1–1012, 77 pls., map.

Boettger, O. 1893. Katalog der Reptilien-Sammlung im Museum der senckenbergischen naturforschenden Gesellschaft in Frankfurt am Main. I Teil. (Rhynchocephalen, Schildkröten, Krokodile, Eidechsen, Chameleons.) Frankfurt a.M. Pp. i–x, 1–140.

Bogert, Charles M. 1930. An annotated list of the amphibians and reptiles of Los Angeles County, California. Bull. South. Calif. Acad. Sci., 24: 3–14.

——. 1937. Note on the growth of the desert tortoise, *Gopherus agassizii*. Copeia, no. 3, pp. 191–192.

——, and Raymond B. Cowles. 1947. Moisture loss in relation to habitat selection in some Floridian reptiles. Amer. Mus. Novitates, no. 1358, pp. 1–34, 23 figs.

——, and James A. Oliver. 1945. A preliminary analysis of the herpetofauna of Sonora. Bull. Amer. Mus. Nat. Hist., 83: 301–425, 13 figs., pls. 31–37, 1 table, 2 maps.

Bonaparte, Carlo Luciano. 1830. Osservazioni sulla seconda edizione del Regno Animale del Barone Cuvier. Ann. Storia Nat., Bologna, 4: 3–26, 159–122, 303–389.

——. 1833. Über Cuviers Thierreich: Schildkröten. Oken, Isis, no. 11, pp. 1041–1099.

Bond, Harley D. 1931. Some reptiles and amphibians of Monogalia County, West Virginia. Copeia, no. 2, pp. 53–54.

Bonnaterre, M. l'A. 1789. Tableau encyclopédique et méthodique des trois règnes de la nature. Erpétologie. Paris. Pp. iii–xxviii, 1–70, 26 pls.

Bosc, L. A. G. 1804. Tortue. Nouv. Dictionnaire Hist. Nat., Paris, 22: 242–271, pl. 8.

Boulenger, G. A. 1889. Catalogue of the chelonians, . . . of the British Museum. New ed. London. Pp. iii–x, 1–311, 6 pls., 73 figs.

——. 1895. On the American box turtles. Ann. Mag. Nat. Hist., ser. 6, 15: 330–331.

——. 1914. Reptiles and batrachians. London, J. M. Dent and Sons; New York, E. P. Dutton and Co. Ch. 2: Chelonia—turtles, terrapins and tortoises, pp. 10–41.

Bourret, R. 1941. Les tortues de l'Indochine. Inst. Oceanogr. l'Indochine, note 38, pp. 1–235, figs., 43 pls.

Boyer, D. A., and A. A. Heinze. 1934. An annotated list of the amphibians and reptiles of Jefferson County, Missouri. Trans. Acad. Sci. St. Louis, 28: 185–200, pls.

Brady, Maurice K. 1924. Muhlenberg's turtle near Washington. Copeia, no. 135, p. 92.

——. 1925. Notes on the herpetology of Hog Island (Va., Northampton Co., E. shore). *Ibid.,* no. 137, pp. 110–111.

——. 1927. Notes on the reptiles and amphibians of the Dismal Swamp. *Ibid.,* no. 162, pp. 26–29.

——. 1937. Natural history of Plummer's Island, Maryland. 6. Reptiles and amphibians. Proc. Biol. Soc. Washington, 50: 137–140.

Breckenridge, W. J. 1944. Reptiles and amphibians of Minnesota. Univ. Minn. Press. Pp. 1–202.

Breder, Charles M., Jr., and R. B. Breder. 1923. A list of fishes, amphibians and reptiles collected in Ashe County, North Carolina. Zoologica, 4: 1–23, 8 figs.

Breder, Ruth Bernice. 1924. Notes on the drinking habits of *Terrapene carolina* (Linnaeus). Copeia, no. 131, pp. 63–64.

——. 1927. Turtle trailing: a new technique for studying the life habits of certain testudinata. Zoologica, 9: 231–243, figs. 278–284.

Brenkelman, John. 1936. A list of the amphibians and reptiles of Chase and Lyons counties, Kansas. Trans. Kansas Acad. Sci., 39: 267–268.

Brennan, L. A. 1934. A check list of the amphibians and reptiles of Ellis County, Kansas. *Ibid.,* 37: 189–191.

——. 1937. A study of the habitat of the reptiles and amphibians of Ellis County, Kansas. *Ibid.,* 40: 341–347.

Brice, J. J. 1897. The fish and fisheries of the coastal waters of Florida. Rept. U.S. Fish Comm. (Senate Doc. no. 100), pp. 1–80.

Brimley, Clement S. 1903. Notes on the reproduction of certain reptiles. Amer. Nat., 37: 261–266.

——. 1904a. Further notes on the reproduction of reptiles. Jour. Elisha Mitch. Sci. Soc., 20: 139–140.

——. 1904b. The box tortoise of southeastern North America. *Ibid.,* 1–8.

——. 1905. Notes on the food and feeding habits of some American reptiles. *Ibid.*, 21: 149–155.

——. 1907. Notes on some turtles of the genus *Pseudemys*. *Ibid.*, June, pp. 76–84.

——. 1909. Some notes on the zoology of Lake Ellis, Craven County, North Carolina, with special reference to herpetology. Proc. Biol. Soc. Washington, 22: 129–138.

——. 1910. Records of some reptiles and batrachians from the southeastern United States. *Ibid.*, 23: 9–18.

——. 1915. List of reptiles and amphibians of North Carolina. Jour. Elisha Mitch. Sci. Soc., 30: 195–206.

——. 1917. Some known changes in the land vertebrate fauna of North Carolina. *Ibid.*, 32: 176–183.

——. 1918a. Brief comparison of the herpetological faunas of North Carolina and Virginia. *Ibid.*, 34: 146–147.

——. 1918b. Eliminations from and additions to the North Carolina list of reptiles and amphibians. *Ibid.*, 148–149.

——. 1920a. The turtles of North Carolina. *Ibid.*, 36: 62–71.

——. 1920b. Notes on *Pseudemys scripta* Schoepff, the yellow-bellied turtle. Copeia, no. 87, pp. 93–94.

——. 1922. Herpetological notes from North Carolina. *Ibid.*, no. 107, pp. 47–48.

——. 1926. Revised key and list of the amphibians and reptiles of North Carolina. Jour. Elisha Mitch. Sci. Soc., 42: 75–93.

——. 1927. Some records of amphibians and reptiles from North Carolina. Copeia, no. 162, pp. 10–12.

——. 1928. Two new terrapins of the genus *Pseudemys* from the southern states. Jour. Elisha Mitch. Sci. Soc., 44: 66–69, 2 pls.

——. 1942–1943. Reptiles and amphibians of North Carolina. The turtles, tortoises or terrapins. Carolina Tips, 1942, 5: 22–23; 1943, 6: 2–3, 6–7, 10–11, 14–15, 18–19.

——, and W. B. Mabee. 1925. Reptiles, amphibians and fishes collected in eastern North Carolina in the autumn of 1923. Copeia, no. 139, pp. 14–16.

——, and Franklin Sherman. 1908. Notes on the life-zones in North Carolina. Jour. Elisha Mitch. Sci. Soc., 24: 14–22.

Britcher, H. W. 1903. Batrachia and Reptilia of Onondaga County. Proc. Onondaga Acad. Sci., 1: 120–122.

Brode, J. Stanley. 1932. Reptilian inhabitants of the school terrarium. Oregon Educ. Jour., 6: 8–9, 26–28, 3 figs.

Brown, Arthur E. 1903. Texas reptiles and their faunal relations. Proc. Acad. Nat. Sci. Philadelphia, 1903, pp. 543–558.

Brown, F. H.   1863.   [*Sphargis coriacea* from off Saco, Maine.]   Proc. Boston Soc. Nat. Hist., 9: 326–327.

Brown, H. H.   1946.   The fisheries of the Windward and Leeward Islands. Development Welfare Bull., no. 20, pp. 1–97, 28 tables.

Brown, J. Roland.   1927.   A Blanding's turtle lays its eggs.   Canadian Field-Nat., 41: 185.

——.   1928.   The herpetology of Hamilton, Ontario and district.   *Ibid.,* 42: 125–127.

Brownell, L. W.   1925.   Posing snakes and other reptiles.   Camera, 31: 362–368, figs.

Brumwell, Malcolm J.   1940.   Notes on the courtship of the turtle, *Terrapene ornata*.   Trans. Kansas Acad. Sci., 43: 391–392.

Bumpus, H. C.   1884–1886.   Reptiles and batrachians of Rhode Island.   Providence, R.I.   1 (1884, nos. 10–12); 2 (1885, nos. 1–12); 3 (1886, nos. 1–2).

Burger, J. Wendell.   1937.   Experimental photoperiodicity in the male turtle, *Pseudemys elegans* (Wied).   Amer. Nat., 71: 481–487.

Burger, W. Leslie, Philip W. Smith, and Hobart M. Smith.   1949.   Notable records of amphibians and reptiles in Oklahoma, Arkansas and Texas.   Jour. Tenn. Acad. Sci., 24: 130–134.

Burne, R. H.   1905.   Notes on the muscular and visceral anatomy of the leathery turtle (*Dermochelys coriacea*).   Proc. Zool. Soc. London, 1905, pp. 291–324.

Burt, Charles E.   1927.   An annotated list of the amphibians and reptiles of Riley County, Kansas.   Occ. Papers Mus. Zool. Univ. Mich., 189: 1–9.

——.   1931.   A report on some amphibians and reptiles from Kansas, Nebraska, and Oklahoma.   Proc. Biol. Soc. Washington, 44: 11–16.

——.   1933.   Some distributional and ecological records of Kansas reptiles. Trans. Kansas Acad. Sci., 36: 186–208.

——.   1935.   Further records of the ecology and distribution of amphibians and reptiles in the Middle West.   Amer. Midland Nat., 16: 311–336.

——.   1937.   Amphibians and reptiles of "Rock City."   Trans. Kansas Acad. Sci., 50: 15.

——.   1944.   An albino false map turtle.   Copeia, 4: 253.

——, and May Danheim Burt.   1929a.   A collection of amphibians and reptiles from the Mississippi Valley, with field observations.   Amer. Mus. Novitates, no. 381, pp. 1–14.

——, and May Danheim Burt.   1929b.   Field notes and locality records on a collection of amphibians and reptiles, chiefly from the western half of the U.S. Jour. Washington Acad. Sci., 19: 428–434, 448–460.

——, and William L. Hoyle.   1934.   Additional records of the reptiles of the central prairie region of the United States.   Trans. Kansas Acad. Sci., 37: 193–216.

Burton, R. W. 1917. Habits of the green turtle (*Chelone mydas*). Jour. Bombay Nat. Hist. Soc., 25: 508.

Butler, Amos W. 1885. Hibernation of the lower vertebrates. Amer. Nat., 19: 37–40.

———. 1887. Herpetology. Jour. Cincinnati Soc. Nat. Hist., 9: 263–265.

———. 1892. Contributions to Indiana herpetology. *Ibid.,* 14: 169–179.

Byrd, Elon E. 1939. Certain aspects of the anatomy of a two-headed turtle. Rept. Reelfoot Lake Biol. Sta. *In* Jour. Tenn. Acad. Sci., 3: 102–109, 1 pl.

Cagle, Fred R. 1937. Egg laying habits of the slider turtle (*Pseudemys troostii*), the painted turtle (*Chrysemys picta*) and the musk turtle (*Sternotherus odoratus*). *Ibid.,* 12: 87–95, 1 fig., 2 tables.

———. 1939. A system of marking turtles for future identification. Copeia, no. 3, pp. 170–173, 5 figs.

———. 1941. Key to the reptiles and amphibians of Illinois. Mus. Nat. Sci. Soc. South. Illinois Normal Univ., pp. 1–32, 66 figs., 3 pls.

———. 1942a. Turtle populations in southern Illinois. Copeia, no. 3, pp. 155–162.

———. 1942b. Herpetological fauna of Jackson and Union counties, Illinois. Amer. Midland Nat., 28: 164–200.

———. 1944a. A technique for obtaining turtle eggs for study. Copeia, no. 1, p. 60.

———. 1944b. Sexual maturity in the female of the turtle *Pseudemys scripta elegans*. *Ibid.,* no. 3, pp. 149–151, 1 table.

———. 1944c. Activity and winter changes of hatchling *Pseudemys*. *Ibid.,* no. 2, pp. 105–109, 2 tables.

———. 1944d. Home range, homing behavior and migration in turtles. Misc. Publ. Mus. Zool. Univ. Mich., 61: 1–34, 2 pls.

———. 1945. Recovery from serious injury in the painted turtle. Copeia, no. 1, p. 45.

———. 1946. The growth of the slider turtle, *Pseudemys scripta elegans*. Amer. Midland Nat., 36: 385–729.

———. 1947. Color abnormalities in *Pseudemys scripta troostii* (Holbrook). Chicago Acad. Sci. Nat. Hist. Miscel., 6: 1–3, 1 fig.

———. 1948a. Sexual maturity in the male turtle, *Pseudemys scripta troostii*. Copeia, no. 2, pp. 108–111, 3 figs.

———. 1948b. The growth of turtles in Lake Glendale, Illinois. *Ibid.,* no. 3, pp. 197–203, 3 figs., 1 table.

———, and Joseph Tihen. 1948. Retention of eggs by the turtle *Deirochelys reticularia*. *Ibid.,* no. 1, pp. 1–66.

Cahn, Alvin R. 1929. The herpetology of Waukesha County, Wisconsin. *Ibid.,* no. 170, pp. 4–8.

——. 1931. *Kinosternon flavescens,* a surprising turtle record from Illinois. *Ibid.,* no. 3, pp. 120–123, 4 figs.

——. 1933. Hibernation of the box turtle. *Ibid.,* no. 1, pp. 13–14.

——. 1937. The turtles of Illinois. Ill. Biol. Mon., 16: 1–218, 15 figs., 31 pls., 20 tables, 20 maps.

——, and Evert Condor. 1932. Mating of the box turtles. Copeia, no. 2, pp. 86–88, 2 figs.

Camp, Charles Lewis. 1913. Notes on the local distribution and habits of the amphibians and reptiles of southeastern California in the vicinity of the Turtle Mountains. Univ. Calif. Publ. Zool., 12: 503–544.

Campbell, Berry. 1934. Report on a collection of reptiles and amphibians made in Arizona during the summer of 1933. Occ. Papers Mus. Zool. Univ. Mich., 289: 1–10.

Carl, Clifford. 1944. The reptiles of British Columbia. Brit. Col. Prov. Mus., Handbook 3: 5–60, 5 figs., 13 text illus.

Carr, Archie F., Jr. 1935. The identity and status of two turtles of the genus *Pseudemys.* Copeia, no. 3, pp. 147–148, 2 figs.

——. 1937a. A new turtle from Florida, with notes on *Pseudemys floridana mobiliensis* (Holbrook). Occ. Papers Mus. Zool. Univ. Mich., 348: 1–7, 2 pls.

——. 1937b. The status of *Pseudemys scripta* (Schoepff) and *Pseudemys troostii* (Holbrook). Herpetologica, 1: 75–77, 1 pl.

——. 1938a. *Pseudemys nelsoni,* a new turtle from Florida. Occ. Papers Boston Soc. Nat. Hist., 8: 305–310, 1 fig.

——. 1938b. A new subspecies of *Pseudemys floridana,* with notes on the *floridana* complex. Copeia, no. 3, pp. 105–109.

——. 1938c. Notes on the *Pseudemys scripta* complex. Herpetologica, 1: 131–135, 1 pl.

——. 1940. A contribution to the herpetology of Florida. Univ. Fla. Biol. Sci. Ser., 3: 1–118.

——. 1942a. The status of *Pseudemys floridana texana,* with notes on parallelism in *Pseudemys.* Proc. New England Zool. Club, 1: 69–76, 2 tables, 1 pl.

——. 1942b. Notes on sea turtles. *Ibid.,* 21: 1–16, 5 pls.

——. 1942c. A new *Pseudemys* from Sonora, Mexico. Amer. Mus. Novitates, no. 1181, pp. 1–4, 3 figs.

——. 1946. Status of the mangrove terrapin. Copeia, no. 3, pp. 170–172, 1 fig.

——. 1948. Sea turtles on a tropical island. Fauna, 10: 50–55, 13 figs.

——. 1949. The identity of *Malacoclemmys kohnii* Baur. Herpetologica, 5: 9–10, 1 fig., 1 map.

——, and Lewis J. Marchand. 1942. A new turtle from the Chipola River, Florida. Proc. New England Zool. Club, 20: 95–100, 1 pl.

Cassell, Richard L. 1945. The ways of a desert tortoise. Desert Mag., Dec., p. 25.

Casteel, D. B. 1911. The discriminative ability of the painted turtle. Jour. Animal Behavior, 1: 1–28.

Cate, J. 1936. Ruckenmarksreflexe bei Schildkröten. Act Brevia Neerland, 6: 63–64.

Catesby, Mark. 1730–1748. The natural history of Carolina, Florida, and the Bahama Islands. London (printed at the expense of the author). Folio 1: i–xii, 1–100, 100 pls., 1 map; 2: 5 + 1–100, i–xliv + 1–20 (appendix) + 6 (index).

Cederstrom, J. A. 1931. A two-headed turtle. Jour. Hered., 22: 137–138, 2 figs.

Chace, Lynnwood. 1945. Fertile turtle. Nat. Hist., 54: 418–421.

Chamberlain, E. B. 1938. The Charleston Museum's list of South Carolina reptiles. Charleston Mus. Leaflet, 4: 1–7.

Chapin, James P. 1906. Specimens exhibited [*Malaclemmys centrata*]. Proc. Staten Island Assoc. Arts Sci., 1: 45.

Chenoweth, W. L. 1949. Comments on some reptiles from the vicinity of Mammoth Cave National Park, Kentucky. Herpetologica, 5: 21–22.

Clark, Austin H. 1930. Records of the wood tortoise (*Clemmys insculpta*), in the vicinity of the District of Columbia. Proc. Biol. Soc. Washington, 43: 13–15.

Clark, H. L. 1904. Notes on the reptiles and batrachians of Eaton County. 4th Ann. Rept. Mich. Acad. Sci., pp. 192–194.

——. 1905. A preliminary list of the Amphibia and Reptilia of Michigan. 7th Ann. Rept. Mich. Acad. Sci., pp. 109–110.

Clark, H. Walton. 1935. On the occurrence of a probable hybrid between the eastern and western box turtles, *Terrapene carolina* and *T. ornata* near Lake Maxinkuckee, Indiana. Copeia, no. 3, pp. 148–150.

——, and John B. Southall. 1920. Fresh water turtles: a source of meat supply. U.S. Bur. Fish. Doc., 889: 3–20, 8 pls.

Clausen, Robert T. 1943. Amphibians and reptiles of Tioga County, New York. Amer. Midland Nat., 29: 360–314.

Cochran, Doris. 1928. Turtles of Galápagos. Nat. Mag., 12: 322–324.

Cockerell, T. D. A. 1896. Reptiles and batrachians of Mesilla Valley, New Mexico. Amer. Nat., 30: 325–327.

——. 1910. Reptiles and amphibians of the University of Colorado Expedition of 1909. Univ. Colo. Studies, 3: 130–131.

——. 1927. Zoology of Colorado. Univ. Colo. Semicentennial Ser., 3: 1–262, illustrated.

Cocteau, J. T., and G. Bibron. 1843. Reptilia. *In* Ramon de la Sagra's Histoire physique, politique et naturelle de l'île de Cuba. Paris. Pp. i–xviii, 1–239, and Atlas, 30 pls.

Coker, R. E. 1905a. Diversity in the horny scutes and bony plates of Chelonia. Science, 21: 384–385 (abstr.)

——. 1950b. Orthogenetic variation? *Ibid.*, 22: 873–875.

——. 1905c. Gadow's hypothesis of "orthogenetic variation" in Chelonia. Johns Hopkins Univ. Circ., 178: 9–24, 7 figs.

——. 1906a. The cultivation of the diamond-back terrapin. N.C. Geol. Surv. Bull., 14: 1–69, 23 pls., 2 figs.

——. 1906b. Other forms of turtles (Chelonia) at Beaufort, N.C. *Ibid.*, 14: 56–67.

——. 1910. Diversity in the scutes of Chelonia. Jour. Morph., 21: 1–76, 17 figs., 14 pls.

——. 1920. The diamond-back terrapin: past, present and future. Sci. Monthly, 11: 171–186, 4 figs.

Coles, R. J. 1914. Egg-laying of the loggerhead turtles. Copeia, no. 4, pp. 3–4.

Collins, R. Lee, and Gardner Lynn. 1936. Fossil turtles from Maryland. Proc. Amer. Phil. Soc., 76: 151–173.

Conant, Roger. 1930. Field notes of a collecting trip. Bull. Antivenin Inst. Amer., 4: 60–64.

——. 1932. A key for the reptiles of Ohio. Pub. Toledo Zool. Soc., pp. 1–6.

——. 1934. Guide to the reptile house of the Toledo Zoological Park. *Ibid.*, pp. 1–24, illustrated.

——. 1938a. The reptiles of Ohio. Amer. Midland Nat., 20: 1–200, 2 pls., 38 maps.

——. 1938b. On the seasonal occurrence of reptiles in Lucas County, Ohio. Herpetologica, 1: 137–144.

——. 1945. An annotated list of the amphibians and reptiles of the Del-Mar-Va Peninsula. Publ. Soc. Nat. Hist. Delaware, pp. 1–8.

——. 1947a. Reptiles and amphibians of the northeastern states. Publ. Zool. Soc. Philadelphia, pp. 1–40, illustrated.

——. 1947b. Reptiles and amphibians in Delaware. *In* Delaware, a history of the first state. New York, Lewis Historical Publ. Co. Pp. 23–25.

——, and Reeve M. Bailey. 1936. Some herpetological records from Monmouth and Ocean counties, New Jersey. Occ. Papers Mus. Zool. Univ. Mich., 328: 1–10.

——, and A. Downs, Jr. 1940. Miscellaneous notes on the eggs and young of reptiles. Zoologica, 25: 33–48.

——, and Coleman J. Goin. 1948. A new subspecies of soft-shelled turtle from the central United States, with comments on the application of the name *Amyda*. Occ. Papers Mus. Zool. Univ. Mich., 510: 1–19, 1 map.

——, and Robert G. Hudson. 1949. Longevity records for reptiles and amphibians in the Philadelphia Zoological Garden. Herpetologica, 5: 1–8.

Cooke, Wells W. 1910. Incubation period of box turtle eggs. Proc. Biol. Soc. Washington, 23: 124.

Cooper, J. G. 1860. Report upon the reptiles collected by the survey (of Washington Territory). Pacific Ry. Reports, 12: 292–306, pls. 12–16, 19–22, 29, 31.

——. 1863. Description of *Xerobates agassizii*. Proc. Calif. Acad. Sci., 2: 118–123.

——. 1870. The fauna of California and its geographic distribution. *Ibid.*, 4: 61–81.

Cooper, John E. 1949. Additional records for *Clemmys muhlenbergii* from Maryland. Herpetologica, 5: 75–76.

Cooper, M. D. 1870. The naturalist in California. Amer. Nat., 3: 182–189.

Cope, Edwin D. 1865. Third contribution to the herpetology of tropical America. Proc. Acad. Nat. Sci. Philadelphia, pp. 185–196.

——. 1866. Reptilia and Batrachia of the Sonoran Province of the Nearctic Region. *Ibid.*, pp. 300–314.

——. 1871. Report on the recent reptiles and fishes of the survey, collected by Campbell Carrington and C. M. Dawes. *In* F. V. Hayden, Preliminary report of the U.S. geological survey of Montana and portions of adjacent territories, being a fifth annual report of progress. Ann. Rept. U.S. Geol. Surv. Terrs., pp. 467–476.

——. 1872. Synopsis of the species of Chelydrinae. Proc. Acad. Nat. Sci. Philadelphia, pp. 22–29.

——. 1875. Check-list of North American Batrachia and Reptilia. Bull. U.S. Nat. Mus., 1: 1–104.

——. 1880. On the zoological position of Texas. *Ibid.*, 17: 1–51.

——. 1887. Catalogue of batrachians and reptiles of Central America and Mexico. *Ibid.*, 32: 1–98.

——. 1888. Catalogue of the Batrachia and Reptilia brought by William Taylor from San Diego, Texas. Proc. U.S. Nat. Mus., 11: 395–398.

——. 1890. Tortoises sold in the markets of Philadelphia. Amer. Nat., 24: 374.

——. 1892. The Batrachia and Reptilia of northwestern Texas. Proc. Acad. Nat. Sci. Philadelphia, pp. 331–336.

——. 1893. On the Batrachia and Reptilia of the plains at latitude 36° 30'. *Ibid.*, pp. 386–387.

——. 1894. [Note on the soft-shell turtle in New Jersey.] Amer. Nat., 28: 889.

——. 1895. Taylor on box turtles. *Ibid.*, 29: 756–757.

——. 1896a. The ancestry of the Testudinata. *Ibid.*, 30: 398–400.

——. 1896b. The geographical distribution of Batrachia and Reptilia in North America. *Ibid.*, 886–902.

Corrington, Julian T. 1927. Field notes on some amphibians and reptiles at Biloxi, Mississippi. Copeia, no. 165, pp. 98–102.

Coues, Elliot. 1871. Notes on the natural history of Fort Macon, N.C. Proc. Acad. Nat. Sci. Philadelphia, 23: 12–49.

——. 1875. Synopsis of the reptiles and batrachians of Arizona. Rept. Geogr. Geol. Exp. Surv. W. 100th Merid., 5: 586–683, pls. 16–25.

——, and H. C. Yarrow. 1878. Notes on the herpetology of Dakota and Montana. Bull. U.S. Geogr. Surv., 4: 259–291.

——, and H. C. Yarrow. 1879. Notes on the natural history of Fort Macon, North Carolina, and vicinity. Proc. Acad. Nat. Sci. Philadelphia, 30: 21–28.

Coventry, A. F. 1931. Amphibia, Reptilia and Mammalia of the Temagami district, Ontario. Canadian Field-Nat., 45: 109–113.

Cowan, Ian McTaggart. 1936. A review of the reptiles and amphibians of British Columbia. Rept. Prov. Mus. Brit. Columbia, pp. 16–25.

——. 1938. Distribution of turtles in coastal British Columbia. Copeia, no. 2, p. 91.

Cowles, Raymond B. 1920. A list and some notes on the lizards and snakes represented in the Pomona College Museum. Jour. Entomol. Zool. Pomona College, 12: 63–66.

——, and Charles M. Bogert. 1936. The herpetology of the Boulder Dam region (Nevada, Arizona, Utah). Herpetologica, 1: 33–42.

Cox, E. T. 1881. The tortoises of Tucson. Amer. Nat., 15: 1003–1004.

Cragin, F. W. 1880a. A preliminary catalogue of Kansas reptiles and batrachians. Trans. Kansas Acad. Sci., 7: 112–119.

——. 1880b. Supplementary list, comprising species now known as extralimital but more or less likely to be found in Kansas. *Ibid.*, 122–123.

——. 1885a. Recent additions to the list of Kansas reptiles and batrachians, with further notes on species previously reported. Bull. Washburn Lab. Nat. Hist., 1: 100–103.

——. 1885b. Second contribution to the herpetology of Kansas, with observations on the Kansas fauna. Trans. Kansas Acad. Sci., 9: 136–140.

——. 1886. Miscellaneous notes. Bull. Washburn Lab. Nat. Hist., 1: 212.

——. 1894. Herpetological notes from Kansas and Texas. Colo. Coll. Studies, 5: 37–39.

Cramer, W. Stuart. 1935. Lancaster County turtles. Penn. Angler, 4: 15.

Creaser, Charles W. 1944. The amphibians and reptiles of the University of Michigan Biological Station area in northern Michigan. Papers Michigan Acad. Sci., Arts, Lett., 29: 229–249.

Creaser, E. P. 1940. A note on the food of the box turtle. Copeia, no. 2, p. 131.

Crenshaw, John W., and George B. Rabb. 1949. Occurrence of the turtle *Graptemys barbouri* in Georgia. *Ibid.*, no. 3, p. 226.

Criddle, Norman. 1919. Additional notes on Manitoba turtles, snakes and batrachians. Ottawa Nat., 32: 135.

Cronise, Titus Fey. 1868. The natural wealth of California. San Francisco, H. H. Bancroft and Co. Pp. i–xvi, 1–699, figs.

Culbertson, Glenn. 1907. Some notes on the habits of the common box turtle (*Cistudo carolina*). Proc. Indiana Acad. Sci., pp. 78–79.

Culver, D. E. 1922. Box tortoise (*T. carolina*) swimming a creek. Copeia, no. 22, pp. 36–37.

Cunningham, Bert. 1923. Some phases in the development of *Chrysemys cinerea*. Jour. Elisha Mitch. Sci. Soc., 38: 51–73.

——, and Elizabeth Huene. 1938. Further studies in water absorption by reptile eggs. Amer. Nat., 72: 380–385.

——, and A. P. Hurwitz. 1936. Water absorption by reptile eggs during incubation. *Ibid.*, 70: 590–595.

Curtis, R. 1924. An account of the natural history of New England and of Nova Scotia and Lower Canada. New York (privately printed). Pp. i–xxi, 22–104.

Cuvier, Georges. 1831. The class Reptilia arranged by the Baron Cuvier with specific descriptions by Edward Griffith and Edward Pidgeon [vol. 9 of Cuvier's Animal Kingdom as revised by Griffith]. London. Pp. 1–481, 55 pls.

Daniel, R. S., and K. V. Smith. 1947. Migration of newly hatched loggerhead turtles toward the sea. Science, 106: 398–399.

Daudin, F. M. 1802. Histoire naturelle des reptiles. Paris. 2: 1–326.

Daugherty, Anita E. 1942. A record of *Graptemys pseudogeographica versa*. Copeia, no. 1, p. 51.

Davis, N. S., and F. L. Rice. 1883a. North American Batrachia and Reptilia found east of the Mississippi River. Bull. Illinois State Lab. Nat. Hist., 5: 3–64.

——, and F. L. Rice. 1883b. List of Batrachia and Reptilia of Illinois. Chicago Acad. Sci., 1: 25–32.

Davis, William T. 1884. The reptiles and batrachians of Staten Island. Proc. Nat. Sci. Assoc. Staten Island, 1: 13.

——. 1887. Color of the eyes as a sexual characteristic in *Cistudo carolina*. Amer. Nat., 21: 88–89.

——. 1897. The common land turtle. Proc. Nat. Sci. Assoc. Staten Island, 6: 20–22.

——. 1908. Notes on New Jersey amphibians and reptiles. Proc. Staten Island Assoc. Arts Sci., 2: 47–52.

——. 1912. Miscellaneous observations on the natural history of Long Island, N.Y. Proc. Staten Island Inst. Arts Sci., 3: 113–115 [common box turtle, *Cistudo carolina*, p. 114].

——. 1922. Natural history records from the meeting of the Staten Island Nature Club. *Ibid.*, 1: 66–80 [record of a large snapping turtle, p. 76].

——. 1923a. Natural history records from the meetings of the Staten Island Nature Club. *Ibid.*, 1: 123–130 [painted turtles observed, p. 129].

——. 1923b. Natural history records from the meetings of the Staten Island Nature Club. *Ibid.*, 2: 139–158 [painted turtles, pp. 151, 153].

——. 1928. Natural history records from the meetings of the Staten Island Nature Club. *Ibid.*, 4: 116–123 [Muhlenberg's turtle, p. 119].

——. 1930. Natural history records from the meetings of the Staten Island Nature Club. *Ibid.*, 5: 26–50 [yellow spotted turtle, *Clemmys guttata*, p. 41; box turtle, p. 46].

——. 1931a. Natural history records from the meetings of the Staten Island Nature Club. *Ibid.*, 163–178 [snapping turtle, p. 168].

——. 1931b. Natural history records from the meetings of the Staten Island Nature Club. *Ibid.*, 6: 28–51 [Kemp's turtle, *Caretta kempi* (Garman) exhibited, p. 35].

Deck, Raymond S. 1927. Longevity of *Terrapene carolina* (Linné). Copeia, no. 160, p. 179.

Deckert, Richard F. 1918. A list of reptiles from Jacksonville, Florida. *Ibid.*, no. 54, pp. 30–33.

DeKay, J. E. 1842. Zoology of New York. Part 3. Reptiles and Amphibians. Pp. i–vii, 1–415, 79 pls. (pls. separate from text).

Dellinger, S. C., and J. D. Black. 1938. Herpetology of Arkansas. Part 1. The Reptiles. Occ. Papers Univ. Arkansas, 1: 1–47.

Deraniyagala, P. E. P. 1930. The Testudinata of Ceylon. Ceylon Jour. Sci., Sec. B, 16: 43–88, pls. 7–13.

——. 1931. Testudinate evolution. Proc. Zool. Soc. London, no. 4, pp. 1057–1070, 3 pls.

——. 1932a. Herpetological notes (from Ceylon). Ceylon Jour. Sci., Sec. B, 17: 44–55.

——. 1932b. Notes on the development of the leathery turtle, *Dermatochelys coriacea*. *Ibid.*, 73–102.

——. 1933. The loggerhead turtles (Carettidae) of Ceylon. *Ibid.*, 18: 61–72, 6 figs., pl. 5.

——. 1934. Relationships among loggerhead turtles (Carettidae). *Ibid.*, 207–209, pl. 18.

——. 1936a. The nesting habit of the leathery turtle *Dermochelys coriacea*. *Ibid.*, 19: 331–336, 3 figs.

——. 1936b. Some postnatal changes in the leathery turtle, *Dermochelys coriacea. Ibid.*, 331–336.

——. 1936c. Further comparative study of *Caretta gigas. Ibid.*, 241–251, 4 figs., 3 tables.

——. 1938. The Mexican loggerhead turtle in Europe. Nature, 142: 540.

——. 1939a. The tetrapod reptiles of Ceylon. Colombo. Pp. ix–xxii, 1–412, 137 figs., 24 pls., 62 tables.

——. 1939b. The distribution of the Mexican loggerhead. Bull. Inst. Oceanogr., Monaco, 772: 1–4, 2 figs.

——. 1939c. The Mexican loggerhead in Europe. Nature, 144: 156.

——. 1943. Subspecies formation in loggerhead turtles. Spolia Zeylanica, 23: 79–92.

——. 1945. Some subspecific characters of the loggerhead *Caretta caretta gigas. Ibid.*, 24: 95–98, 2 figs.

——. 1946. Marginal scutes in races of the brown-red loggerhead, *Caretta caretta* Linné. *Ibid.*, 195–196, pl. 25.

DeSola, Ralph. 1928. Extirpation of certain neck growths in pond turtles. Copeia, no. 169, pp. 104–106.

——. 1931a. Sex determination in a species of the Kinosternidae, with notes on sound production in reptiles. *Ibid.*, no. 3, pp. 124–125.

——. 1931b. The turtles of the northeastern states. Bull. N.Y. Zool. Soc., 34: 131–160, 19 figs.

——. 1932. Observations on the use of the sucking-fish or remora, *Echeneis naucrates,* for catching turtles in Cuban and Columbian waters. Copeia, no. 2, pp. 45–52, 4 figs.

——. 1935. Herpetological notes from southeastern Florida. *Ibid.*, no. 1, pp. 44–45.

——, and Fredrica Abrams. 1933. Testudinata from southeastern Georgia, including the Okefinokee Swamp. *Ibid.*, pp. 10–12.

Despott, G. 1914. Our reptiles. Jour. Malta Hist. Sci. Soc., 2: 13–16, 93–95.

——. 1930. Cattura di due esimplara di *Chelone mydas* nei mari di Malta. Naturalista Siciliana, 7: 73–75, 2 figs.

Dexter, Ralph W. 1948. More records of reptiles from Portage County, Ohio. Copeia, no. 2, p. 139.

Dice, Lee R. 1916. Distribution of the land vertebrates of southeastern Washington. Univ. Calif. Publ. Zool., 16: 293–348, pls. 24–26.

——. 1923. Notes on the communities of the vertebrates of Riley County, Kansas, with special reference to amphibians, reptiles and mammals. Ecology, 4: 40–53.

Ditmars, Raymond L. 1905. The reptiles of the vicinity of New York City. Amer. Mus. Jour., 5: 93–140, 47 figs.

——. 1907. The reptile book. New York, Doubleday, Page and Co. Pp. i–xxxii, 1–472, 136 pls.

——. 1910. Reptiles of the world. New York, Sturgis and Walton Co. Pp. i–xi, 1–373, 89 pls.

——. 1934. A review of the box turtles. Zoologica, 42: 1–44, 41 figs.

——. 1936. The reptiles of North America. New York, Doubleday, Doran and Co. Pp. i–xvi, 1–476, 135 pls.

Dodge, Ernest S. 1944. Status of the ridley turtle in Massachusetts waters. Copeia, no. 2, pp. 120–121.

Dolley, John S. 1933. Preliminary notes on the biology of the St. Joseph River. Amer. Midland Nat., 14: 193–227, 2 pls., tables, 3 maps.

Driver, E. C. 1946. Delayed hatching in the box turtle. Copeia, no. 3, pp. 173–174.

Drowne, F. P. 1905. Reptiles and amphibians of Rhode Island. Roger Williams Park Mus., Providence, R.I., Monograph 15: 1–24.

Duellman, William E. 1947. Herpetological records from Logan County, Ohio. Copeia, no. 3, p. 208.

Duméril, André M. C., and G. Bibron. 1834–1854. Erpétologie générale ou histoire naturelle complète des reptiles. Paris, Librairie Encyclopédique de Roret. Vols. 1–9, and atlas. Turtles: vol. 1 (1834), pp. i–xxiv, 1–439, 12 pls.; vol. 2 (1835), pp. 1–680, 24 pls.

——, and Auguste Duméril. 1851. Catalogue méthodique de la collection des reptiles du Muséum d'Histoire Naturelle. Paris, Gide and Boudry. Pp. i–iv, 1–224.

Duméril, Aug. 1854–1855. Notice historique sur la ménagerie des reptiles du Muséum d'Histoire Naturelle et observations qui y ont été recueillies. Arch. Mus. Hist. Nat., 7: 193–320.

Duncan, David D. 1943. Capturing giant turtles in the Caribbean. Nat. Geogr., 84: 177–190, 13 illus., map.

Duncan, P. Martin (ed.). 1884. Class Reptilia. The reptiles. In Cassell's Natural history. London. 4: 241–380.

Dunn, E. R. 1915a. Some amphibians and reptiles of Delaware County, Pennsylvania. Copeia, no. 16, pp. 2–4.

——. 1915b. List of amphibians and reptiles observed in the summers of 1912, 1913 and 1914 in Nelson County, Virginia. Ibid., no. 18, pp. 5–7.

——. 1915c. List of amphibians and reptiles from Clark County, Virginia. Ibid., no. 25, pp. 62–63.

——. 1917. Reptile and amphibian collections from the North Carolina mountains, with especial reference to salamanders. Bull. Amer. Mus. Nat. Hist., 37: 624–626.

——. 1918a. A preliminary list of the reptiles and amphibians of Virginia. Copeia, no. 53, pp. 16–27.

——. 1918b. *Caretta kempii* in Jamaica. *Ibid.*, no. 59, pp. 75–76.

——. 1919. *Clemmys muhlenbergii* at Lake George, New York. *Ibid.*, no. 72, p. 68.

——. 1920a. Some reptiles and amphibians from Virginia, North Carolina, Tennessee and Alabama. Proc. Biol. Soc. Washington, 33: 129–135.

——. 1920b. Notes on *Melanemys* Shufeldt. Copeia, no. 77, pp. 7–8.

——. 1920c. Two new Virginia records. *Ibid.*, p. 8.

——. 1930. Reptiles and amphibians of Northampton and vicinity. Bull. Boston Soc. Nat. Hist., 57: 3–8.

——. 1936. List of Virginia amphibians and reptiles. Haverford, Penna., Haverford College. Pp. 5 (mimeographed).

——. 1945. Los géneros de anfibios y reptiles de Colombia. IV. Reptiles, órdenes Testudineos y Crocodilinos. Caldasia, 3: 306–335.

Dury, Ralph. 1932a. Recent acquisitions to the department of herpetology. Proc. Junior Soc. Nat. Sci., 3: 26–28.

——. 1932b. Notes on amphibians and reptiles from Clifty Falls State Park, Jefferson County, Indiana. *Ibid.*, 23–26.

——, and William Gessing, Jr. 1940. Additions to the herpetofauna of Kentucky. Herpetologica, 2: 31–32.

——, and Raymond S. Williams. 1933. Notes on some Kentucky amphibians and reptiles. Bull. Baker-Hunt Found. Mus., Williams Nat. Hist. Coll. (Covington, Ky.), 1: 1–22.

Eaton, Theodore H., Jr. 1945. Herpetological notes from Allegany State Park, New York. Copeia, no. 2, p. 115.

——. 1947. Box turtle eating a horned toad. *Ibid.*, no. 4, p. 270.

Eckel, Edwin C., and Frederick C. Paulmier. 1902. Catalogue of New York reptiles and batrachians. N.Y. State Mus. Bull., 51: 353–414, 24 figs., 1 pl.

Edgren, Richard A., Jr. 1942. Amphibians and reptiles from Van Buren County, Michigan. Copeia, no. 3, p. 180.

——. 1943. *Pseudemys scripta troostii* in Michigan. *Ibid.*, no. 4, p. 249.

——. 1948. Some additional notes on Michigan *Pseudemys*. Chicago Acad. Sci. Nat. Hist. Miscel., 22: 2.

——. 1949. Variation in the sizes of eggs of the turtles, *Chelydra s. serpentina* (Linné) and *Sternotherus odoratus* (Latreille). *Ibid.*, 53: 1.

——, and W. T. Stille. 1948. Check list of Chicago area amphibians and reptiles. *Ibid.*, 26: 1–7.

Edney, J. M. 1949. *Haemogregarina stepanowi* Danilewsky (1885) in middle Tennessee turtles. Jour. Tenn. Acad. Sci., 24: 220–222.

Eigenmann, C. H. 1896. The inhabitants of Turkey Lake. Testudinata. Proc. Indiana Acad. Sci., 5: 262–264.

——. 1918. The aquatic vertebrates. *In* H. B. Ward and G. C. Whipple,

Freshwater biology. New York, John Wiley and Sons. Pp. i–ix, 1–111, 1547 figs.

Eights, James. 1835. Papers on natural history. The naturalist's every day book. [Clippings from the *Zodiac*.] June, Thursday, 18th, and July, May 6th.

Ellis, Max M. 1917a. Amphibians and reptiles from the Pecos Valley. Copeia, no. 43, pp. 39–40.

——. 1917b. Amphibians and reptiles of the Douglas Lake Region. 19th Ann. Rept. Mich. Acad. Sci., 45–63.

——, and Junius Henderson. 1913. The Amphibia and Reptilia of Colorado. Part I. Univ. Colo. Studies, 10: 39–129, 8 pls.

——, and Junius Henderson. 1915. Amphibia and Reptilia of Colorado. Part II. Univ. Colorado Studies, 15: 253–263.

Engelhardt, George P. 1912. The turtle market at Key West. Mus. News, Brooklyn Inst. Arts Sci., 7: 73–74.

——. 1916. Burrowing habits of the box turtle. Copeia, no. 31, pp. 42–43.

Engels, William L. 1942. Vertebrate fauna of North Carolina coastal islands. Amer. Midland Nat., 28: 273–304, 11 figs.

Ennis, Jacob. 1861. On the young of *Kalemys muhlenbergii*. Proc. Acad. Nat. Sci. Philadelphia, 13: 124–125.

Eschscholtz, J. F. 1829. Zoologischer Atlas, enthaltend Abbildungen und Beschreibungen neuer Thierarten während des Flottcapitains v. Kotzebue zweiter Reise um die Welt, auf den russisch-kaiserlichen Kriegsschlupp Predpriaetie in den Jahren 1823–1826. Berlin, G. Reimer. 1: 1–28, 25 pls.

Evans, Arthur T. 1916. A collection of amphibians and reptiles from Gogebic County, Michigan. Proc. U.S. Nat. Mus., 49: 351–354.

Evans, Howard. 1947a. Herpetology of Crystal Lake, Sullivan County, New York. Herpetologica, 4: 19–22.

——. 1947b. Notes on Panamanian reptiles and amphibians. Copeia, no. 3, pp. 166–170.

Evans, L. T. 1949. The reproductive behavior of the giant tortoises, *T. vicina* and *T. vandenburghii*. Anat. Rec., 105: 579 (abstr.).

——, and J. Quaranta. 1949. Patterns of cooperative behavior in a herd of 14 giant tortoises at the Bronx Zoo. *Ibid.,* 506 (abstr.)

Evenden, Fred G. 1948. Distribution of the turtles of western Oregon. Herpetologica, 4: 201–204.

Evermann, Barton W. 1900. General report on the investigations in Porto Rico of the United States Fish Commission Steamer Fish Hawk in 1899. Bull. U.S. Fish Comm., 20: 3–26.

——. 1918. Notes on some amphibians and reptiles of Pike County, Pa. Copeia, no. 58, pp. 66–67.

——, and Howard Walton Clark. 1916. The turtles and batrachians of the Lake Maxinkuckee Region. Proc. Indiana Acad. Sci., 1916, pp. 472–518.

——, and Howard Walton Clark. 1920. Lake Maxinkuckee: a physical and biological survey. The turtles. Publ. Indiana Dept. Cons., 7: 591–620.

Ewing, H. E. 1926. The common box turtle, a natural host for chiggers. Proc. Biol. Soc. Washington, 39: 19–20.

——. 1933. Reproduction in the eastern box turtle, *Terrapene carolina carolina*. Copeia, no. 2, pp. 95–96.

——. 1935. Further notes on reproduction of the eastern box turtle. *Ibid.,* pp. 102–103.

——. 1937. Notes on a Florida box turtle, *Terrapene bauri* Taylor, kept under Maryland conditions. *Ibid.,* p. 141, 1 fig.

——. 1939. Growth in the eastern box turtle, with special reference to the dermal shields of the carapace. *Ibid.,* pp. 87–92, 3 figs.

——. 1943. Continued fertility in female box turtles following mating. *Ibid.,* pp. 112–114.

Faouzi, Hussein. 1936. On the occurrence of the leathery turtle *Dermochelys coriacea* Linn. in Egyptian Mediterranean waters. Proc. Zool. Soc. London, 2: 1175.

Fichter, George S. 1947. Preliminary list of the reptiles of Butler County, Ohio. Herpetologica, 4: 71–73.

Finneran, Leo Charles. 1947. A large clutch of eggs of *Chelydra serpentina serpentina* (Linnaeus). *Ibid.,* 3: 182.

——. 1948a. Diamond-back terrapin in Connecticut. Copeia, no. 2, p. 138.

——. 1948b. Reptiles in Branford, Connecticut. Herpetologica, 4: 123–126.

Fisher, A. K. 1887. Muhlenberg's tortoise (*Chelopus muhlenbergii* Schweigger) at Lake George, New York. Amer. Nat., 21: 672–673.

Fisher, G. Clyde. 1917. "Gopher pulling" in Florida. Amer. Mus. Jour., 17: 291–293, 2 figs.

——. 1945. Early spring mating of the wood turtle. Copeia, no. 3, pp. 175–176.

Fitch, Henry S. 1936. Amphibians and reptiles of the Rogue River Basin, Oregon. Amer. Midland Nat., 17: 634–652.

Fitzinger, L. J. 1835. Entruf einer systematischen Anordnung der Schildkröten nach den Grundsätzen der naturlichen Methode. Ann. Mus. Wien, 1: 103–128.

——. 1843. Systema reptilium. Vindobonae, Braumüller et Seidel, Bibliopolas. Pp. i–vi, 1–106.

Fleming, John. 1822. The philosophy of zoology. Edinburgh, Archibald Constable and Co. 2: 1–618.

Fletcher, William B. 1900. The Florida gopher. Proc. Indiana Acad. Sci., 1899, pp. 46–52.

Flower, Stanley S. 1925. Contributions to our knowledge of the duration of life in vertebrate animals. No. III. Reptiles. Proc. Zool. Soc. London, pp. 911–981.

——. 1937. Further notes on the duration of life in animals. *Ibid.*, 107: 1–39.

Fogg, B. F. 1862a. List of reptiles and amphibians found in the state of Maine. Maine Board Agric., 7th Ann. Rept. Secy., pp. 141–142.

——. 1862b. List of reptiles and amphibians found in the state of Maine. Proc. Portland Soc. Nat. Hist., 1: 86.

Forbes, T. R. 1940. A note on reptilian sex ratios. Copeia, no. 2, p. 132.

Force, Edith R. 1925. Notes on reptiles and amphibians of Ocmulgee County, Oklahoma. *Ibid.*, no. 141, pp. 25–27.

——. 1928. A preliminary list of the amphibians and reptiles of Tulsa County, Oklahoma. Proc. Oklahoma Acad. Sci., 8: 78–79.

——. 1930. The amphibians and reptiles of Tulsa County, Oklahoma and vicinity. Copeia, no. 2, pp. 25–39.

Ford, John. 1879. The leather turtle. Amer. Nat., 13: 633–637.

Fowler, H. W. 1906a. Some cold blooded vertebrates from the Florida Keys. Proc. Acad. Nat. Sci. Philadelphia, 59: 77–113, 13 figs., pls. 3–4.

——. 1906b. Notes on Muhlenberg's turtle. Amer. Nat., 40: 596.

——. 1907. The amphibians and reptiles of New Jersey. Ann. Rept. New Jersey State Mus., 1906, pp. 23–250, 69 pls.

——. 1908. A supplementary account of New Jersey amphibians and reptiles. Ann. Rept. New Jersey State Mus., 1907–1908, pp. 190–202.

——. 1909. Notes on New Jersey amphibians and reptiles. Ann. Rept. New Jersey State Mus., 1908, pp. 393–408.

——. 1913. An interesting form of the snapping turtle. Copeia, no. 1, pp. 1–2.

——. 1914. Amphibians and reptiles at Jennings, Maryland. *Ibid.*, no. 2, p. 2.

——. 1915. Some amphibians and reptiles of Cecil County, Maryland. *Ibid.*, no. 22, pp. 37–40.

——. 1916. Some amphibians and reptiles from Bucks County, Pennsylvania. *Ibid.*, no. 40, pp. 14–15.

——. 1925a. Records of amphibians and reptiles for Delaware, Maryland and Virginia. I. Delaware. *Ibid.*, no. 145, pp. 57–61.

——. 1925b. Records of amphibians and reptiles for Delaware, Maryland and Virginia. II. Maryland. *Ibid.*, pp. 61–64.

——. 1925c. Records of amphibians and reptiles for Delaware, Maryland and Virginia. III. Virginia. *Ibid.*, no. 146, pp. 65–67.

Fowler, J. A. 1942. Herpetological notes from Lake Cobbosseecontee and vicinity, Kennebec Co., Maine. *Ibid.*, no. 3, pp. 185–186.

——. 1945. The amphibians and reptiles of the national capitol parks and the District of Columbia region. U.S. Dept. Int., Nat. Park Serv. Publ., 194: 1–4.

Fowler, R. L. 1934. Some amphibians and reptiles of the district around High River, Alberta. Canadian Field-Nat., 48: 139–140.

Fox, Herbert. 1941. Kyphoscoliosis in a painted tortoise. Rept. Penrose Res. Lab., p. 16.

Freiberg, Marcos A. 1938. Catálogo sistemático y descriptivo de las tortugas argentinas. Mem. Mus. Entre Rios, 9: 1–23, 8 pls.

Frothingham, Langdon. 1936. Observations on young box turtles. Bull. Boston Soc. Nat. Hist., 78: 3–8, 1 fig.

Frum, W. Gene. 1947. *Graptemys geographica* in West Virginia. Copeia, no. 3, p. 211.

Fry, D. B. 1913. On the status of *Chelonia depressa* Garman. Rec. Austr. Mus., 10: 159–185, figs. 38–49, pls. 19–22.

Funkhouser, W. D. 1925. Wild life in Kentucky. Kentucky Geol. Surv., Frankfort, ser. 6, 16: 1–385, 88 figs.

Gadow, Hans. 1899. Orthogenetic variation in the shells of Chelonia. Wiley Zool. Results. Cambridge Univ. Press, 3: 207–222, pls. 24–25.

——. 1901. Amphibians and reptiles. New York and London, Macmillan and Co. Pp. i–xiii, 1–668, 181 figs. (2d ed., 1920.)

——. 1905a. Distribution of Mexican amphibians and reptiles. Proc. Zool. Soc. London, 2: 191–244, figs. 29–32, pls. 6–7.

——. 1905b. Orthogenetic variation. Science, 22: 637–640.

Gage, S. H. 1884. Pharyngeal respiration in the soft-shelled turtle (*Aspidonectes spinifer*). Proc. Amer. Assoc. Adv. Sci., 32: 316.

——, and S. P. Gage. 1886. Aquatic respiration in soft-shelled turtles: a contribution to the physiology of respiration in vertebrates. Amer. Nat., 20: 233–236.

Gage, Susanna Phelps. 1895. Comparative morphology of the brain of the soft-shelled turtle (*Amyda mutica*) and the English sparrow (*Passer domestica*). Trans. Amer. Microsc. Soc., 17: 185–238, 5 pls.

Gaige, Helen Thompson. 1914. A list of the amphibians and reptiles observed in Richland Co., Illinois in May, 1913. Copeia, no. 11, p. 4.

——. 1937. Some amphibians and reptiles from Tamaulipas. Univ. Mich. Studies Sci. Ser., 12: 301–304.

Gammons, Frank. 1871. How the sculptured turtle (*Glyptemys insculpta* Ag.) deposits her eggs. Amer. Nat., 4: 53.

Garman, Harrison. 1889. A preliminary report on the animals of the waters of the Mississippi bottoms, near Quincy, Illinois in August, 1888. Bull. Illinois State Lab. Nat. Hist., 3: 123–184.

——. 1890a. Notes on reptiles and amphibians, including several species not before recorded from the northern states. *Ibid.*, 185–190.

——. 1890b. The differences between the geographic turtles. Bull. Essex Inst., 12: 1–14.

——. 1890c. On geographic turtles. *Ibid.*, 22: 1–14, 1 pl.

——. 1892. A synopsis of the reptiles and amphibians of Illinois. Bull. Illinois State Lab. Nat. Hist., 3: 215–385, pls. 9–15.

———. 1894. A preliminary list of the vertebrate animals of Kentucky. Bull. Essex Inst., 26: 1–63.

Garman, Samuel. 1880. On certain species of Chelonioidae. Bull. Mus. Comp. Zool., 6: 123–126.

———. 1883. The reptiles and batrachians of North America. Introduction— Testudinata. Mem. Mus. Comp. Zool., 8: 4–8.

———. 1884a. The North American reptiles and batrachians. Bull. Essex Inst., 16: 1–46.

———. 1884b. The reptiles of Bermuda. U.S. Nat. Mus. Bull., 25: 285–303.

———. 1887. Reptiles and batrachians from Texas and Mexico. Bull. Essex Inst., 19: 1–20.

———. 1888. Reptiles and batrachians from the Caymans and the Bahamas. Ibid., 20: 1–13.

———. 1891. On a tortoise found in Florida and Cuba. Bull. Essex Inst. 23(7–9): 1–4.

———. 1892. On Texas reptiles. Ibid., 24: 1–12.

———. 1917. The Galápagos tortoises. Mem. Mus. Comp. Zool., 30: 261–296, 42 pls.

Garneir, J. H. 1881. List of reptiles of Ontario. Canadian Sportsman Nat., 1: 37–39.

Gee, N. Gist. 1929–1930. A contribution towards a preliminary list of reptiles recorded from China. Bull. Dept. Biol., Yenching Univ., 1: 53–84.

Gentry, Glenn. 1941. Herpetological collections from counties in the vicinity of the Obey River drainage of Tennessee. Jour. Tenn. Acad. Sci., 16: 329–332.

———. 1944. Some predators of the Flintville State Fish Hatchery. Ibid., 19: 265–267.

Geoffrey–St. Hilaire, H. 1909. Sur les tortues molles nouveau genre sous le nom Trionyx, et sur la formation des carapaces. Ann. Mus. Hist. Nat. Paris, 14: 1–20, 5 pls.

Gibbs, Morris. 1895. The reptiles and batrachians of Michigan. Amer. Field, 44: 6–7, 30–31, 55, 79–80, 106, 132, 155–156.

———, F. N. Notestein, and H. L. Clark. 1905. A preliminary list of the Amphibia and Reptilia of Michigan. 7th Rept. Mich. Acad. Sci., pp. 109–110.

Gilmore, C. W. 1923. A new fossil turtle, Kinosternon arizonense, from Arizona. Proc. U.S. Nat. Mus., 62: 1–8, 7 figs., 4 pls.

———. 1927. On fossil turtles from the Pleistocene of Florida. Ibid., 71: 1–10, 4 figs., 5 pls.

———. 1937. A new marine turtle from the Miocene of California. Proc. Calif. Acad. Sci., 23: 171–174, pl. 14.

———, and Doris M. Cochran. 1930. Cold-blooded vertebrates. Chapter X. Turtles. Smithsonian Sci. Ser., 8: 306–319, pls. 67–68.

Giral, Francisco, and Andre Marquez. 1948. Mexican turtle oils. III. *Caretta caretta* Linn. Arch. Biochem., 16: 187–189.

Giral, Jose, Francisco Giral, and Maria Luisa Giral. 1948. Mexican turtle oils. II. *Chelone mydas* Linn. *Ibid.*, 181–186.

Girard, Charles. 1858. United States exploring expedition during the years 1838–1842. Under the command of Charles Wilkes, U.S.N. Philadelphia, J. B. Lippincott and Co. Herpetology, 20: i–xvii, 1–496.

Glaesner, L. 1924. Ueber drei Doppeldildungen von *Chelone mydas*. Zool. Anz., 60: 185–194, 5 figs.

Glass, Bryan P. 1949. Records of *Macrochelys temminckii* in Oklahoma. Copeia, no. 2, pp. 138–141.

Gloyd, Howard K. 1928. The amphibians and reptiles of Franklin County, Kansas. Trans. Kansas Acad. Sci., 31: 115–141.

——. 1932. The herpetological fauna of the Pigeon Lake region, Miami County, Kansas. Papers Mich. Acad. Sci., Arts, Lett., 15: 389–409, pls. 30–32, map 3.

——. 1937. A herpetological consideration of faunal areas in southern Arizona. Bull. Chicago Acad. Sci., 5: 79–136, 22 figs.

Gmelin, Jean-Frederic. 1789. Systema Naturae. (13th ed., vol. 1.)

Goette, A. 1899. Über die Entwicklung des knockernen Rückenschildes (Carapax) der Schildkröten. Zeitschr. Wiss. Zool., 66: 407–434, 3 pls.

Goff, C. C., and Dorothy S. Goff. 1932. Egg laying and incubation of *Pseudemys floridana*. Copeia, no. 2, pp. 92–94, 1 fig.

Goff, Dorothy S., and C. C. Goff. 1935. On the incubation of clutch of eggs of *Amyda ferox* (Schneider). *Ibid.*, no. 3, p. 156.

Goin, Coleman J. 1943. The lower vertebrate fauna of the water hyacinth community in northern Florida. Proc. Florida Acad. Sci., 6: 143–154.

——. 1948. The occurrence of *Amyda spinifera aspera* in Florida. Copeia, no. 4, p. 304.

——, and C. C. Goff. 1941. Notes on the growth rate of the gopher turtle, *Gopherus polyphemus*. Herpetologica, 2: 66–68.

Goode, G. Brown. 1876. Classification of the animal resources of the United States. Bull. U.S. Nat. Mus., 6: iv–xiii, 1–126.

Gordon, Kenneth. 1931. The Amphibia and Reptilia of Oregon. Oregon State Mon. Studies Zool., 1: 1–82, 54 figs., 7 pls.

Gosse, P. H. 1851. A naturalist's sojourn in Jamaica. London, Longman. Pp. i–xxiv, 1–508.

Graf, William, Stanley G. Jewett, Jr., and Kenneth L. Gordon. 1939. Records of amphibians and reptiles from Oregon. Copeia, no. 2, pp. 101–104.

Grant, Chaffee. 1936. Breeding of *Pseudemys elegans* in California and notes on other captive reptiles. *Ibid.*, no. 1, pp. 112–113.

Grant, Chapman. 1911. Key West record turtles. Zool. Soc. Bull., pp. 506–508.

——. 1927a. Autoist, spare that turtle. Bull. N.Y. Zool. Soc., 29: 215–216.

——. 1927b. Notes on sea turtles. Copeia, no. 164, p. 69.

——. 1935a. Secondary sexual differences and notes on the mud turtle, *Kinosternon subrubrum* in northern Indiana. Amer. Midland Nat., 16: 798–800.

——. 1935b. Notes on the spotted turtle in northern Indiana. Proc. Indiana Acad. Sci., 44: 244–247.

——. 1935c. The "eastward migration" of *Terrapene ornata*. Copeia, no. 4, pp. 186–188.

——. 1936a. Herpetological notes from northern Indiana. Proc. Indiana Acad. Sci., 45: 323–333.

——. 1936b. The southwestern desert tortoise, *Gopherus agassizii*. Zoologica, 21: 225–229.

——. 1937a. Orthogenetic variations. Proc. Indiana Acad. Sci., 46: 240–245, 5 figs.

——. 1937b. The "midventral keel" in Testudinata. *Ibid.*, 246–252, 5 figs.

——. 1946. Identification of *Lepidochelys kempii* (Garman). Herpetologica, 3: 39.

Graper, Ludwig. 1932. Das Zungenbein und die Zunge bewegenden Muskeln des Schildkröten. I und II. Jenaische Zeitschr. Naturwiss., 66: 169–198.

Gray, I. E. 1941. Amphibians and reptiles of the Duke Forest and vicinity. Amer. Midland Nat., 25: 652–658.

Gray, John Edward. 1831. Synopsis reptilium. Part I. Cataphracta. Tortoises, crocodiles, and enaliosaurians. London, Treuttel, Wurtz and Co. Pp. i–vii, 1–78, 11 pls.

——. 1844. Catalogue of the tortoises, crocodiles and amphisbaenians in the collection of the British Museum. London. Pp. i–viii, 1–80.

——. 1855. Catalogue of the shield reptiles in the collection of the British Museum. Part I. Testudinata (tortoises). London. Pp. 1–79, 42 pls.

——. 1863. Notes on American *Emydidae,* and Professor Agassiz's observations on my catalog of them. Ann. Mag. Nat. Hist., 12: 176–183.

——. 1869. Notes on the families and genera of tortoises (Testudinata) and on the classification afforded by the study of their skulls. Proc. Zool. Soc. London, 1869, pp. 165–225.

——. 1870. Supplement to the catalogue of shield reptiles in the collection of the British Museum. Part I. Testudinata (tortoises). London. Pp. i–x, 1–120, 40 figs.

——. 1872. Appendix to the catalogue of shield reptiles in the collection of the British Museum. Part I. Testudinata (tortoises). London. Pp. 1–28.

——. 1873. Hand list of the specimens of shield reptiles in the British Museum. London. Pp. 1–124.

Greaves, J. B. 1933. Nesting of *Caretta olivacea*. Jour. Bombay Nat. Hist. Soc., 37: 494.

Green, Harold T. 1923. Notes on middle states amphibians and reptiles. Copeia, no. 122, pp. 99–100.

Green, N. Bayard. 1937a. The herpetological status of Randolph County, West Virginia. Mag. Hist. Biography, 9: 61–69.

——. 1937b. The amphibians and reptiles of Randolph County, West Virginia. Herpetologica, 1: 113–116.

——. 1941. Amphibians and reptiles of the Huntington region. Marshall Rev., 4: 33–40.

Grimpe, G. 1928. Merkwürdige Todesursache einer Elefantenschildkröte. Zool. Garten. Leipzig, 1: 225–226.

Grinnell, Joseph, and C. L. Camp. 1917. A distributional list of the amphibians and reptiles of California. *Ibid.,* 17: 127–208, 14 figs.

——, J. Dixon, and J. M. Linsdale. 1930. Vertebrate natural history of a section of northern California through our Lassen Peak region. Univ. Calif. Publ. Zool., 35: i–v, 1–594, 181 figs.

——, and H. W. Grinnell. 1907. Reptiles of Los Angeles County, California. Throop. Inst. Bull., 35: 1–35, 23 figs.

——, and T. I. Storer. 1924. Animal life in the Yosemite. Berkeley, Calif., Univ. Calif. Press. Pp. i–xviii, 1–752, 62 pls., 64 text figs.

Gudger, E. W. 1919. On the use of the sucking fish for catching fish and turtles: studies in Echeneis or Remora. Amer. Nat., 43: 289–311, 446–467, 515–525, 5 figs., 2 pls.

Guérin-Méneville, M. F. E. 1829–1844. [Reptiles.] *In* Iconographie du Règne Animal de G. Cuvier. Paris, J. B. Baillière, Libraire Acad. Roy. Med. 1: 1–30.

Gundlach, Juan. 1867. Revista y catálogo de los reptiles cubanos. *In* Felipe Poey, Repert. físico-nat. de Cuba, 2: 102–122.

——. 1875. Catálogo de los reptiles cubanos. An. Soc. Esp. Hist. Nat., 4: 347–368.

——. 1880. Contribución a la herpetología cubana. Havana, G. Montiel y Cía. Pp. 1–99.

——. 1881. Apuntas para la fauna puerto-riquena. An. Soc. Esp. Hist. Nat., 10: 305–317.

Gunter, Gordon. 1945. The northern range of Berlandier's tortoise. Copeia, no. 3, p. 175.

Günther, Albert C. L. G. 1864. The reptiles of British India. London. Pp. i–xxvii, 1–452, 4 figs., 26 pls.

——. 1885. Biología centrali-americana. Reptilia and Batrachia. R. H. Porter and Dulan and Co. Pp. i–xx, 1–326, 76 pls.

Hadsall, Leo F. 1935. Snakes, lizards and turtles. Sci. Guide Elementary Schools, Calif. State Dept. Educ., 1: 1–30, 13 figs., 1 pl.

Hahn, Walter L. 1908. Notes on mammals and cold blooded vertebrates of the Indiana University Farm, Mitchell, Indiana. Proc. U.S. Nat. Mus., 35: 545–581.

Hall, Harvey M., and Joseph Grinnell. 1919. Life-zone indicators in California. Proc. Calif. Acad. Sci., 9: 37–67.

Hall, Henry H., and Hobart M. Smith. 1947. Selected records of reptiles and amphibians from southeastern Kansas. Trans. Kansas Acad. Sci., 49: 447–454.

Hallinan, Thomas. 1923. Observations made in Duval County, northern Florida, on the gopher tortoise (*Gopherus polyphemus*). Copeia, no. 115, pp. 11–20, 4 figs.

Hallowell, Edward. 1854. Descriptions of some new reptiles from California. Proc. Acad. Nat. Sci. Philadelphia, pp. 91–97.

——. 1856. [Notes on Kinosternidae.] *Ibid.*, 7: 105–108.

——. 1857a. Notes on a collection of reptiles from Kansas and Nebraska, presented to the Academy of Natural Sciences, by Dr. Hammond, U.S.A. *Ibid.*, 8: 238–253.

——. 1857b. Notes on the reptiles in the collection of the Academy of Natural Sciences of Philadelphia. *Ibid.*, 221–238.

——. 1857c. Notes on the collection of reptiles from the neighborhood of San Antonio, Texas, recently presented to the Academy of Natural Sciences by Dr. Heermann. *Ibid.*, 306–310.

——. 1859. Report upon the reptiles collected on the survey [Williamson's Route]. Rept. Pacific R.R. Surv., 10: 1–27, 10 pls.

——. 1860. Report upon the Reptilia of the North Pacific exploring expedition under the command of Captain John Rogers. Proc. Acad. Nat. Sci. Philadelphia, 12: 480–510.

Haltom, William L. 1931. Alabama reptiles. Alabama Mus. Nat. Hist., Paper no. 11, pp. 1–145, 57 figs., 39 pls.

Hamilton, Rodgers D. 1944. Notes on mating and migrations in Berlandier's turtle. Copeia, no. 1, p. 62.

——. 1947. The range of *Pseudemys scripta gaigeae*. *Ibid.*, pp. 65–66.

Hamilton, W. J., Jr. 1940. Observations on the reproductive behavior of the snapping turtle. Copeia, no. 2, pp. 124–126, 1 fig.

——. 1947. Egg laying of *Trionyx ferox*. *Ibid.*, no. 3, p. 209.

Hammann, Cecil B. 1939. A note on the occurrence of the spotted turtle, *Clemmys guttata* (Schneider) in Indiana. Proc. Indiana Acad. Sci., 48: 227.

Hankinson, T. L. 1908. Biological survey of Walnut Lake, Michigan. Rept. Geol. Surv. Michigan, 1907, pp. 157–288, pls. 14–75.

——. 1915. The vertebrate life of certain prairie and forest regions near Charleston, Illinois. Bull. Illinois State Lab. Nat. Hist., 11: 283–303, pls. 64–79.

——. 1917. Amphibians and reptiles of the Charleston region. Trans. Illinois Acad. Sci., 10: 322–330.

——. 1929. Fishes of North Dakota. Papers Mich. Acad. Sci., Arts, Lett., 10: 439–460.

Hansemann, D. von. 1915. Die Lungenatmung der Schildkröten. Sitzber. Akad. Wiss. Berlin, 2: 661–672, pls. 3–4.

Hansen, Ira B. 1941. The breathing mechanism of turtles. Science, 94: 64.

——. 1943. Hermaphroditism in a turtle of the genus *Pseudemys*. Copeia, no. 1, pp. 7–9.

Hardy, Ross, and Lee Lamoreaux. 1945. Emory's turtle in the Virgin River drainage of northwestern Arizona. *Ibid.*, no. 3, p. 168.

Harlan, Richard. 1826–1827. Genera of North American Reptilia and a synopsis of the species. Jour. Acad. Nat. Sci. Philadelphia, 5: 317–372.

——. 1829. Genera of North American Reptilia and a synopsis of the species [and catalogue of North American Reptilia]. *Ibid.*, 6: 7–38.

——. 1835. Genera of North American reptiles, and a synopsis of the species. Medical and Physical Researches. Philadelphia, Lydia R. Bailey. Pp. 84–160.

——. 1837. Description of a new species of freshwater tortoise, inhabiting the Columbia River. Amer. Jour. Sci., 31: 382–383.

Harper, F. 1926. Tales of the Okefinokee. American Speech, 1: 407–420.

——. 1940. Some works of Bartram, Daudin, Latreille and Sonnini and their bearing upon North American herpetological nomenclature. Amer. Midland Nat., 23: 692–723.

Harris, Kilroy. 1931. The green turtle of Australia. Nat. Mag., 18: 96.

Hartweg, Norman. 1938. *Kinosternon flavescens stejnegeri*, a new turtle from northern Mexico. Occ. Papers Mus. Zool. Univ. Mich., 371: 1–5.

——. 1939a. Further notes on the *Pseudemys scripta* complex. Copeia, no. 1, p. 55.

——. 1939b. A new American Pseudemys. Occ. Papers Mus. Zool. Univ. Mich., 397: 1–4.

——. 1944. Spring emergence of painted turtle hatchlings. Copeia, no. 1, pp. 20–22.

——. 1946. Confirmation of overwintering in painted turtle hatchlings. *Ibid.*, no. 4, p. 255.

Harwood, P. D. 1932. The helminths parasitic in the Amphibia and Reptilia of Houston, Texas and vicinity. Proc. U.S. Nat. Mus., 81: 1–71.

Hatt, R. T. 1923. The land vertebrate communities of western Leelanare

County, Michigan, with an annotated list of the mammals of the county. Papers Mich. Acad. Sci., Arts, Lett., 3: 369–402, pls. 24–26.

Hay, O. P. 1887a. A preliminary catalogue of the Amphibia and Reptilia of the state of Indiana. Jour. Cincinnati Soc. Nat. Hist., pp. 59–69.

——. 1887b. The amphibians and reptiles of Indiana. Ann. Rept. Indiana State Board of Agric., 28: 201–223.

——. 1892a. Some observations on the turtles of the genus *Malaclemys*. Proc. U.S. Nat. Mus., 15: 379–384.

——. 1892b. The batrachians and reptiles of the state of Indiana. Ann. Rept. Indiana Dept. Geol. Nat. Res., 17: 412–602, 3 pls.

——. 1898. On Protostega . . . and the morphology of the chelonian carapace and plastron. Amer. Nat., 32: 929–948, 3 figs.

——. 1901. The composition of the shell of turtles. Ann. N.Y. Acad. Sci., 14: 110.

——. 1903. On existing genera of the Trionychidae. Proc. Amer. Phil. Soc., 42: 268–274.

——. 1908a. The fossil turtles of North America. Carnegie Inst. Washington Publ., no. 75, pp. iii–iv, 1–568, 704 figs., 113 pls.

——. 1908b. On three existing species of sea-turtles, one of them (*Caretta remivaga*) new. Proc. U.S. Nat. Mus., 34: 183–198, pls. 6–11.

——. 1916. Descriptions of some Floridian fossil vertebrates, belonging mostly to the Pleistocene. 8th Ann. Rept. Florida State Geol. Surv., pp. 39–76, 9 pls.

——. 1922. On the phylogeny of the shell of the Testudinata and the relationships of *Dermochelys*. Jour. Morph., 36: 421–441, 1 fig., 2 pls.

——. 1923. Oligocene sea turtles of South Carolina. Pan Amer. Geologist, 40: 29–31, pls. 2–3.

——. 1928. Further consideration of the shell of *Chelys* and of the constitution of the armor of turtles in general. Proc. U.S. Nat. Mus., 73: 1–12, 2 pls.

Hay, W. P. 1902. A list of batrachians and reptiles of the District of Columbia and vicinity. Proc. Biol. Soc. Washington, 15: 121–145, 3 figs.

——. 1904. A revision of *Malaclemmys*, a genus of turtles. Bull. U.S. Bureau Fish., 24: 1–20, 12 pls.

——, and H. D. Aller. 1913. Artificial propagation of the diamond-back terrapin. U.S. Bureau Fish. Econ. Circ., 5: 1–14, 3 figs.

Haycroft, J. B. 1891. The development of the carapace of Chelonia. Trans. Roy. Soc. Sci. Edinburgh, 36: 335–342.

Hayden, F. V. 1863. On the geology and natural history of the upper Missouri. Trans. Amer. Phil. Soc., 12: 1–218, 9 figs., 1 map.

——. 1875. Catalogue of the collections in geology and natural history. *In* Preliminary rept. explorations in Nebraska and Dakota. Washington. Pp. 59–125.

Hecht, Max. 1943. A new record for Blanding's turtle in eastern New York. Copeia, no. 3, pp. 196–197.

Heller, Edmund. 1903. Papers of the Hopkins Stanford Galápagos expedition, 1898–1899. XIV. Reptiles. Proc. Washington Acad. Sci., 5: 39–98.

Henning, Willard L. 1938. Amphibians and reptiles of a 2200 acre tract in central Missouri. Copeia, no. 2, p. 91.

Henshaw, S. 1904. Fauna of New England. List of the Reptilia. Occ. Papers Boston Soc. Nat. Hist., 7(1): 1–13.

Herrera, Alfonso L. 1895. Catálogo de la colección de reptiles batracios del Museo Nacional. No 3. México, Imprenta del Museo Nacional. Pp. 1–66.

Heude, M. 1880. Mémoire sur les *Trionyx*. Mém. Hist. Nat. Emp. Chinois, 1: v–viii, 1–38.

Hibbard, Claude W. 1936. The amphibians and reptiles of Mammoth Cave National Park (proposed). Trans. Kansas Acad. Sci., 39: 277–281.

Higley, W. K. 1889. Reptilia and Batrachia of Wisconsin. Trans. Wisconsin Acad. Sci., Arts, Lett., 7: 155–176.

Hildebrand, Samuel F. 1929. Review of experiments on artificial culture of diamond-back terrapin. Bull. U.S. Bureau Fish., 45: 25–70, 14 figs., 36 tables.

——. 1930. Duplicity and other abnormalities in diamond-back terrapins. Jour. Elisha Mitch. Sci. Soc., 46: 41–53.

——. 1932. Growth of diamond-back terrapins. Size attained, sex ratio and longevity. Zoologica, 9: 551–563, figs. 383–384, 4 tables.

——. 1933. Hybridizing diamond-back terrapins. Jour. Hered., 24: 231–238, figs. 6–7, 4 tables.

——, and C. Hatsel. 1926. Diamond-back culture at Beaufort, North Carolina. U.S. Bureau Fish. Econ. Circ., 60: 1–20.

——, and C. Hatsel. 1927. On the growth, care and age of loggerhead turtles in captivity. Proc. Nat. Acad. Sci., 13: 374–377.

Hilgendorf, F. 1880. Bemerkungen über die von ihm in Japan gesammelten Amphibien nebst Beschreibung zweier neuen Schlangenarten. Sitz. Ber. Ges. Naturf. Freunde Berlin, no. 8, pp. 111–121, 10 figs.

Hingston, R. W. G. 1933. The meaning of animal color and adornment. London, Edward Arnold and Co. Pp. 1–411, illustrated.

Holbrook, John Edwards. 1842. North American herpetology; or a description of the reptiles inhabiting the United States. Ed. 2. Philadelphia, J. Dobson. 1: i–xv, 1–152, 24 pls. [First edition, vol. 1, 1836]

——. 1849. Catalogue of the reptiles of Georgia, White's statistics of the state of Georgia. Savannah, Thorne Williams. Pp. 13–15.

Holt, Ernest G. 1925. Additional records for the Alabama herpetological catalogue. Copeia, no. 136, pp. 100–101.

Honigmann, H. 1921. Zur Biologie der Schildkröten. Biologisches Centralb., 41: 241–250.

Hooker, Davenport. 1908a. The breeding habits of the loggerhead turtle and some early instincts of the young. Science, 27: 490–491.

——. 1908b. Preliminary observation on the behavior of some newly hatched loggerhead turtles (*Thallasochelys caretta*). Yearbook Carnegie Inst. Washington, 6: 111–112.

——. 1909. Report on the instincts and habits of newly hatched loggerhead turtles. *Ibid.*, 7: 124.

——. 1911. Certain reactions to color in the young loggerhead turtle. Papers Tortugas Lab., Carnegie Inst. Washington, 3: 69–76.

Hoopes, Isabel. 1936. Reptiles in the home zoo. New England Mus. Nat. Hist., spec. publ., 1: 1–64, illus.

Hornell, James. 1927. The turtle fisheries of the Seychelles Islands. H. M. Stationery Office, pp. 1–55.

Hough, Franklin B. 1852. Catalogue of reptiles and fishes from St. Lawrence County. Ann. Rept. State Cab. Nat. Hist., N.Y., 5: 23–28.

Householder, Victor H. 1916. The lizards and turtles of Kansas, with notes on their distribution and habits. Dissertation, Univ. Kansas. Pp. 1–110, illustrated.

Howe, R. Heber, Jr. 1911. The second record for Blanding's turtle in Concord, Massachusetts. Science, 34: 272.

Hoy, P. R. 1883. Catalogue of the cold-blooded vertebrates of Wisconsin. Geol. Wisconsin, 1: 422–426.

Hoyt, J. Southgate. 1941. The incubation period of the snapping turtle. Copeia, no. 3, p. 180.

Hubbard, Henry G. 1893. The Florida land tortoise-gopher, *Gopherus polyphemus*. Science, 22: 57–58.

——. 1894. The insect guests of the Florida land tortoise. Insect Life, 6: 302–315.

——. 1896. Additional notes on the insect guests of the Florida land tortoise. Proc. Ent. Soc. Washington, 3: 299–302.

Hubbs, Carl L. 1924. The growth of the painted turtle. Ecology, 5: 108.

Hudson, George E. 1942. The amphibians and reptiles of Nebraska. Nebraska Conserv. Bull., no. 24, pp. 1–146.

Hughes, Edward. 1886. Preliminary list of reptiles and batrachians of Franklin County. Bull. Brookville Soc. Nat. Hist., no. 2, pp. 40–45.

Hurley, Frank. 1924. Pearls and savages. New York, G. P. Putnam's Sons. Pp. iii–xiii, 1–414, 42 illus.

Hurter, Julius. 1883. Catalogue of reptiles and batrachians collected in the state of Missouri. Privately printed.

——. 1893. Catalogue of reptiles and batrachians found in the vicinity of St. Louis, Missouri. Trans. Acad. Sci. St. Louis, 6: 251–261.

——. 1903. Second contribution of the herpetology of Missouri. *Ibid.,* 13: 77–86.

——. 1911. Herpetology of Missouri. *Ibid.,* 20: 59–274, pls. 18–24.

——, and John K. Strecker, Jr. 1909. The amphibians and reptiles of Arkansas. *Ibid.,* 11–27.

Huse, W. H. 1901. The Testudinata of New Hampshire. Proc. Manchester Inst. Arts Sci., 2: 47–51.

Ingle, Robert M., and F. G. Walton Smith. 1949. Sea turtles and the turtle industry. Special Publ. Univ. Miami, pp. 6–107, 2 figs.

Ives, J. E. 1891. Reptiles and batrachians from northern Yucatan and Mexico. Proc. Acad. Nat. Sci. Philadelphia, pp. 458–463.

Jackson, V. W. 1934. A manual of vertebrates of Manitoba. Winnipeg, Canada. Pp. 1–48, 178 illus., 7 maps.

Jaekel, O. 1910. Über die Paratheria, eine neue Klasse von Wirbeltieren. Zool. Anz., 34: 113–124, 5 figs.

——. 1916. Die Wirbeltierfunde aus dem Keuper von Halberstadt. Paleont. Zeit., 1: 155–214.

James, J. F. 1887. Catalogue of the mammals, birds, reptiles, batrachians and fishes in the collection of the Cincinnati Society of Natural History. Jour. Cincinnati Soc. Nat. Hist., 10: 34–48.

Jesup, A. M. 1891. Flora and fauna within thirty miles of Hanover, New Hampshire. Privately printed. Pp. i–vii, 1–91, map.

Jopson, Harry G. M. 1940. Reptiles and amphibians from Georgetown County, South Carolina. Herpetologica, 2: 39–43.

Johnson, C. E. 1925. The muskrat in New York: its natural history and economics. 13. Relations to associated birds and reptiles. Roosevelt Wild Life Bull., 3: 294–299, figs. 85–86.

Johnson, Charles W. 1894. Trionyches in the Delaware drainage. Amer. Nat., 28: 889.

Johnson, F. F. [n.d.]. Markets for fresh-water turtles. Fishery Industries Leaflet, no. 2004 (mimeographed).

Johnson, Murray L. 1942. A distributional check-list of the reptiles of Washington. Copeia, no. 1, pp. 15–18.

Jones, A. W., B. W. Mounts, and G. B. Wolcott. 1945. A pronocephalid fluke from the *Pseudemys.* Trans. Texas Acad. Sci., 28: 92–93.

Jones, J. M. 1865. Contributions to the natural history of Nova Scotia. Reptilia. Proc. Trans. Nova Scotian Inst. Nat. Sci., 1: 128–144.

Jordan, David Starr. 1929. Manual of the vertebrate animals of the northeastern United States. 13th ed. New York, World Book Co. Pp. i–xxi, 1–446, 15 figs., map.

Josselyn, John. 1612. New England rarities discovered in birds, beasts, fishes, serpents, and plants of that country. . . . London. Pp. 1–113.

Kanberg, Hans. 1931. Schildkröten. Wochenschr. Aquar. Terrar., 46: 747–750.

Kennicott, Robert. 1855. Catalogue of animals observed in Cook County, Illinois. Trans. Illinois State Agric. Soc., 1: 577–595.

Kermode, F. 1909. Visitor's guide to the Provincial Museum of Natural History and Ethnology. Victoria, B.C. Pp. 1–92, illustrated.

——. 1932. A remarkable capture of leatherback turtles off Bajo Reef near Nootka Sound, west coast of Vancouver Island, British Columbia. An. Rept. Prov. Mus. Nat. Hist., 1931, pp. B6–B7.

Kerr, Walter C. 1895. A large turtle. Proc. Nat. Sci. Assoc. Staten Island, 4: 75.

Khalil, Fouad. 1947. Excretion in reptiles. I. Non-protein nitrogen constituents of the urine of the sea turtle *Chelone mydas* L. Jour. Biol. Chem., 171: 611–616.

King, F. Willis. 1932. Herpetological records and notes from the vicinity of Tucson, Arizona, July and August, 1930. Copeia, no. 4, pp. 175–177.

——. 1939. A survey of the herpetology of Great Smoky Mountains National Park. Amer. Midland Nat., 21: 531–582.

Kingsley, John Sterling. 1885. The standard natural history. Vol. III. Lower vertebrates. Boston, S. E. Cassio and Co. Pp. i–vi, 1–478, 270 figs., 16 pls.

Kirsch, P. H. 1895a. Report on investigations in the Maumee Basin during the summer of 1893. Bull. U.S. Fish Comm., 14: 315–337.

——. 1895b. A report upon explorations made in Eel River basin in the northeastern part of Indiana in the summer of 1892. *Ibid.*, 31–41.

Kirtland, Jared P. 1838. Reptiles [of Ohio]. Ann. Rept. Ohio Geol. Surv., 2: 158, 167, 188–189.

Kitchin, Irwin C. 1949. Nesting habits of diamond-back terrapins. Anat. Rec., 105: 550 (abstr.).

Klauber, L. M. 1928. A list of the amphibians and reptiles of San Diego County, California. Bull. Zool. Soc. San Diego, 5: 1–8.

——. 1930. A list of the amphibians of San Diego County, California. 2d ed. *Ibid.*, 1–8.

——. 1932. Amphibians and reptiles observed en route to Hoover Dam. Copeia, no. 3, pp. 118–128.

——. 1934. Annotated list of the amphibians and reptiles of the southern border of California. Bull. Zool. Soc. San Diego, 11: 1–28, 8 figs.

Klingelhöffer, W. 1931. Terrarienkunde. Stuttgart, Julius E. G. Wegner. Pp. i–x, 1–576, 559 figs.

Knight, C. F. 1871. [Remarks on the Florida turtles.] Proc. Boston Soc. Nat. Hist., 14: 16–18.

Knoll, C. M. 1935. Shield variation, reduction, and age in box terrapin, *Terrapene carolina*. Copeia, no. 2, p. 100, 1 fig.

Knowlton, Josephine Gibson. 1943. My turtles. Privately printed. Pp. i–xxxvii, 39–222, 22 figs.

Kopstein, F. 1926. Reptilien von den Molukken und den Benachbarten Inseln. Zool. Meded. Leiden, 9: 71–112.

Koster, William J. 1946. Records of the snapping turtle from New Mexico. Copeia, no. 3, p. 173.

Krefft, Paul. 1928. Ueber die Dosenschildkrötengattung *Terrapene* Merr. Blatt. Aquar. Terrar., 39: 315–320, 337–342.

Lacépède, B. G. E. 1788. Histoire naturelle des quadrupèdes ovipares et des serpens. Paris, 1: 18 + 651, 41 pls.; (1789), 2: 20 + 527, 22 pls., 1 table.

Lagler, Karl F. 1940a. A turtle loss. Amer. Wildlife, 29: 41–44, 3 figs.

——. 1940b. Turtles, friends or foes of fish culture. Prog. Fish Culturist, 50: 14–18.

——. 1941a. Predatory animals and game fish. Amer. Wildlife, 30: 87–90, 8 figs.

——. 1941b. Fall mating and courtship of the musk turtle. Copeia, no. 4, p. 268.

——. 1942. Here's your meat. Sports Afield, 103: 20–21, 55.

——. 1943a. Methods of collecting freshwater turtles. Copeia, no. 1, pp. 21–25.

——. 1943b. Food habits and economic relations of the turtles of Michigan with special reference to game management. Amer. Midland Nat., 29: 257–312, 9 figs.

——. 1943c. Turtle: An unrationed ration. Michigan Conserv., 12: 6–7, 3 figs.

——, and Vernon C. Applegate. 1943. Relationship between the length and the weight in the snapping turtle *Chelydra serpentina* Linnaeus. Amer. Nat., 77: 476–478.

——, and Mary Jane Lagler. 1944. Natural enemies of crayfishes in Michigan. Papers Mich. Acad. Sci., Arts, Lett., 29: 293–303.

Lampe, E. 1901. Catalog der Reptilien Sammlung (Schildkröten, Crocodile, Eidechsen, und Chamaeleons) des Naturhistorischen Museums zu Wiesbaden. Jahrb. Nassau. Ver. Naturk., 54: 177–222, pl. 3.

Lamson, George Herbert. 1935. The reptiles of Connecticut. State Geol. Nat. Hist. Surv. Bull., 57: 1–35, 12 pls.

Lapham, I. A. 1852. Fauna and flora of Wisconsin. Trans. Wisconsin State Agric. Soc., 2: 337–445.

La Rivers, Ira. 1942. Some new amphibian and reptile records from Nevada. Jour. Ent. Zool., 34: 52–68.

Latham, Roy. 1916a. Studying the box turtle. Copeia, no. 40, pp. 15–16.

——. 1916b. Notes on *Cistudo carolina* from Orient, Long Island. *Ibid.*, no. 34, pp. 65–67.

Latreille, see Sonnini and Latreille.

Le Conte, John. 1836. Description of the species of North American tortoises. Ann. Lyceum Nat. Hist., N.Y., 3: 91–131.

——. 1854. Description of four new species of *Kinosternum*. Proc. Acad. Nat. Sci. Philadelphia, 7: 180–190.

——. 1856. Notice of American animals, formerly known but now forgotten or lost. *Ibid.*, 8: 8–14.

——. 1859. Description of two new species of tortoises. *Ibid.*, 11: 4–7.

Lederer, Gustàv. 1931. Aus meinem Tagebuch. Wie lange können Aquarien- und Terrarientiere hungern? Wochenschr. Aquar. Terrar., 28: 588–589.

Leffingwell, D. J. 1926. Vertebrates. *In* A preliminary biological survey of the Lloyd-Cornell reservation. Lloyd Library Botany, Pharm. Mat. Med. Bull., 27: 71–82.

Lesson, R. P. 1834. Reptiles. *In* Belanger's Voyage aux Indes-Orientales . . . Paris, A. Bertrand. Zoology, pp. 291–336; atlas, reptiles, pls. 1–7.

Le Sueur, C. A. 1817. An account of the American species of tortoise, not noticed in the systems. Jour. Acad. Nat. Sci. Philadelphia, 1: 86–88.

——. 1827. Note sur deux espèces de tortues du genre *Trionyx*. Gffr. St. II. Mem. Mus. Hist. Nat. Paris, 15: 257–268.

Lewis, C. Bernard. 1940. The Cayman Islands and marine turtle. Bull. Inst. Jamaica, 2 (appendix): 56–65.

Lindholm, W. A. 1929. Revidiertes Verzeichnis der Gattungen der regenten Schildkröten nebst Notizen zur Nomenklatur eineger Arten. Zool. Anz., 81: 275–295.

Linné, Carolus von. 1758. Systema naturae. Editio Reforma decima, Reformata. vol. 1.

——. 1766. Systema naturae. 12th ed.

Linsdale, Jean M. 1927. Amphibians and reptiles of Doniphan County, Kansas. Copeia, no. 164, pp. 75–81.

——. 1932. Amphibians and reptiles from Lower California. Univ. Calif. Publ. Zool., 38: 346–386.

——. 1940. Amphibians and reptiles in Nevada. Proc. Amer. Acad. Arts Sci., 73: 197–257, 29 figs.

——, and J. Linsley Gressit. 1937. Soft-shelled turtles in the Colorado River basin. Copeia, no. 4, pp. 222–225, 3 figs.

Linsley, J. H. 1844. A catalogue of the reptiles of Connecticut arranged according to their natural families. Amer. Jour. Sci. Arts, Ser. 1, 46: 35–51.

Little, Elbert L., Jr. 1940. Amphibians and reptiles of the Roosevelt Reservoir area, Arizona. Copeia, no. 4, pp. 260–265.

——, and J. G. Keller. 1937. Amphibians of the Jornada Experimental Range, New Mexico. *Ibid.*, pp. 216–222.

Llewellyn, Leonard M. 1940. The amphibians and reptiles of Mineral County, West Virginia. Proc. West Virginia Acad. Sci., 14: 14–150.

Lockington, W. N. 1879. Notes on some reptiles and batrachia of the Pacific coast. Amer. Nat., 13: 780–781.

Lockley, A. S. 1948. Note on the three-toed box turtle. Copeia, no. 2, p. 132.

Löding, H. P. 1922. A preliminary catalogue of Alabama amphibians and reptiles. Papers Alabama Mus. Nat. Hist., 5: 1–59.

Logier, E. B. S. 1925. Notes on the herpetology of Point Pelee, Ontario. Canadian Field-Nat., 39: 91–95.

——. 1928. The amphibians and reptiles of the Lake Nipigon region. Trans. Roy. Canadian Inst., 16: 290.

——. 1930. A faunal investigation of King Township, York County, Ontario. IV. The amphibians and reptiles of King Township. *Ibid.*, 17(2): 203–208.

——. 1931. A faunal investigation of Long Point and vicinity, Norfolk County, Ontario. The amphibians and reptiles of Long Point. *Ibid.*, 18: 229–236.

——. 1932. Some account of the amphibians and reptiles of British Columbia. *Ibid.*, 311–336.

——. 1939. The reptiles of Ontario. Roy. Ontario Mus. Zool. Handbook, 4: 5–63, 8 pls.

——. 1941. The amphibians and reptiles of Prince Edward County, Ontario. Univ. Toronto Studies, no. 48, pp. 93–106.

——, and G. C. Toner. 1942. Amphibians and reptiles of Canada. Canadian Field-Nat., 56: 15–16.

Long, Edward. 1774. The history of Jamaica, or general survey of the ancient and modern state of that island. London, T. Loundes. 3 vols.

Lönnberg, Einar. 1894. Reptiles and batrachians collected in Florida in 1892 and 1893. Proc. U.S. Nat. Mus., 17: 317–339, 1 fig.

——. 1896. Is the Florida box turtle a distinct species? *Ibid.*, 19: 253–254.

Loomis, R. B. 1948. Notes on the herpetology of Adams County, Iowa. Herpetologica, 4: 121–122.

——, and J. Knox Jones. 1948a. New herpetological records from Central Minnesota. *Ibid.*, 213–214.

——, and J. Knox Jones. 1948b. An additional record of *Graptemys p. pseudogeographica* (Gray), from Nebraska. *Ibid.*, 147.

Lord, J. K. 1866. The naturalist in Vancouver Island and British Columbia. London, George Bently. 2 vols. 1: 1–358; 2: 1–375, illustrated.

Loveridge, Arthur. 1928. Field notes on vertebrates collected by the Smithsonian Chrysler East African Expedition of 1926. Proc. U.S. Nat. Mus., 73: 1–69, 4 pls.

——. 1931. Some herpetological records from Vermont. Bull. Boston Soc. Nat. Hist., no. 61, pp. 15–16.

——. 1934. Australian reptiles in the Museum of Comparative Zoology, Cambridge, Massachusetts. Bull. Mus. Comp. Zool., 77: 243–383, 1 pl.

——. 1944. Pancake tortoise. Fauna, 6: 13–15, illustrated.

——. 1945. Reptiles of the Pacific world. New York, Macmillan Co. Pp. i–xii, 1–259, 2 pls.

——. 1947. Bone making material for turtles. Copeia, no. 2, p. 136.

——. 1948. New Guinean reptiles and amphibians in the Museum of Comparative Zoology and United States Natural Museum. Bull. Mus. Comp. Zool., 101: 305–430.

——, and Benjamin Shreve. 1947. The "New Guinea" snapping turtle. Copeia, no. 2, pp. 120–123, 1 fig.

Lucas, F. A. 1916. Occurrence of *Pseudemys* at Plymouth, Mass. *Ibid.*, no. 38, pp. 98–100.

——. 1922. Historic tortoise and other aged animals. Nat. Hist., 22: 301–305.

Ludicke, M. 1936. Über die Atmung von *Emys orbicularis* L. Zool. Jahrb. Abt. Allg. Zool. Physiol., 56: 82–106, 9 figs., 1 table.

Luederwaldt, Hermann. 1926. Os Chelonios brasileiros. Rev. Mus. Paulista, 14: 1–66, 4 figs., 11 pls.

Lumsden, T 1924. Chelonian respiration. Jour. Physiol., 58: 259–266.

Lynn, W. Gardner. 1936. Reptile records from Stafford County, Virginia. Copeia, no. 3, pp. 169–171.

——. 1937. Variation in scutes and plates in the box turtle, *Terrapene carolina*. Amer. Nat., 71: 421–426.

——, and Theodor von Brand. 1945. Studies on the oxygen consumption and water metabolism of turtle embryos. Biol. Bull., 88: 112–125.

"M.M.W." [n.d.]. Notes on egg laying habits of the turtle. Turtox News, 3: 37–38.

McAtee, W. L. 1907. A list of the mammals, reptiles and batrachians of Monroe County, Indiana. Proc. Biol. Soc. Washington, 20: 1–16.

——. 1918. A sketch of the natural history of the District of Columbia. Bull. Biol. Soc. Washington, no. 1, pp. 1–142, maps.

McCauley, Robert Henry, Jr. 1940. A distributional study of the reptiles of Maryland and the District of Columbia. Abstr. Theses, Cornell Univ., 1940, pp. 267–269.

——. 1945. The reptiles of Maryland and District of Columbia. Hagerstown, Maryland, published by the author. Pp. 1–194, 48 figs., 46 maps.

——, and Charles S. East. 1940. Amphibians and reptiles from Garrett County, Maryland. Copeia, no. 2, pp. 120–123.

MacCoy, Clinton. 1931. Key for the identification of New England amphibians and reptiles. Bull. Boston Soc. Nat. Hist., 59: 25–33.

——. 1932. Herpetological notes from Tucson, Arizona. Occ. Papers Boston Soc. Nat. Hist., 8: 11–28.

McCulloch, Allan R. 1908. A new genus and species of turtle from North Australia. Rec. Australian Mus., 7: 126–128, pls. 26–27.

McCutcheon, F. H. 1941. The breathing mechanism of turtles. Science, 94: 609.

——. 1943. The respiratory mechanism in turtles. Physiol. Zool., 16: 255–269.

——. 1947. Specific oxygen affinity of hemoglobin in elasmobranchs and turtles. Jour. Cell. Comp. Physiol., 29: 333–334. 3 figs.

MacKay, A. H. 1898. Batrachia and Reptilia of Nova Scotia. Proc. Trans. Nova Scotian Inst. Sci., 9: 41–43, pl. 43.

McLain, Robert Baird. 1899a. Critical notes on a collection of reptiles from the western coast of the United States. Wheeling, West Virginia, privately published. Pp. 1–13.

——. 1899b. Notes on a collection of reptiles made by Mr. C. J. Pierson at Fort Smith, Arkansas, with remarks on other eastern reptiles. Wheeling, West Virginia, privately printed. Pp. 1–5.

Macnamara, C. 1919. Notes on turtles. Ottawa Nat., 32: 135.

Maluquer, Joaquim, and Nicolan Maluquer. 1919. Les tortugues de Catalunya. Treb. Mus. Cien. Nat. Barcelona, 2: 93–159.

Mansuete, Romeo. 1941. A descriptive catalogue of the reptiles found in and around Baltimore City, Maryland. Proc. Nat. Hist. Soc. Maryland, 7: 56, 2 pls., 1 map (mimeographed).

Manville, R. H. 1939. Notes on the herpetology of Mt. Desert Island, Maine. Copeia, no. 3, p. 174.

Marcellin, P. 1926. Two marine turtles. Bull. Soc. d'Etude Sci. Nat. Nîmes, 44: 151–152.

Marchand, Lewis J. 1942. A contribution to a knowledge of the natural history of certain freshwater turtles. Master's thesis, University of Florida. Pp. 1–83, 12 pls., 7 graphs, 4 tables.

——. 1944. Notes on the courtship of a Florida terrapin. Copeia, no. 3, pp. 191–192.

——. 1945a. Water goggling: a new method for the study of turtles. *Ibid.*, no. 1, pp. 37–40.

——. 1945b. The individual range of some Florida turtles. *Ibid.*, no. 2, pp. 75–77.

Marr, John C. 1944. Notes on amphibians and reptiles from the central United States. Amer. Midland Nat., 32: 478–490.

Mast, S. O. 1911. Behavior of the loggerhead turtle in depositing its eggs. Papers Tortugas Lab., Carnegie Inst. Washington, 3: 63–67.

Mattern, E. S., and W. I. Mattern. 1917. Amphibians and reptiles of Lehigh County, Pennsylvania. Copeia, no. 46, pp. 64–66.

Mattox, Norman T. 1936. Annular rings in the long bones of turtles and their correlation with size. Trans. Illinois State Acad. Sci., 28: 225–226.

Mawson, N. 1921. Breeding habits of the green turtle, *Chelonia mydas*. Jour. Bombay Nat. Hist. Soc., 27: 956–957.

Mearns, E. A. 1897. A study of the vertebrate fauna of the Hudson highlands. Bull. Amer. Mus. Nat. Hist., 10: 303–352.

——. 1907. Mammals of the Mexican boundary of the United States. Bull. U.S. Nat. Mus., 56: i–xi, 1–530, 126 figs., 13 pls.

Medsger, Oliver P. 1919. Notes on the first turtle I ever saw. Copeia, no. 69, p. 29.

Meek, S. E. 1905. An annotated list of a collection of reptiles from southern California and northern Lower California. Field Columbian Mus. Zool. Ser., 7: 3–19, 3 pls., 1 table.

Mellen, Ida M. 1924. Box tortoises over 35 years old. Bull. N.Y. Zool. Soc., 1924, p. 98.

Merrem, Blasius. 1820. Tentamen systematis amphibiorum. Marburg. Pp. i–xv, 1–191.

Mertens, Robert. 1928. Über die Einwirkung der Kulturlandschaft auf die Verbreitung der Amphibien und Reptilien. Zool. Garten Leipzig, 1: 195–203.

——. 1936. Schildkröten-Beobachtungen in Freiland-Terrarium. Blät. Aquar. Terrar., 47: 253–275, 268–272.

——. 1939. Herpetologische Ergebnisse einer Reise nach der Insel Hispaniola, Westindien. Abh. Senck. Naturf. Ges., 449: 1–84, 10 pls.

——. 1940. Der Knochenpanzer einer kyphotischen Weichschildkröte. Senckenbergiana, 22: 236–243, 3 figs.

——. 1943. Über die Umwandlung der Antillen-schildkröte *Pseudemys palustris* in *Pseudemys rugosa*. *Ibid.*, 26: 313–319, 6 figs.

——, and Lorenz Müller. 1928. Liste der Amphibien und Reptilien Europas. Abh. Senck. Naturf. Ges., 41: 1–62.

——, and Lorenz Müller. 1940. Die Amphibien und Reptilien Europas. *Ibid.*, 451: 1–56.

Miles, M. A. 1861. A catalogue of the mammals, birds, reptiles and mollusks of Michigan. 1st Bienn. Rept. Geol. Surv. Mich., pp. 213–241.

Miller, E. Morton. 1937. A gopher turtle of unusual coloration. Copeia, no. 4, pp. 230–231.

Miller, Loye. 1932. Notes on the the desert tortoise (*Testudo agassizii*). Trans. San Diego Soc. Nat. Hist., 7: 187–208.

——. 1942. A Pleistocene tortoise from the McKittrick asphalt. *Ibid.*, 9: 439–442.

Miller, Newton. 1917. A method for killing turtles. Science, 45: 408–409.

Miller, Robert R. 1946. The probable origin of the soft-shelled turtle in the Colorado River basin. Copeia, no. 1, p. 46.

Mills, R. Colin. 1948. A check list of the reptiles and amphibians of Canada. Herpetologica, 2d suppl., 4: 1-15.

Mitchell, S. Weir, and George R. Morehouse. 1863. Researches upon the anatomy and physiology of respiration in the Chelonia. Smithsonian Contrib., 13: 1-42.

Mitsukuri, K. 1893. On the process of gastrulation in Chelonia. (Contrib. to Embryol. Rept. IV.) Jour. Coll. Sci. Tokyo, 6: 227-277, pls. 6-8.

——. 1895a. How many times does the snapping turtle lay eggs in one season? Zool. Mag. Tokyo, 6: 143-147.

——. 1895b. On the fate of the blastopore, the relations of the primitive streak, and the formation of the posterior end of the embryo in Chelonia, together with remarks on the nature of mesoblastic ova in vertebrates. (Contrib. to Embryol. Rept. V.) Jour. Coll. Sci. Tokyo, 10: 1-118, 11 pls.

——. 1905. The cultivation of marine and fresh-water animals in Japan. Bull. Bureau Fish. Washington, 24: 257-289.

——, and C. Ishikawa. 1886. On the formation of the germinal layers in Chelonia. Quart. Jour. Micros. Sci. n.s., 27: 17-48, pls. 2-5.

Mittleman, M. B. 1944. The status of *Testudo terrapin* (Schoepff). Copeia, no. 4, pp. 245-250.

——. 1945a. Type localities of two American turtles. *Ibid.*, no. 3, p. 171.

——. 1945b. Additional notes on the name *Testudo terrapin* Schoepff. *Ibid.*, no. 4, pp. 233-234.

——. 1947. The allocation of *Testudo rugosa* Shaw. Herpetologica, 3: 173-176.

——, and Bryce C. Brown. 1947. Notes on *Gopherus berlandieri* (Agassiz). Copeia, no. 3, p. 211.

Mocquard, F. 1899. Contribution à la faune herpétologique de la Basse-Californie. Nouv. Arch. Mus. d'Hist. Nat. Paris, ser. 4, 1: 297-344, pls. 11-13.

Moore, George A., and Carl C. Rigney. 1941. Notes on the herpetology of Payne County, Oklahoma. Proc. Oklahoma Acad. Sci., 1941, pp. 77-80.

Moorhouse, F. W. 1933. Notes on the green turtle (*Chelonia mydas*). Rept. Great Barrier Reef Comm., 4: 1-22.

Morris, Mary B. 1912. Turtles and tortoises. Mus. News Brooklyn Inst. Arts Sci., 7: 100-104.

Morse, Max. 1901. Ohio reptiles in the Ohio State University Zoological Museum. Ohio Nat., 1: 126-128.

——. 1903. Ohio reptiles and batrachians. *Ibid.*, 3: 360-361.

——. 1904. The batrachians and reptiles of Ohio. Proc. Ohio Acad. Sci., 4: 93-144, 2 pls.

Mosauer, Walter. 1935. The identity of the turtles from San Ignacio, Lower California. Copeia, no. 4, p. 191.

Mowbray, Louis L. 1922. Certain citizens of the warm sea. Nat. Geogr., 41: 27–62, pls.

Muller, J. F. 1921. Notes on the habits of the soft-shell turtle *Amyda mutica*. Amer. Midland Nat., 7: 180–184.

Munroe, Ralph W. 1898. The green turtle, and the possibilities of its protection and increase on the Florida coast. Bull. U.S. Bureau Fish., 17: 273–275.

Murphy, Robert C. 1914. *Thalassochelys caretta* in the South Atlantic. Copeia, no. 2, p. 4.

———. 1916. Long Island turtles. *Ibid.,* no. 33, pp. 56–60.

Murray, Leo T. 1937. System of indices for the dermal skeleton of the Testudinata. Proc. Texas Acad. Sci., 20: 16.

———. 1938. Basic research problems in out-door herpetology in Texas. *Ibid.,* 22: 18–19.

Musgrave, A., and C. P. Whitley. 1926. From sea to soup. An account of the turtles of Northwest Inlet. Australian Mus. Mag., 2: 331–336, 11 figs.

Myers, George S. 1926. A synopsis for the identification of the amphibians and reptiles of Indiana. Proc. Indiana Acad. Sci., 35: 277–294, 1 fig.

———. 1927. Notes on Indiana amphibians and reptiles. *Ibid.,* 36: 337–340.

———. 1933. Two records of the leatherback turtle on the California coast. Copeia, no. 1, p. 44.

Nash, C. W. 1905. Check list of the vertebrates of Ontario, and catalogue of the specimens of the biological section of the Provincial Museum. Batrachians, reptiles and mammals. Toronto Dept. Educ. Pp. 1–32.

———. 1908. Batrachians and reptiles of Ontario. *In* Vertebrates of Ontario. Toronto Dept. Educ. Pp. 5–18.

Necker, Walter L. 1934. Contribution to the herpetology of the Smoky Mountains of Tennessee. Bull. Chicago Acad. Sci., 5: 1–4.

———. 1935. Reptiles and amphibians of the dunes. *In* The dunes of the Chicago region. Progr. Activ. Chicago Acad. Sci., 6: 22–23.

———. 1938. Check list of reptiles and amphibians of the Chicago region. Chicago Acad. Sci., leaflet, 1: 1–4, 1 map.

———. 1939. Revised check list of reptiles and amphibians of the Chicago region. *Ibid.,* 11: 1–4.

———. 1940. Hump-backed turtles. Chicago Nat., 3: 62, 1 fig.

Neill, Wilfred T. 1948a. Hibernation of amphibians and reptiles in Richmond County, Georgia. Herpetologica, 4: 107–114.

———. 1948b. Use of scent glands by prenatal *Sternotherus minor*. *Ibid.,* 148.

———. 1948c. Odor of young box turtles. Copeia, no. 2, p. 130.

———. 1948d. The musk turtles of Georgia. Herpetologica, 4: 181–183.

——. 1951. The taxonomy of North American soft-shelled turtles, Genus *Amyda*. Publ. Research Div. Ross Allen's Rept. Inst., 1(2): 7–24, fig. 1.

Nelson, Edward W. 1921. Lower California and its natural resources. Mem. Nat. Acad. Sci., 16: 1–194, 34 pls.

Nelson, Julius. 1890. Descriptive catalogue of the vertebrates of New Jersey. Geol. Surv. New Jersey; Final Rept. State Geol., 2: 489–824.

Netting, M. Graham. 1927a. Muhlenberg's turtle in western Pennsylvania. Ann. Carnegie Mus., 17: 403–408, pl. 36.

——. 1927b. Amphibians and reptiles in relation to birds. Cardinal, 2: 1–6.

——. 1929. A note on the egg-laying of *Pseudemys floridana* (Le Conte). Copeia, no. 170, pp. 24–25.

——. 1932. Blanding's turtle, *Emys blandingii* (Holbrook), in Pennsylvania. *Ibid.,* no. 4, pp. 173–174.

——. 1935. A non-technical key to the amphibians and reptiles of western Pennsylvania. Nawakwa Fireside, n.s., 3–4: 34–49.

——. 1936a. Hibernation and migration of the spotted turtle, *Clemmys guttata* (Schneider). Copeia, no. 2, p. 112.

——. 1936b. Hand list of the amphibians and reptiles of Pennsylvania. Carnegie Mus. Herp. Leaflet, 1: 1–4.

——. 1936c. The amphibians and reptiles of Indiana County, Pennsylvania. Proc. Penn. Acad. Sci., 10: 25–28.

——. 1939. Hand list of the amphibians and reptiles of Pennsylvania. Bienn. Rept. Penn. Fish Comm., pp. 109–112.

——. 1940. The spotted turtle, *Clemmys guttata* (Schneider). Proc. West Virginia Acad. Sci., 14: 146–147.

——. 1944. The spineless soft-shelled turtle, *Amyda mutica* (Le Sueur), in Pennsylvania. Ann. Carnegie Mus., 30: 85–88.

——. 1946. The amphibians and reptiles of Pennsylvania. Pennsylvania, Board Fish Comm. Pp. 1–29.

Newcombe, W. A. 1932. Accession notes. Rept. Prov. Mus. Nat. Hist. for 1931, Victoria, British Columbia, pp. B8–B16.

Newman, H. H. 1906a. The habits of certain tortoises. Jour. Comp. Neurol., 16: 126–152.

——. 1906b. The significance of scute and plate "abnormalities" in Chelonia. Biol. Bull., 10: 68–114, 58 figs.

Nichols, J. T. 1917. Stray notes on *Terrapene carolina*. Copeia, no. 46, pp. 66–68.

——. 1920. Turtle eggs. Forest and Stream, July, pp. 379, 396–397.

——. 1921. The snapping turtle's egg. *Ibid.,* Sept., p. 401.

——. 1933. Further notes on painted turtles. Copeia, no. 1, pp. 41–42.

——. 1939a. Range and homing of individual box turtles. *Ibid.,* no. 3, pp. 125–127.

——. 1939b. Disposition in the box turtle. *Ibid.,* no. 2, p. 107.

——. 1939c. Data on size, growth and age in the box turtle, *Terrapene carolina. Ibid.,* no. 1, pp. 14–20, 2 figs., 2 tables.

——. 1947. Notes on the mud turtle. Herpetologica, 3: 147–148.

——, and Ralph DeSola. 1933. The probable size maximum of *Chrysemys picta picta* (Schneider). Copeia, no. 3, p. 151.

Nick, L. 1912. Das Kopfskelet von *Dermochelys coriacea* L. Zool. Jahrb., Abth. Anat., 33: 1–238.

Nixon, William, and Hobart M. Smith. 1949. The occurrence of kyphosis in turtles. Turtox News, 27: 1–2, 3 figs.

Noble, G. K. 1923. *Chelys* and the phylogeny of the turtle. Amer. Nat., 57: 377–379.

——. 1929. Distributional list of the reptiles and amphibians of the New York City region. Amer. Mus. Nat. Hist., Guide Leaflet Ser., 69: 1–16.

——, and A. Braslovsky. 1935. The sensory mechanisms involved in the migration of newly hatched fresh-water turtles. Anat. Rec. (suppl.) 64: 88.

——, and A. M. Breslau. 1938. The senses involved in the migration of young fresh-water turtles after hatching. Jour. Comp. Psychol., 25: 175–193.

Nopsca, Baron Francis. 1922. A case of secondary adaptation in a tortoise. Ann. Mag. Nat. Hist., ser. 9, 10: 155–157.

Norris-Elye, L. T. S. 1949. The common snapping turtle (*Chelydra serpentina*) in Manitoba. Canadian Field-Nat., 63: 145–147.

Norton, A. H. 1929. Notes on the history of herpetology in Maine. Maine Nat., 9: 53–68.

Obrecht, Carl P. 1946. Notes on South Carolina reptiles and amphibians. Copeia, no. 2, pp. 71–74.

Ogilby, J. D. 1905. Catalogue of the emydosaurian and testudinian reptiles of New Guinea. Proc. Royal Soc. Queensland, 19: 1–31.

Okada, S. 1891. Catalogue of vertebrated animals of Japan. Tokyo. Pp. 1–125.

Oliver, James A. 1937. Notes on a collection of amphibians and reptiles from Colima, Mexico. Occ. Papers Mus. Zool. Univ. Mich., 360: 1–28, 1 fig.

——. 1946. An aggregation of Pacific sea turtles. Copeia, no. 2, p. 103.

——, and Joseph R. Bailey. 1939. Amphibians and reptiles of New Hampshire. Biol. Surv. Conn. Watershed Rept., no. 4, pp. 195–217, 19 figs.

Ortenburger, A. I. 1921. A list of Amphibia and Reptilia collected in Indiana. Copeia, no. 99, pp. 73–76.

——. 1926a. Reptiles and amphibians collected in the Arbuckle Mts., Murray Co., Oklahoma. *Ibid.,* no. 156, pp. 145–146.

——. 1926b. Reptiles and amphibians collected in the Wichita Mts., Comanche County, Oklahoma. *Ibid.,* no. 155, pp. 137–138.

——. 1927a. Report on the amphibians and reptiles of Oklahoma. Proc. Oklahoma Acad. Sci., 6: 89–100.

——. 1927b. A list of reptiles and amphibians from the Oklahoma panhandle. Copeia, no. 163, pp. 46–48.

——. 1929a. Reptiles and amphibians from northeastern Oklahoma. *Ibid.*, no. 170, pp. 26–28.

——. 1929b. Reptiles and amphibians from southeastern Oklahoma and southwestern Arkansas. *Ibid.*, pp. 8–12.

——, and Beryl Freeman. 1930. Notes on some reptiles and amphibians from western Oklahoma. Oklahoma Univ. Biol. Surv. Publ., 2: 175–188.

——, and R. D. Ortenburger. 1927. Field observations on some amphibians and reptiles of Pima County, Arizona. Proc. Oklahoma Acad. Sci., 6: 101–121.

Osbeck, Peter. 1771. A voyage to China and the East Indies. London, Benjamin White. 2: 1–367 and index.

Otto, C. L. 1914. Young snapping turtles. Brooklyn Inst. Arts Sci., Children's Mus. News, 1: 29.

Over, William H. 1923. Amphibians and reptiles of South Dakota. South Dakota Geol. Nat. Hist. Surv. Ser., 23: 5–34, 18 pls.

Overton, Frank. 1916. Aquatic habits of the box turtle. Copeia, no. 26, pp. 4–5.

Owen, R. 1849. On the development and homologies of the carapace and the plastron of the Chelonia reptiles. Phil. Trans. Roy. Soc. London, 139: 151–171.

Owen, Robert P. 1940. A list of the reptiles of Washington. Copeia, no. 3, pp. 169–172.

Owens, David W. 1941. Some amphibians and reptiles from southern Illinois. *Ibid.*, p. 184.

Owens, Virgil. 1949. New snake records and notes, Morgan County, Missouri. Herpetologica, 5: 49–50.

Ozuma, K. 1937. Studies on sauropsid chromosomes. III. The chromosomes of the soft-shelled turtle, *Amyda japonica* (Temminck and Schlegel), as additional proof of heterogamity in the Reptilia. Jour. Genetics, 34: 247–264.

Packard, A. S. 1881. Probable cause of the longevity of turtles. Amer. Nat., 15: 738–739.

Palmer, E. Laurence. 1922. Amphibia and Reptilia. Cornell Rural School Leaflet, 15: 326–340.

Park, Orlando, W. C. Allee, and V. E. Shelford. 1939. A laboratory introduction to animal ecology and taxonomy. Univ. Chicago Press. Pp. i–x, 1–272, 17 pls.

Parker, G. H. 1901. Correlated abnormalities in the scutes and bony plates of the carapace of the sculptured tortoise. Amer. Nat., 35: 17–24, 5 figs.

——. 1922a. The crawling of young loggerheads toward the sea. Jour. Exp. Zool., 36: 323–331.

——. 1922b. The instinctive locomotor reactions of the loggerhead turtle in relation to its senses. Jour. Comp. Psychol., 2: 425–429, 2 figs.

——. 1925. The time of submergence necessary to drown alligators and turtles. Occ. Papers Boston Soc. Nat. Hist., 5: 157–159.

——. 1926. The growth of turtles. Proc. Nat. Acad. Sci., 12: 422–424.

——. 1928. Ciliary currents in oviducts of turtles in relation to transportation of spermatozoa. Proc. Soc. Exp. Biol. Med., 26: 52.

——. 1929. The growth of the loggerhead turtle. Amer. Nat., 63: 367–373, 1 table.

Parker, H. W. 1939a. Marine turtles as current indicators. Nature, 143: 121.

——. 1939b. [Short note on sea turtles in European waters.] *Ibid.*, 144: 156–157.

——. 1939c. Turtles stranded on the British coast. Proc. Linn. Soc. London, 151: 127–129.

Parker, Malcolm V. 1937. Some amphibians and reptiles from Reelfoot Lake, Tennessee. Jour. Tenn. Acad. Sci., 12: 60–86.

——. 1939. The amphibians and reptiles of Reelfoot Lake and vicinity, with a key for the separation of species and subspecies. *Ibid.*, 14: 72–101, 14 figs.

——. 1948. A contribution to the herpetology of western Tennessee. *Ibid.*, 23: 20–30.

Parks, Hal B., Frank Archibald, and Marl Caldwell. 1939. Amphibians and reptiles of east Texas pine belt. Tech. Bull. Stephen F. Austin State Teachers College, 1: 1–4.

——, and V L. Cory. 1936. Biological survey of the east Texas big thicket area. Texas Acad. Sci. Special Publ., pp. 1–51, illustrated.

Parmenter, C. S. 1897. Fossil turtle cast from the Dakota epoch. Trans. Kansas Acad. Sci., 16: 67.

Parschin, A. N. 1929. Bedingte Reflexe bei Schildkröten. Pfluger's Arch. Ges. Physiol., 222: 328–333.

Patch, Clyde L. 1918. A list of amphibians and reptiles of the Ottawa, Ontario, district. Ottawa Nat., 32: 53.

——. 1919. A rattlesnake, melano garter snakes and other reptiles from Point Pelee, Ontario. Canadian Field-Nat., 33: 60–61.

——. 1925. *Graptemys geographica* in Canada. Copeia, no. 149, pp. 95–96.

Pawling, R. Oldt. 1939. The amphibians and reptiles of Union County, Pennsylvania. Herpetologica, 1: 165–169.

Pearse, A. S. 1923a. The abundance and migration of turtles. Ecology, 4: 24–28, figs. 1–2, 1 table.

——. 1923b. The growth of the painted turtle. Biol. Bull., 45: 145–148, 2 tables.

——, S. Lepkowsky, and Laura Hintze. 1925. The growth and chemical composition of three species of turtles fed on rations of pure foods. Jour. Morph., 41: 191–216, 13 figs.

Pell, S. Morris. 1940. Notes on the food habits of the common snapping turtle. Copeia, no. 2, p. 131.

Pellegrin, J. 1926. La longévité chez les reptiles en captivité. Rev. Gen. Sci., Paris, 37: 47–49.

Penn, George H. 1940. Notes on the summer herpetology of De Kalb County, Alabama. Jour. Tenn. Acad. Sci., 15: 352–355.

——. 1943. Herpetological notes from Cameron Parish, Louisiana. Copeia, no. 1, pp. 58–59.

——, and Karl E. Pottharst. 1940. The reproduction and dormancy of *Terrapene major* in New Orleans. Herpetologica, 2: 25–29.

Pennant, Thomas. 1769. Indian zoology. London [no title page]. 14 pp., 12 pls.

Perkins, C. B. 1947. A note on longevity of amphibians and reptiles in captivity. Copeia, no. 2, p. 144.

Peters, J. A. 1942. Reptiles and amphibians of Cumberland County, Illinois. *Ibid.,* no. 3, pp. 182–183.

——. 1946. Reptiles and amphibians of Sam A. Baker State Park, Wayne County, Missouri. *Ibid.,* no. 1, p. 44.

Philippi, R. A. 1899. Las tortugas chilenas. An. Univ. Chile, 104: 727–736, 3 figs.

Pickens, A. L. 1927. Reptiles of upper South Carolina. Copeia, no. 165, pp. 110–113.

——. 1940. Observations on growth in the box terrapin. Neighborhood Res., 4: 7.

Pickwell, Gayle. 1947. Amphibians and reptiles of the Pacific states. Stanford Univ. Press. Pp. v–xiv, 1–236, 65 pls., 20 figs.

Piers, Harry. 1897. Notes on Nova Scotian zoology, No. 4. Proc. Trans. Nova Scotian Inst. Sci., 9: 255–267.

Pilsbry, Henry A. 1910. *Stomatolepas,* a barnacle commensal in the throat of the loggerhead turtle. Amer. Nat., 44: 304–306, 1 fig.

Pirnie, Miles D. 1935. Michigan waterfowl management. Mich. Dept. Conserv., pp. 1–318, 212 text figs.

Pope, Clifford H. 1934. List of Chinese turtles, crocodilians, and snakes, with keys. Amer. Mus. Novitates, no. 733, pp. 1–29.

——. 1939a. Turtles of the United States and Canada. New York, Alfred A. Knopf. Pp. i–xvii, 1–343, 99 figs., table.

——. 1939b. Structure and habits of turtles. School Nat. League Bull., ser. 9, no. 8, pp. 1–3, 6 figs.

——. 1944. Amphibians and reptiles of the Chicago area. Chicago Nat. Hist. Mus. Press. Pp. 1–275, 50 figs., frontispiece.

——. 1945. Turtles. School Nat. League Bull., ser. 16, no. 2, 4 pp. (unnumbered), 6 figs.

Pope, T. E. B. 1928. Wisconsin herpetological notes. Year Book Mus. City Milwaukee, 8: 177–184.

——. 1931. Wisconsin herpetological notes. Trans. Wisconsin Acad. Sci., Arts, Lett., 26: 321–329.

——, and W. E. Dickinson. 1928. The amphibians and reptiles of Wisconsin. Bull. Publ. Mus. City Milwaukee, 8: 1–138, 28 figs., 22 pls.

Potter, Doreen. 1920a. Reptiles and amphibians taken in central Michigan in 1919. Copeia, no. 82, pp. 39–41.

——. 1920b. Reptiles and amphibians collected in northern Mississippi in 1919. Ibid., no. 86, pp. 82–83.

Pratt, H. S. 1923. A manual of land and fresh water vertebrate animals of the United States. Philadelphia, P. Blakiston's Son. Pp. i–xv, 1–422, 184 figs.

Proctor, Joan B. 1922. A study of the remarkable tortoise, Testudo loveridgii Blgr., and the morphogeny of the chelonian carapace. Proc. Zool. Soc. London, 34: 483–526, 21 figs., 3 pls.

Provancher, L'Abbé. 1874–75. Fauna canadienne. Les reptiles. Naturaliste Canadien, 6: 273–278; 7: 10–20, 42–46, 65–73, 289–298, 321–330, 353–370.

Pycraft, W. P. 1914. The courtship of animals. New York. Pp. 1–318, illustrated.

Quaranta, John V. 1949. The color discrimination of Testudo vicina. Anat. Rec., 105: 510–511.

——, and L. T. Evans. 1949. The visual learning of Testudo vicina. Ibid., 580 (abstr.).

Rafinesque, C. S. 1832. Description of two new genera of soft shell turtles of North America. Atlantic Jour. and Friend of Knowledge, Philadelphia, 1: 64–65.

Raj, B. S. 1927. The vertebrate fauna of Krusadai Island. Bull. Madras Govt. Mus., n.s. Nat. Hist., sec. 1, pp. 182–183.

Ramsey, E. R. 1901. The cold blooded vertebrates of Winona Lake and vicinity. Proc. Indiana Acad. Sci., 1900, pp. 218–224.

Randall, W. C., D. E. Stullken, and W. A. Hiestand. 1944. Respiration of reptiles as influenced by the composition of the inspired air. Copeia, no. 3, pp. 136–144, 4 figs.

Raney, E. C., and E. A. Lachner. 1942. Summer food of Chrysemys picta marginata in Chautauqua Lake, New York. Ibid., no. 2, pp. 83–85.

Rathke, H. 1848. Ueber die Entwicklung der Schildkröten. Braunschweig. Pp. 1–268.

Reed, Hugh, and A. H. Wright. 1909. The vertebrates of the Cayuga Lake basin, New York. Proc. Amer. Phil. Soc., 48: 370–459, pls. 17–20.

Reese, A. M. 1917. Reptiles as food. Sci. Monthly, pp. 545–550.

Reinhardt, J., and C. F. Lütken. 1862. Bidrag til det vest indiske Origes og naunligen til de dansk-vestindiske Oers Herpetologie. Videnskabelige Meddelelser fra den naturhistoriske Forening i Kobenhavn for Aaret, 1862, pp. 153–291.

Rhoads, Samuel N. 1895. Contributions to the zoology of Tennessee. No. I. Reptiles and amphibians. Proc. Acad. Nat. Sci. Philadelphia, 47: 376–406.

Richmond, Neil D. 1936. Seventeen-year locust in the diet of the snapping turtle. Herpetologica, 1: 8.

——. 1945. Nesting habits of the mud turtle. Copeia, no. 4, pp. 217–219, 2 tables.

——, and Coleman J. Goin. 1938. Notes on a collection of amphibians and reptiles from New Kent County, Virginia. Ann. Carnegie Mus., 27: 301–310.

Ridley, H. N. 1899. The habits of Malay reptiles. Jour. Straits Branch R. Asiatic Soc., 32: 185–210.

Risley, Paul L. 1930. Anatomical differences in the sexes of the musk turtle, *Sternotherus odoratus* (Latreille). Papers Mich. Acad. Sci., Arts, Lett., 11: 445–464, figs. 19–30.

——. 1933. Observations on the natural history of the common musk turtle, *Sternotherus odoratus* (Latreille). *Ibid.*, 17: 685–711, 2 figs., 1 table.

——. 1941. Some observations on hermaphroditism in turtles. Jour. Morph., 68: 101–121, 1 fig., 2 pls.

Rivers, J. J. 1889. Description of a new turtle from the Sacramento River belonging to the family of Trionychidae. Proc. Calif. Acad. Sci., 2: 333–336.

Robertson, H. C. 1939. A preliminary report on the reptiles and amphibians of Catoctin National Park, Maryland. Bull. Nat. Hist. Soc. Maryland, 9: 88–93, 5 figs.

Roddy, Harry Justin. 1928. Reptiles of Lancaster County and the state of Pennsylvania. Lancaster Sci. Press. Pp. 1–53, illustrated.

Rodeck, Hugo G. 1949. Notes on box turtles in Colorado. Copeia, no. 1, pp. 32–34.

Rodgers, Thomas L., and William L. Jellison. 1942. A collection of reptiles and amphibians from western Montana. *Ibid.*, pp. 10–13.

Roemier, Ferdinand. 1849. Texas. Bonn. Pp. i–xiv, 1–464, map.

Rogers, Charles H. 1917. Notes on three common New Jersey turtles. Copeia, no. 48, pp. 74–76.

Rollinat, R. 1899. Sur l'accouplement automnal de la cistudo d'Europe. Bull. Soc. Zool. France, 24: 103–106.

Rooij, Nelly de. 1915. The reptiles of the Indo-Australian archipelago. Leiden, E. J. Brill. Vol. 1 Pp. i–xiv, 1–382, 132 figs.

Roosevelt, Theodore. 1917. Notes on Florida turtles. Amer. Mus. Jour., 17: 289–291, 3 figs.

Rosenberger, Randle C. 1916. Interesting facts about turtles. Forest and Stream, 86: 764–765.

——. 1936. Notes on some habits of *Terrapene carolina* (Linné). Copeia, no. 3, p. 177.

Ruckes, Herbert. 1929a. Studies in chelonian osteology. Part II. The morphological relationships between the girdles, ribs, and carapace. Ann. New York Acad. Sci., 31: 81–120, 30 figs.

——. 1929b. Studies in chelonian osteology. Part I. Truss and arch analogies in chelonian pelves. *Ibid.*, 31–80, pls. 4–7.

——. 1937. The lateral arcades of certain emydids and testudinids. Herpetologica, 1: 97–119.

Russel, F. S. 1939. Turtles in the English Channel. Nature, 143: 206.

Rust, Hans Theodor. 1934. Systematische Liste der lebenden Schildkröten. Blät. Aquar. Terrar., 45: 42–45, 59–68.

——. 1936a. Verzeichnis der bisher gepflegten Schildkröten. Taschenkalender Aquar. Terrar., 28: 159–208.

——. 1936b. Ergänzung zum "Verzeichnis der bisher gepflegten Schildkröten." Wochenschr. Aquar. Terrar., 47: 163–165.

——. 1937. Interessante Schildkröten. III. Wochenschr. Aquar. Terrar., 34: 637–639.

Rüthling, Paul D. R. 1915. Hibernation of reptiles in southern California. Copeia, no. 19, pp. 10–11.

Ruthven, Alexander G. 1904. Notes on the molluscs, reptiles and amphibians of Ontonagon County, Michigan. Ann. Rept. Mich. Acad. Sci., 6: 188–192.

——. 1906. The cold-blooded vertebrates of the Porcupine Mountains and Isle Royale, Michigan. Rept. Geol. Surv. Mich., 1905, pp. 107–112.

——. 1907. A collection of reptiles and amphibians from southern New Mexico and Arizona. Bull. Amer. Mus. Nat. Hist., 23: 483–603, 22 figs.

——. 1909a. Notes on Michigan reptiles and amphibians. 11th Ann. Rept. Mich. Acad. Sci., pp. 116–117.

——. 1909b. The cold-blooded vertebrates of Isle Royale. *In* An ecological survey of Isle Royale, Lake Superior. Rept. Geol. Surv. Mich., 1908, pp. 329–333.

——. 1910a. Contributions to herpetology of Iowa. Proc. Iowa Acad. Sci., 17: 198–209, 2 figs.

——. 1910b. Notes on Michigan reptiles and amphibians. II. 12th Ann. Rept. Mich. Acad. Sci., p. 59.

——. 1911a. Notes on Michigan reptiles and amphibians. III. *Ibid.*, 13: 114–115, 1 pl.

——. 1911b. A biological survey of the sand dunes region on the south shore of Saginaw Bay: amphibians and reptiles. Mich. Geol. Biol. Surv., publ. no. 4, Biol. Ser., 2: 257–272.

——. 1912a. Directions for collecting and preserving specimens of reptiles and amphibians for museum purposes. Ann. Rept. Mich. Acad. Sci., 13: 165–176.

——. 1912b. Contributions to the herpetology of Iowa. II. Proc. Iowa Acad. Sci., 19: 207.

——. 1919. Contributions to the herpetology of Iowa. III. Occ. Papers Mus. Zool. Univ. Mich., 66: 1–3.

——. 1924. A check list of North American amphibians and reptiles. By Leonhard Stejneger and Thomas Barbour. [Review.] Science, 59: 339–340.

——. 1927. A large Blanding's turtle from Michigan. Copeia, no. 164, pp. 86–87.

——, and Crystal Thompson. 1915. On the occurrence of *Clemmys insculpta* (Le Conte) in Michigan. Occ. Papers Mus. Zool. Univ. Mich., 12: 1–2.

——, Crystal Thompson, and Helen T. Gaige. 1912. The herpetology of Michigan. Mich. Geol. Biol. Surv., publ. no. 10, Biol. Ser., 3: 1–166, 55 figs., 20 pls.

——, Crystal Thompson, and Helen T. Gaige. 1928. The herpetology of Michigan. Univ. Mich. Handbook, ser. 3, pp. i–ix, 1–228, 52 figs., 19 pls., frontispiece.

Sagar, Abram. 1839. Catalogue of mammals, birds, reptiles, amphibians, fishes and molluscs of Michigan. Senate Doc. State Mich., pp. 294–305.

Saville-Kent, W. 1897. The naturalist in Australia. London, Chapman and Hall. Pp. v–viii, 1–302, 104 figs., 50 pls., 9 color pls.

Say, Thomas. 1825. On the fresh water and land tortoises of the United States. Jour. Acad. Nat. Sci. Philadelphia, 4: 203–219.

Schinz, H. R. 1833. Naturgeschichte und Abbildungen der Reptilien. Leipzig. Pp. i–iv, 1–392, 102 pls.

Schmidt, F. J. W. 1926. List of amphibians and reptiles of Worden Township, Clark County, Wisconsin. Copeia, no. 154, pp. 131–132.

Schmidt, Francis V. 1948. Evaluation of wildlife populations on the Tuckahoe-Corbin City area. Pittman-Robertson Quart., 8: 70.

Schmidt, J. 1916. Marking experiments with turtles in the Danish West Indies. Meddelelser Kommissionen Havundersogelser, Ser: Fiskeri, 5: 1–26, 11 figs., 5 tables.

Schmidt, Karl P. 1916. Notes on the herpetology of North Carolina. Jour. Elisha Mitch. Sci. Soc., 32: 32–37.

——. 1922. The amphibians and reptiles of Lower California and the neighboring islands. Bull. Amer. Mus. Nat. Hist., 46: 607–707, pls. 47–57, 13 figs.

——. 1924a. Emory's soft-shelled turtle in Arizona. Copeia, no. 131, p. 64.

——. 1924b. A list of amphibians and reptiles collected near Charleston, South Carolina. *Ibid.*, no. 132, pp. 67–69.

——. 1938. Turtles of the Chicago area. Field Mus. Nat. Hist., Zool. Leaflet, 14: 1–24, 14 figs., 2 pls.

——. 1939. [Review of] Einführung in die vergleichende biologische Anatomie der Wirbeltiere. Erster Band. By Hans Boker. Jena. Copeia, no. 2, pp. 116–117.

——. 1940. A new turtle of the genus *Podocnemis* from the Cretaceous of Arkansas. Field Mus. Nat. Hist. Geol. Ser., 8: 1–12, 5 figs.

——. 1941. The amphibians and reptiles of British Honduras. Field Mus. Nat. Hist. Zool. Ser., 22: 475–510, 1 fig.

——. 1944. Two new thalassemyd turtles from the Cretaceous of Arkansas. Field Mus. Nat. Hist. Geol. Ser., 8: 63–74, 5 figs.

——. 1945a. A new turtle from the Paleocene of Colorado. Fieldiana-Geology, 10: 1–4, 1 fig.

——. 1945b. Problems in the distribution of marine turtles. Marine Life, 1: 7–10.

——. 1946. Turtles collected by the Smithsonian Biological Survey of the Panama Canal Zone. Smithsonian Miscel. Coll., 106: 1–9.

——, and Emmett R. Dunn. 1917. Notes on *Colpochelys kempi* Garman. Copeia, no. 44, pp. 50–52.

——, and Walter L. Necker. 1935. Amphibians and reptiles of the Chicago region. Bull. Chicago Acad. Sci., 5: 57–77.

——, and David W. Owens. 1944. Amphibians and reptiles of northern Coahuila, Mexico. Field Mus. Nat. Hist. Zool. Ser., 29: 97–115.

——, and T. F. Smith. 1944. Amphibians and reptiles of the Big Bend region. *Ibid.*, 75–96.

Schneck, J. 1886. Longevity of turtles. Amer. Nat., 20: 897.

Schnee, ——. 1898. Seeschildkröten. Natur und Hans, 6: 325–329.

——. 1899a. Einige Notizen über Weichschildkröten. Zeitschr. Naturwiss., 72: 197–208.

——. 1899b. Über Landschildkröten. Zool. Garten, 40: 119–122.

Schneider, J. G. 1792. Beschreibung und Abbildung einer neuen Art von Wasserschildkröte. Schriften Ges. Naturf. Freunde Berlin, 10: 259–283.

Schoepff, Johann David. 1792a. Historia testudinium. Erlangae, Io. Iac. Palmii. Pp. i–xii, 1–136, 31 pls.

——. 1792b. Naturgeschichte der Schildkröten. Erlangen, Johann Jacob Palm. Pp. 1–160, 31 pls.

Schoffman, Robert J. 1949. Turtling for the market at Reelfoot Lake. Jour. Tenn. Acad. Sci., 24: 143–145.

Schoonhoven, J. J. 1911. Blanding's turtle [from Long Island]. Science, n.s., 34: 917.

Schreitmuller, Wilhelm. 1931. *Caretta caretta* Linné (dickkopfige See-oder

Karettschildkröte) und andere. Wochenschr. Aquar. Terrar., 28: 216–217.

Schroeder, William C. 1924. Fisheries of Key West and the clam industry of southern Florida. U.S. Bureau Fish. Doc., 962: 1–74, 29 figs.

——. 1931. The turtle industry of Key West, Florida. U.S. Bureau Fish. Mem. S-239, pp. 1–4 (mimeographed).

Schweigger, A. F. 1812. Prodomus monographiae cheloniorum. Königsberger Archiv. Naturwiss. Math., 1: 271–405, 406–458.

Seale, A. 1911. The fisheries resources of the Philippine Islands. Part IV. Miscellaneous marine products. Philippine Jour. Sci., 6: 283–320, 12 pls.

Sears, J. H. 1886. *Dermatochelys coriacea,* trunk back or leathery turtle. Bull. Essex Inst., 18: 87–94.

Seba, Albertus. 1734. Locupletissimi rerum naturalium thesauri accurata descriptio et iconibus artificiosissimus expressio, per universam physices historiam. Amsterdam. 4 vols., 1734–1765. Turtles, 1: 126–130, pls. 79–80.

Seeliger, L. M. 1945. Variation in the Pacific mud turtle. Copeia, no. 3, pp. 150–159, 5 figs.

Self, J. Teague. 1938. Note on the food habits of *Chelydra serpentina. Ibid.,* no. 4, p. 200.

Sergeev, A. 1937. Some materials to the problem of reptilian post embryonic growth. Zool. Jour. Moscow, 16: 723–735.

Seton, E. T. 1918. A list of the turtles, snakes and batrachians of Manitoba. Ottawa Nat., 32: 79–83.

Shannon, Frederick A., and Hobart M. Smith. 1949. Herpetological results of the University of Illinois Field Expedition, Spring, 1949. Trans. Kansas Acad. Sci., 52: 494–509.

Sharp, D. L. 1911. Turtle eggs for Agassiz. *In* The face of the fields. Boston. Pp. 1–250.

Shaw, Charles E. 1946. An anomalous Pacific loggerhead turtle from the northwestern coast of Baja California. Herpetologica, 3: 123–124, 1 pl.

——. 1947. First records of the red-brown loggerhead turtle from the eastern Pacific. *Ibid.,* 4: 55–56.

Shaw, George. 1802. General zoology, or systematic natural history. London, G. Kearsley. 3: i–vi, 1–615, 140 pls.

Shaw, R. J., and Francis Marsh Baldwin. 1935. The mechanics of respiration in turtles. Copeia, no. 1, pp. 2–15, 2 figs.

Shimek, B. 1924. Holbrook's North American herpetology. 1st ed. Iowa Acad. Sci., 31: 427–430.

Shockley, Clarence H. 1949. Fish and invertebrate populations of an Indiana bass stream. Invest. Indiana Lakes Streams, 3: 247–270.

Shoup, C. S., J. H. Peyton, and Glenn Gentry. 1941. A limited biological survey of the Obey River and adjacent streams in Tennessee. Tenn. Dept. Conserv. Div. Game Fish Miscel. Publ., 3: 48–77.

Shreve, Benjamin. 1947. On Venezuelan reptiles and amphibians collected by Dr. H. G. Kugler. Bull. Mus. Comp. Zool., 99: 519–537.

Shufeldt, R. W. 1893. Scientific taxidermy for museums. Rept. U.S. Nat. Mus., 1892, pp. 369–436, pls. 15–96.

——. 1921. Observations on the cervical region of the spine in chelonians. Jour. Morph., 35: 213–227.

Siebenrock, F. 1902. Zur Systematik der Schildkrötenfamilie Trionychidae Bell, nebst der Beschreibung einer neuen Cyclanorbis-Art. Sitsber. Akad. Wiss. Wien, 91: 1–40, 18 figs.

——. 1906. Schildkröten aus Südmexico. Zool. Anz., 30: 3–4.

——. 1907. Die Schildkrötenfamilie Cinosternidae. Sitsber. Akad. Wiss. Wien, 116: 1–73, 7 figs., 2 pls., 2 maps.

——. 1909. Synopsis der rezenten Schildkröten. Zool. Jahrb., Suppl., 10: 427–618.

——. 1923. Die nearktischen Trionychidae. Verh. Zool. Bot. Ges. Wien, 73: 180–194.

Simpson, George Gaylord. 1943. Turtles and the origin of the fauna of Latin America. Amer. Jour. Sci., 241: 413–429.

Skinner, Alanson. 1909. [*Terrapene carolina* recorded at meeting of March 20, 1909.] Proc. Staten Island Assoc. Arts Sci., 2: 209.

Slater, James R. 1939. *Clemmys marmorata* in the state of Washington. Occ. Papers Dept. Biol. Coll. Puget Sound, no. 5, p. 32.

——. 1941. The distribution of reptiles and amphibians in Idaho. *Ibid.*, 14, pp. 78–108.

Slevin, Joseph R. 1926. Expedition to the Revillagigedo Islands, Mexico, in 1925. III. Notes on a collection of reptiles and amphibians from the Tres Marias and Revillagigedo Islands, and west coast of Mexico, with description of a new species of *Tantilla*. Proc. Calif. Acad. Sci. San Francisco, 15: 195–207, pl. 22.

——. 1927. An additional record of the leather-back turtle on the coast of California. Copeia, no. 159, pp. 172–173.

——. 1931. Range extension of certain western species of reptiles and amphibians. *Ibid.*, no. 3, pp. 140–141.

——. 1934. A handbook of reptiles and amphibians of the Pacific States. Calif. Acad. Sci. Special Publ., pp. 1–73, 9 figs., 11 pls.

Smith, Eugene. 1899. The turtles and lizards found in the vicinity of New York City. Proc. Linn. Soc. N.Y., 11: 11–32.

Smith, Hobart M. 1938. Notes on reptiles and amphibians from Yucatan, and Campeche, Mexico. Occ. Papers Mus. Zool. Univ. Mich., 388: 1–22, 1 pl.

——. 1939a. An annotated list of the Mexican amphibians and reptiles in the Carnegie Museum. Ann. Carnegie Mus., 27: 311–320.

——. 1939b. Notes on Mexican amphibians and reptiles. Field Mus. Nat. Hist. Zool. Ser., 24: 15–35, 1 fig.

——. 1946. The map turtles of Texas. Proc. Trans. Texas Acad. Sci., 30: 60.

——. 1947. Kyphosis and other variations in soft-shelled turtles. Mus. Nat. Hist. Univ. Kansas Publ., 1: 117–124, 3 figs.

——, and Sidney O. Brown. 1946. A hitherto neglected integumentary gland in the Texas tortoise. Proc. Trans. Texas Acad. Sci., 30: 59.

——, and Bryan P. Glass. 1947. A new musk turtle from the southern United States. Jour. Washington Acad. Sci., 37: 22–24.

——, and Arthur B. Leonard. 1934. Distributional records of reptiles and amphibians in Oklahoma. Amer. Midland Nat., 15: 190–196.

——, C. W. Nixon, and S. A. Minton, Jr. 1949. Observations on constancy and color pattern in soft-shelled turtles. Trans. Kansas Acad. Sci., 52: 92–98, 3 figs.

Smith, Hugh M. 1894. Notes on a reconnaissance of the fisheries of the Pacific Coast of the United States in 1894. Bull. U.S. Fish Comm., 14: 223–288.

——. 1904. Notes on the breeding habits of the yellow-bellied terrapin. Smithsonian Miscel. Coll., 45: 252–253.

Smith, Karl V., and Robert S. Daniel. 1947. Observation of behavioral development in the loggerhead turtle (*Caretta caretta*). Science, 104: 154–156.

Smith, Malcolm A. 1931. The fauna of British India, including Ceylon and Burma. Reptilia and Amphibia. Vol. 1, Loricata, Testudines. Pp. v–xxviii, 1–185, 42 figs., 2 pls.

Smith, Philip W. 1947. The reptiles and amphibians of eastern central Illinois. Bull. Chicago Acad. Sci., 8: 21–40.

——. 1948. Noteworthy herpetological records from Illinois. Chicago Acad. Sci. Nat. Hist. Miscel., 33: 1–4.

Smith, W. H. 1879. Catalogue of the Reptilia and Amphibia of Michigan. Science News, 1: i–viii (supplement).

——. 1882. Report on the reptiles and amphibians of Ohio. Rept. Geol. Surv. Ohio, 4: 629–734, 8 figs.

Snyder, L. L. 1921. Some observations on Blanding's turtle. Canadian Field-Nat., 35: 17–18.

——, E. B. S. Logier, and T. B. Kurata. 1942. A faunal investigation of the Sault Ste. Marie region, Ontario. Trans. Roy. Canadian Inst., 24: 99–165.

Somes, M. P. 1911. Notes on some Iowa reptiles. Proc. Iowa Acad. Sci., 18: 149–154.

Sonnini, C. S., and P. A. Latreille. 1801. Histoire naturelle des reptiles avec figures dessinées d'après nature. Paris, Deterville. 1: i–xx, 1–280, 14 pls.; 2: 1–332, 21 pls.; 3: 1–335, 6 pls.; 4: 1–410, 13 pls.

Sowerby, J., and E. Lear. 1872. Tortoises, terrapins and turtles. London. Pp. 1–16, 60 pls. (Introductory text by J. E. Gray.)

Speck, Frank G. 1943. Turtle music. Fauna, 5: 82–84, illustrated.

Springer, Stewart. 1928. A list of reptiles and amphibians taken in Marion County, Indiana in 1924–1927. Proc. Indiana Acad. Sci., 37: 491–492.

Steindachner, Franz. 1891. Ueber die Reptilien und Batrachier der westlichen und östlichen Gruppe der canarischen Inseln. Ann. K. K. Naturhist. Hofmus., 6: 287–306.

Stejneger, Leonhard. 1893. Annotated list of the reptiles and batrachians collected by the Death Valley Expedition in 1891, with descriptions of new species. North Amer. Fauna, no. 7, pp. 159–228.

——. 1902a. The reptiles of the Huachuca Mountains of Arizona. Proc. U.S. Nat. Mus., 25: 149–158.

——. 1902b. Some generic names of turtles. Proc. Biol. Soc. Washington, 15: 235–238.

——. 1904. The herpetology of Porto Rico. Rept. U.S. Nat. Mus., 1902, pp. 553–724, 197 figs., 1 pl.

——. 1905. Generic names of soft-shelled turtles. Science, n.s., 21: 228–229.

——. 1907. Herpetology of Japan and adjacent territory. Bull. U.S. Nat. Mus., 58: i–xx, 1–577, 528 figs., 35 pls.

——. 1914. On the systematic names of the snapping turtles. Copeia, no. 6, pp. 3–4.

——. 1918. Description of a new lizard and a new snapping turtle from Florida. Proc. Biol. Soc. Washington, 31: 89–92.

——. 1923. Rehabilitation of a hitherto overlooked species of musk turtle of the southern states. Proc. U.S. Nat. Mus., 62: 1–3.

——. 1925. New species and subspecies of American turtles. Jour. Washington Acad. Sci., 15: 462–463.

——. 1936. The correct name for the northern diamond-back terrapin. Copeia, no. 2, p. 115.

——. 1938. Restitution of the name *Ptychemys hoyi* Agassiz for a western river tortoise. Proc. Biol. Soc. Washington, 51: 173–176.

——. 1944. Notes on the American soft shelled turtles with special reference to *Amyda agassizii*. Bull. Mus. Comp. Zool., 94: 1–75, 30 pls., 10 tables.

——, and Thomas Barbour. 1917. A check list of North American amphibians and reptiles. Cambridge, Mass., Harvard Univ. Press. Pp. i–iv, 1–125.

——, and Thomas Barbour. 1933. A check list of North American amphibians and reptiles. Cambridge, Mass., Harvard Univ. Press. 3d ed. Pp. i–xiv, 1–185.

——, and Thomas Barbour. 1939. A check list of North American amphibians and reptiles. Cambridge, Mass., Harvard Univ. Press. 4th ed. Pp. i–xvi, 1–207.

——, and Thomas Barbour. 1943. A check list of North American amphibians and reptiles. Bull. Mus. Comp. Zool., 5th ed., 93: i–xix, 1–260.

Stephens, Frank. 1921. An annotated list of the amphibians and reptiles of San Diego County, California. Trans. San Diego Soc. Nat. Hist., 3: 57–69.

Stewart, George D. 1947. A record for Muhlenberg's turtle. Copeia, no. 1, p. 68.

Stewart, N. H. 1928. Some rare vertebrates of the Susquehanna Valley, Pennsylvania. Proc. Penn. Acad. Sci., 2: 21–24.

Stickel, Elizabeth L. 1949. Population and home range relationships of the box turtle, *Terrapene carolina* (Linnaeus). Microfilm Abstr., 9: 195–196.

Stickel, Lucille F. 1948. Observations on the effect of flood on animals. Ecology, 29: 505–507.

Stille, W. T. 1947. *Kinosternon subrubrum subrubrum* in the Chicago area. Copeia, no. 2, p. 143.

——, and R. A. Edgren, Jr. 1948. New records for amphibians and reptiles in the Chicago area. Bull. Chicago Acad. Sci., 8: 195–202.

Stockwell, A. G. 1888. Notes upon soft shelled turtles and the anatomical vagaries of *Aspidonectes spinifer*. Jour. Comp. Med. Surg., 9: 28–42.

Stone, Witmer. 1903. A collection of reptiles and batrachians from Arkansas, Indian Territory and western Texas. Proc. Acad. Nat. Sci. Philadelphia, pp. 538–542.

——. 1906. Notes on reptiles and batrachians of Pennsylvania, New Jersey and Delaware. Amer. Nat., 40: 159–170.

Storer, D. H. 1839. Reptiles of Massachusetts. *In* Report on the fishes, reptiles and birds of Mass. Boston. Pp. 202–253.

Storer, H. R. 1851. Miscellaneous remarks. Proc. Boston Soc. Nat. Hist., 4: 147.

Storer, Tracy I. 1930. Notes on the range and life-history of the Pacific freshwater turtle, *Clemmys marmorata*. Univ. Calif. Publ. Zool., 32: 429–441.

——. 1932. The western limit of range for *Chrysemys picta bellii*. Copeia, no. 1, pp. 9–11.

——. 1937. Further notes on the turtles of the Pacific coast of North America. *Ibid.*, pp. 66–67.

Strader, L. D. 1936. Herpetology of the eastern panhandle of West Virginia. Proc. West Virginia Acad. Sci., 9: 32–35.

Strauch, A. 1862. Chelonologische Studien. Mem. Acad. Sci. St. Petersbourg, Ser. 7, 5: 1–196, 1 pl.

——. 1865. Die Vertheilung der Schildkröten über den Erdball. Mem. Acad. Sci. St. Petersbourg, 3: 1–207.

——. 1890. Bemerkungen über die Schildkrötensammlung im zoologischen Museum der Kaiserlichen Akademie der Wissenschaften zu St. Petersbourg. *Ibid.*, 38: 1–127, 4 pls.

Strecker, John K. 1902. A preliminary report on the reptiles and batrachians of McLennan County, Texas. Trans. Texas Acad. Sci., 4: 1–7.

——. 1908a. Notes on the habits of two Arkansas salamanders and a list of the batrachians and reptiles collected at Hot Springs. Proc. Biol. Soc. Washington, 21: 85–90.

——. 1908b. The reptiles and batrachians of Victoria and Refugio County, Texas. *Ibid.*, 45–52.

——. 1908c. The reptiles and batrachians of McLennan County, Texas. *Ibid.*, 69–84.

——. 1909a. Reptiles and amphibians collected in Brewster County, Texas. Baylor Univ. Bull., 12: 11–15.

——. 1909b. Notes on the herpetology of Burnett County, Texas. *Ibid.*, 1–9.

——. 1909c. Contributions to Texan herpetology. *Ibid.*, 1–20.

——. 1910a. Notes on the fauna of a portion of the Canyon region of northwestern Texas. *Ibid.*, 13: 1–31.

——. 1910b. Description of a new solitary spadefoot (*Scaphiopus hurterii*) from Texas, with other herpetological notes. Proc. Biol. Soc. Washington, 23: 115–122.

——. 1915. Reptiles and amphibians of Texas. Baylor Univ. Bull., 18: 1–82.

——. 1922. An annotated catalogue of the amphibians and reptiles of Bexar County, Texas. Bull. Sci. Soc. San Antonio, no. 4, pp. 1–31, 4 pls.

——. 1924. Notes on the herpetology of Hot Springs, Arkansas. Bayor Univ. Bull., 27: 29–47.

——. 1926a. Amphibians and reptiles collected in Somervell County, Texas. Contrib. Baylor Univ. Mus., no. 2, pp. 1–2.

——. 1926b. Notes on the herpetology of the east Texas timber belt. I. Liberty County amphibians and reptiles. *Ibid.*, no. 3, pp. 1–3.

——. 1926c. A list of the reptiles and amphibians collected by Louis Garni in the vicinity of Boerne, Texas. *Ibid.*, no. 6, pp. 1–9.

——. 1926d. Notes on the herpetology of the east Texas timber belt. II. Henderson County amphibians and reptiles. *Ibid.*, no. 7, pp. 1–3.

——. 1927a. Chapters from the life histories of Texas reptiles and amphibians. *Ibid.*, no. 10, pp. 1–14.

——. 1927b. Notes on a specimen of *Gopherus berlandieri* (Agassiz). Copeia, no. 161, pp. 189–190.

——. 1927c. Observations on the food habits of Texas amphibians and reptiles. *Ibid.*, no. 162, pp. 6–9.

——. 1928a. Amphibians and reptiles collected at Harlingen, Texas. Contrib. Baylor Univ. Mus., no. 15, pp. 7–8.

——. 1928b. Common and English folk names for Texas amphibians and reptiles. *Ibid.*, no. 16, pp. 1–21.

——. 1929a. Further studies in the folk-lore of reptiles. Baylor Univ. Contrib. Folk-lore, 1: 1–16, 2 pls.

——. 1929b. The eggs of *Gopherus berlandieri* Agassiz. Contrib. Baylor Univ. Mus., no. 18, p. 6.

——. 1929c. Animals and streams: a contribution to the study of Texas folk names. Baylor Univ. Contrib. Folk-lore, no. 2, pp. 1–23.

——. 1929d. Random notes on the zoology of Texas. Contrib. Baylor Univ. Mus., no. 18, pp. 1–12.

——. 1929e. A preliminary list of the amphibians and reptiles of Tarrant County, Texas. *Ibid.,* no. 19, pp. 10–15.

——. 1929f. Field notes on the herpetology of Wilbarger County, Texas. *Ibid.,* pp. 3–9.

——. 1931. A catalogue of the amphibians and reptiles of Travis County, Texas. *Ibid.,* no. 23, pp. 1–16.

——. 1935. The reptiles of West Frio, Real County, Texas. *In* W. J. Williams, Notes on the zoology of Texas. Baylor Univ. Bull., 38: 1–69.

——, and L. S. Frierson, Jr. 1926. The herpetology of Caddo and De Soto parishes, Louisiana. Contrib. Baylor Univ. Mus., no. 5, pp. 1–10.

——, and W. J. Williams. 1927. Herpetological records from the vicinity of San Marcos, Texas, with distributional data on the amphibians and reptiles of Edwards Plateau and central Texas. *Ibid.,* no. 12, pp. 1–16.

——, and W. J. Williams. 1928. Field notes on the herpetology of Bowie County, Texas. *Ibid.,* no. 17, pp. 1–19.

Street, J. F. 1914. Amphibians and reptiles observed at Beverly, New Jersey. Copeia, no. 4, p. 2.

Stromsten, F. A. 1917. The development of the musk glands in the logger-head turtle. Proc. Iowa Acad. Sci., 24: 311–313.

——. 1923. Nest digging and egg laying habits of Bell's turtle, *Chrysemys marginata bellii* (Gray). Univ. Iowa Studies Nat. Hist., 10: 67–70.

Stuart, L. C. 1935. A contribution to the knowledge of the herpetology of a portion of the savanna region of central Peten, Guatemala. Univ. Mich. Mus. Zool. Misc. Publ., 29: 1–56, 4 pls., 1 map.

Sumner, Francis B., Raymond C. Osburn, and Leon J. Cole. 1911. A biological survey of the waters of Woods Hole and vicinity. Section III. A catalogue of the marine fauna. Bull. Bureau Fish., 31: 545–860.

Surface, H. A. 1908. First report on the economic features of the turtles of Pennsylvania. Zool. Bull. Div. Zool. Penn. Dept. Agric., 6: 105–196, 16 figs., pls. 4–12.

Svihla, Arthur, and Ruth Dowell Svihla. 1933. Amphibians and reptiles of Whitman County, Washington. Copeia, no. 3, pp. 125–128.

Swanson, Paul L. 1939. Herpetological notes from Indiana. Amer. Midland Nat., 22: 684–695.

Taft, A. C. 1944. Diamond-back terrapin introduced into California. Calif. Fish Game, 30: 101–102, 1 fig.

Tanner, Vasco M. 1927. Distributional list of the amphibians and reptiles of Utah. Copeia, no. 163, pp. 54–58.

Taylor, Edward H. 1920. Philippine turtles. Philippine Jour. Sci., 16: 111–144, 7 pls.

——. 1929. List of reptiles and batrachians of Morton County, Kansas, reporting species new to the state fauna. Univ. Kansas Sci. Bull., 19: 63–65.

——. 1933. Observations on the courtship of turtles. *Ibid.*, 21: 269–271.

——. 1935. Arkansas amphibians and reptiles in the Kansas University Museum. *Ibid.*, 22: 207–218.

——. 1936. Notes on the herpetological fauna of the Mexican state of Sonora. *Ibid.*, 24: 475–503.

——. 1943. An extinct turtle of the genus *Emys* from the Pleistocene of Kansas. *Ibid.*, 29: 249–254, 1 fig.

Taylor, W. E. 1895. The box turtles of North America. Proc. U.S. Nat. Mus., 17: 573–588, 7 figs.

Telford, H. S., and O. A. Stevens. 1942. Uses and management of ponds and lakes. North Dakota Agric. Exp. Station Bull., 313: 1–40.

Temminck, C. J., and H. Schlegel. 1835–1838. Fauna japonica. Reptilia, pp. i–xxi, 1–444, map, pls. 1–9.

Terron, C. C. 1919. Las tortugas mas importantes y su pesca. Bol. Dirección Estudios Biológicos [Mexico], 2: 395.

——. 1921. Datos para una monografía de la fauna erpetologica de la Peninsula de la Baja California. Mem. Rev. Soc. Cient. "Antonio Alzate," Mexico, 39: 165.

Test, F. C. 1894. Annotated list of the reptiles and batrachians collected in Missouri and Texas in the fall of 1891. Bull. U.S. Fish. Comm., 1892, 12: 121–122.

Thacker, T. L. 1924. Notes on Bell's painted turtles (*Chrysemys marginata bellii*) in British Columbia. Canadian Field-Nat., 38: 164–167, 1 fig.

Theobald, W. 1868. Catalogue of the reptiles of British Burma, embracing the provinces of Pegu, Martaban and Tenasserim, with descriptions of new or little known species. Jour. Linn. Soc. London, 10: 4–68.

Thomas, E. S., and Milton B. Trautman. 1937. Segregated hibernation of *Sternotherus odoratus* (Latreille). Copeia, no. 4, p. 231.

Thompson, C. Wyville. 1878. Voyage of the Challenger. The Atlantis. New York, Harper and Bros. 2 vols.

Thompson, Crystal. 1911. Notes on the amphibians and reptiles of Cass County, Michigan. Rept. Mich. Acad. Sci., 13: 105–107.

——. 1915a. The reptiles and amphibians of Manistee County, Michigan. Occ. Papers Mus. Zool. Univ. Mich., 18: 1–6.

——. 1915b. The reptiles and amphibians of Munroe County, Michigan. Mich. Geol. Biol. Surv., publ. no. 20, Biol. Ser., 4: 61–63.

——, and Helen Thompson. 1912a. Results of the Merson expedition to the Charity Islands, Lake Huron. Rept. Mich. Acad. Sci., 14: 156–158.

——, and Helen Thompson. 1912b. Results of the Shiras expedition to White-fish Point, Michigan. *Ibid.,* 215–217.

Thompson, D'Arcy Wentworth. 1945. On growth and form. American ed. New York, Macmillan Co. Pp. 1–1116, 554 figs., 2 pls.

Thompson, J. Stuart. 1932. The anatomy of the tortoise. Sci. Proc. Roy. Dublin Soc., 20: 359–461.

Thompson, Joseph C. 1889. [Reference made to snapping turtle at meeting of May 9, 1889.] Proc. Nat. Sci. Assoc. Staten Island, 2: 11.

——. 1890. [Reference made to snapping turtle and eggs at meeting of Sept. 11, 1890.] *Ibid.,* 61.

Thompson, Zadock. 1842. History of Vermont, natural, civil, and statistical. Pt. 1. Natural history. Burlington, Chauncey Goodrich. Pp. 1–224.

Tihen, Joe A. 1937. Additional records of amphibians and reptiles in Kansas counties. Trans. Kansas Acad. Sci., 40: 401–409.

——, and James M. Sprague. 1939. Amphibians, reptiles and mammals of the Meade County State Park. *Ibid.,* 42: 499–512, 5 figs.

Tinklepaugh, O. L. 1932. Maze learning of a turtle. Jour. Comp. Psychol., 13: 201–206, 3 figs.

Toner, G. C. 1933. Over winter eggs of the snapping turtle. Copeia, no. 4, pp. 221–222.

——. 1936. Notes on the turtles of Leeds and Frontenac counties, Ontario. *Ibid.,* p. 236.

——, and W. E. Edwards. 1938. Cold-blooded vertebrates of Grippen Lake, Leeds County, Ontario. Canadian Field-Nat., 52: 40–43.

Tornier, Gustav. 1902. Die Crocodile, Schildkröten und Eidechsen in Kamerun. Zool. Jahrb., 15: 663–677, pl. 35.

Townsend, C. H. 1887. Field notes on the mammals, birds, and reptiles of northern California. Proc. U.S. Nat. Mus., 10: 159–241, pl. 5.

——. 1904a. The collection of sea-turtles. Bull. N.Y. Zool. Soc., no. 13, pp. 143–144.

——. 1904b. Collection of freshwater turtles. *Ibid.,* p. 145.

——. 1906a. A note on Muhlenberg's turtle. *Ibid.,* 22: 289.

——. 1906b. Growth of confined hawksbill turtles. *Ibid.,* 291.

——. 1911. Blanding's turtle [and *Graptemys geographica* in Orange County, New York]. Science, no. 873, p. 381.

——. 1916. Voyage of the "Albatross" to the Gulf of California in 1916. Bull. Amer. Mus. Nat. Hist., 35: 399–476, 45 figs.

——. 1926. Items of interest. Bull. N.Y. Zool. Soc., 24: 217–218.

——. 1937. Growth of Galápagos tortoises, *Testudo vicina,* from 1928 to 1937. Zoologica, 22: 289–292.

Trapido, Harold, and Robert T. Clausen. 1938. The amphibians and reptiles of eastern Quebec. Copeia, no. 3, pp. 117–125.

Trautman, Milton B. 1931. List of turtles of Ohio. Bull. Ohio Dept. Agric., Bureau Sci. Res., 53: 1 p.

Tressler, Donald K. 1923. Marine products of commerce. New York, The Chemical Catalog Company. Pp. 762, 257 figs., 90 tables.

Troost, Gerard. 1844. List of reptiles inhabiting the state of Tennessee. Geol. Rept. State Tennessee, 7: 39–42.

Trowbridge, A. H. 1937. Ecological observations on amphibians and reptiles collected in southeastern Oklahoma during the summer of 1934. Amer. Midland Nat., 18: 285–303, 1 map.

True, Frederick W. 1882. On the North American land tortoises of the genus *Xerobates*. Proc. U.S. Nat. Mus., 4: 434–448, 3 figs.

——. 1883. A list of the vertebrate animals of South Carolina. *In* Harry Hammond, South Carolina. Charleston, S.C., State Board of Agric. Pp. 209–264.

——. 1884. The useful aquatic reptiles and batrachians. Fish. Indiana U.S. Sect. 1, Part II, pp. 137–162.

Tuge, Hideomi. 1931. Early behavior of embryos of the turtle Terrapene carolina (L.). Proc. Soc. Exp. Biol., 29: 52–53.

Vaillant, L. 1894. Essai sur la classification générale des chéloniens. Ann. Soc. Sci. Nat. Zool., 16: 331–345.

Van Denburgh, John. 1895. A review of the herpetology of Lower California. Part I. Reptiles. Proc. Calif. Acad. Sci., 2: 77–162, 14 pls.

——. 1897. The reptiles of the Pacific coast and the Great Basin. Occ. Papers Calif. Acad. Sci., 5: 1–236, text figs.

——. 1905. On the occurrence of the leather-back turtle, *Dermochelys,* on the coast of California. Proc. Calif. Acad. Sci., 4: 51–55, pls. 9–11.

——. 1912. Notes on a collection of reptiles from southern California and Arizona. *Ibid.,* 147–156.

——. 1922. The reptiles of western North America. Vol. II. Snakes and turtles. Occ. Papers Calif. Acad. Sci., 10: 623–1028, pls. 58–128.

——. 1924a. A fifth record of the Pacific leatherback turtle on the coast of California. Copeia, no. 130, p. 53.

——. 1924b. Notes on the herpetology of New Mexico, with a list of species known from that state. Proc. Calif. Acad. Sci., ser. 4, 13: 189–230.

——, and J. R. Slevin. 1913. List of the amphibians and reptiles of Arizona, with notes on the species in the collection of the Academy. *Ibid.,* 3: 391–454, pls. 17–28.

——, and J. R. Slevin. 1921. A list of the amphibians and reptiles of the peninsula of Lower California with notes on the species in the collection of the Academy. *Ibid.,* ser. 4, 11: 49–72.

Van Hyning, O. C. 1931. Reproduction of some Florida snakes. Copeia, no. 2, pp. 59–60.

——. 1933. Batrachia and Reptilia of Alachua County, Florida. *Ibid.*, no. 1, pp. 3–7.

Verrill, A. E. 1863. Catalogue of the reptiles and batrachians found in the vicinity of Norway, Oxford County, Maine. Proc. Boston Soc. Nat. Hist., 9: 195–199.

Versluys, J. 1913. On the phylogeny of the carapace and the affinities of *Dermochelys cariacea*. Rept. 83rd Meeting Brit. Assoc. Adv. Sci., Birmingham, pp. 791–807.

——. 1914. Über die Phylogenie des Panzers der Schildkröten und über die Verwandtschaft der Lederschildkröten (*Dermochelys coriacea*). Paleont. Zeitschr. Berlin, 1: 321–347.

Viosca, Percy, Jr. 1923. An ecological study of the cold-blooded vertebrates of southeastern Louisiana. Copeia, no. 115, pp. 35–44.

——. 1926. Distribution problems of the cold-blooded vertebrates of the Gulf coastal plain. Ecology, 7: 307–314.

——. 1927. Notes on *Gopherus berlandieri* in Louisiana. Copeia, no. 164, pp. 83–84.

——. 1931. Amphibians and reptiles of Louisiana. Southern Biol. Supply Co., Price List, 20: 3–11.

——. 1933. The *Pseudemys troostii-elegans* complex, a case of sexual dimorphism. Copeia, no. 4, pp. 208–210.

Visher, S. S. 1914. A preliminary list of the reptiles and amphibians of Harding County. South Dakota Geol. Nat. Hist. Surv. Bull., no. 6, pp. 92–93.

Volker, H. 1913. Über das Stamm-Gliedmassen und Hautskelet vom *Dermochelys coriacea* L. Zool. Jahrb., 33: 431–552.

Vollbrecht, John L. 1947. Skeeter turtle hunt. Florida Outdoors, June, pp. 6–7.

Von Bloeker, Jack C., Jr. 1942. Fauna and flora of the El Segundo sand dunes. 13. Amphibians and reptiles of the dunes. Bull. South. Calif. Acad. Sci., 41: 29–38.

Vorhies, Charles T. 1945. Water requirements of desert animals in the Southwest. Univ. Ariz. Tech. Bull., 107: 485–525.

Wagler, J. 1830. Natürliches System der Amphibien. München, Stuttgart und Tübigen. Pp. 1–354.

Wagner, George. 1922. A wood turtle from Wisconsin. Copeia, no. 107, pp. 43–44.

Wailes, Benjamin L. C. 1854. Report on the agriculture and geology of Mississippi, embracing a sketch of the social and natural history of the state. Agric. Geol. Surv. Mississippi, pp. i–xx, 1–371, pls. 14–17.

Waite, Edgar R. 1925. Field notes on some Australian reptiles and a batrachian. Rec. S. Australian Mus., 3: 17–32, 15 figs.

Walker, Charles F. 1931. Notes on reptiles in the collection of the Ohio State Museum. Copeia, no. 1, pp. 9–13.

Walker, Warren F. 1947. The development of the shoulder region of the turtle, *Chrysemys picta marginata,* with special reference to the primary musculature. Jour. Morph., 80: 195–250, 28 figs.

Walls, G. L. 1934. The reptilian retina. I. A new concept of visual-ceil evolution. Amer. Jour. Ophthalmology, 17: 892–915.

Ward, Henry Baldwin, and George Chandler Whipple. 1918. Fresh-water biology. New York, John Wiley and Sons. Pp. i–ix, 1–1111, 1547 figs. [Reptiles, pp. 1026–1028.]

Watson, D. M. S. 1914. *Eunotosaurus africanus* Seeley, and the ancestors of Chelonia. Proc. Zool. Soc. London, 2: 1011–1020.

Weber, Jay A. 1928. Herpetological observations in the Adirondack Mountains, New York. Copeia, no. 169, pp. 106–112.

Weed, Alfred C. 1922. Reptile notes. Copeia, no. 112, pp. 84–87.

——. 1923. Notes on reptiles and batrachians of central Illinois. *Ibid.,* no. 116, pp. 45–50.

Wegner, Th. 1918. *Chelonia gwinneri* Wegner aus dem Rupelton von Flörsheim a.M. Abh. Senckenberg. Naturf. Gesell., 36: 359–372, 1 fig., pls. 29–30.

Weller, W. H. 1930. Records of some reptiles and amphibians from Chimney Rock Camp, Chimney Rock, North Carolina, and vicinity. Proc. Cincinnati Soc. Nat. Sci., 1: 9–12.

——. [n.d.]. Guide to the exhibition of amphibians and reptiles in the Cincinnati Society of Natural History. Compiled by Jun. Soc. Nat. Hist. Cincinnati.

Welter, Wilfred A., and Katherine Carr. 1939. Amphibians and reptiles of northeastern Kentucky. Copeia, no. 3, pp. 128–130.

Werner, F. 1902. Beiträge zur Biologie der Reptilien und Batrachier. Biol. Zentralbl., 22: 737–758.

Wetmore, Alexander. 1920. Observations on the hibernation of the box turtle. Copeia, no. 77, pp. 3–5.

——. 1925. Bird life among lava rock and coral sand. Nat. Geogr., 48: 77–108.

——, and Francis Harper. 1917. A note on the hibernation of *Kinosternon pennsylvanicum.* Copeia, no. 45, pp. 56–59.

Wheeler, George C. 1947. The amphibians and reptiles of North Dakota. Amer. Midland Nat., 38: 162–190.

Wheeler, W. M. 1899. George Baur's life and writings. Amer. Nat., 33: 15–30.

White, Gilbert. 1901. *In* Rashleigh Holt-White, The life and letters of Gilbert White of Selborne. New York, E. P. Dutton and Co.

White, Theodore. 1928. The osteology of the recent turtles of central North America. Master's thesis, University of Kansas. Pp. 1–307, 12 pls.

——. 1935. Adaptive evolution of the pelvic musculature of the turtles of the genera *Clemmys, Emys* and Terrapene. Doctor's dissertation, University of Michigan. Pp. 1–34, 5 pls.

Wickham, M. M. 1922. Notes on the migration of *Macrochelys lacertina*. Oklahoma Acad. Sci., n.s., 247, Univ. Studies, 15: 20–22.

Wied, Maximilian zu. 1838. Reise von Bethlehem nach Pittsburgh über die Alleghanys, vom 17. September bis zum 7. October. *In* Reise in das innere Nord-America in den Jahren 1832 bis 1834. Coblenz, J. Hoelscher. Pp. 121–142.

——. 1865. Verzeichnis der Reptilien welche auf einer Reise im nördlichen America beobachtet wurden. Nova. Act. Acad. Leopold Carol. Nat. Curios, 32: i–vii, 1–143, 7 pls.

Wieland, George R. 1897. Variability of external sutures in the skull of *Chelone mydas*. Amer. Nat., 31: 446.

——. 1903. Notes on the marine turtle *Archelon*. Amer. Jour. Sci., 15: 211–216.

——. 1909. Revision of the Protosteigidae. *Ibid.,* 27: 101–130, 12 figs.

Wilcox, LeRoy. 1933. Incubation of a painted turtle's eggs. Copeia, no. 1, p. 41.

Williamson, William. 1832. History of the state of Maine. Masters and Co., Hallowell. Vol. 1, pp. 165–166.

Williston, S. W. 1899. A new turtle from the Kansas Cretaceous. Trans. Kansas Acad. Sci., 17: 195–199.

——, and W. K. Gregory. 1925. Osteology of the reptiles. Harvard Univ. Press. Pp. 1–298.

Wojtusiak, Roman J. 1932. Über den Farbensinn der Schildkröten. Zeitschr. Wiss. Biol. Abt. C., Zeitschr. Vergleich. Physiol., 18: 393–436.

——. 1934. Über den Formensinn der Schildkröten. Bull. Internat. Acad. Polanaise Sci. Lett., Sci. Math., Nat. Ser. B, Sci. Nat. no. 11 (Zoology), no. 8/10, pp. 349–373, 19 figs.

Wolf, Siegfried. 1933. Zur Kenntnis von Bau und Funktion der Reptilien Lunge. Zool. Jahrb. Abt. Anat. Ont., 57: 139–190.

Wood, John Thornton. 1946. *Graptemys geographica* (Le Sueur) added to the herpetofaunal list of Great Smoky Mountains National Park. Copeia, no. 3, p. 168.

——, and W E. Duellman. 1947. Preliminary herpetological survey of Montgomery County, Ohio. Herpetologica, 4: 3–6.

Woodbury, Angus M. 1931. The reptiles of Utah. Bull. Univ. Utah, 21: 1–129, 58 figs.

——, and Ross Hardy. 1940. The dens and behavior of the desert tortoise. Science, 92: 529.

——, and Ross Hardy. 1948. Studies of the desert tortoise, *Gopherus agassizii*. Ecol. Monogr., 18: 145–200, 25 figs.

Woodward, A. Smith. 1887. Leathery turtles recent and fossil. Proc. Geol. Assoc., 10: 2–14.

Wright, A. H. 1918a. Notes on the Muhlenberg's turtle. Copeia, no. 52, pp. 5–7.

——. 1918b. Notes on *Clemmys*. Proc. Biol. Soc. Washington, 31: 51–57, 1 pl.

——. 1919. The turtles and the lizards of Monroe and Wayne counties, New York. Copeia, no. 66, pp. 6–8.

——. 1926. The vertebrate life of the Okefinokee Swamp in relation to the Atlantic coastal plain. Ecology, 7: 77–85, pls. 2–6.

——. 1935. Some rare amphibians and reptiles of the United States. Proc. Nat. Acad. Sci., 21: 340–345.

——, and W. D. Funkhouser. 1915. A biological reconnaissance of the Okefinokee Swamp in Georgia. The reptiles. Part I. Turtles, lizards and alligators. Proc. Nat. Acad. Sci. Philadelphia, 67: 108–139, 4 figs.

——, and S. E. R. Simpson. 1920. The vertebrates of the Otter Lake region, Dorset, Ontario. III. The batrachians and reptiles. Canadian Field-Nat., 34: 141–145.

Wyman, J. 1851. [The shell and sternum of *Trionyx ferox*.] Proc. Boston Soc. Nat. Hist., 4: 10.

Yarrow, Henry C. 1875. Report upon the collections of batrachians and reptiles made in portions of Nevada, Utah, California, Colorado, New Mexico and Arizona during the years 1871, 1872, 1873 and 1874. Rept. Geogr. Geol. Expl. Surv. W. 100th Merid., 5: 509–584.

——. 1882. Check-list of North American Reptilia and Batrachia, with a catalogue of specimens in the U.S. National Museum. U.S. Nat. Mus. Bull., no. 24, pp. 1–249. [Turtle section by F. W. True, pp. 26–38.]

Yerkes, R. M. 1901. The formation of habits in the turtle. Pop. Sci. Monthly, 58: 519–529, 6 figs.

——. 1904. Space perception of tortoises. Jour. Comp. Neurol., 2: 15–26.

——. 1905. The color pattern of *Nanemys guttata* Schneider. Science, n.s., 21: 386.

Young, F. N., and C. C. Goff. 1939. An annotated list of the arthropods found in the burrows of the Florida gopher tortoise. Florida Entomol., 22: 53–62.

Young, Robert T. 1924. The biology of North Dakota. Quart. Jour. Univ. North Dakota, 15: 53–68.

Zangerl, Rainer. 1939. Homology of the shell elements in turtles. Jour. Morph., 65: 383–410.

——. 1941. A series of lateral organs found in embryos of the snapping turtle (*Chelydra serpentina*). Papers Mich. Acad. Sci., Arts, Lett., 26: 339–341, 1 fig., 1 pl.

——. 1945. Fossil specimens of *Macrochelys* from the Tertiary of the plains. Fieldiana-Geology, Chicago Nat. Hist. Mus., 10: 5–12, 4 figs.

REFERENCES FOR STATES AND PROVINCES

(For complete references see general bibliography.)

United States of America

ALABAMA
Barkalow, 1940
Baur, 1893c
Carr, 1937a
Dunn, 1920a
Haltom, 1931
Holt, 1925
Löding, 1922
Penn, 1940

ARIZONA
Campbell, 1934
Cope, 1866
Coues, 1875
Cowles and Bogert, 1936
Cox, 1881
Gilmore, 1923
Gloyd, 1937
Hardy and Lamoreaux, 1945
King, 1932
Little, 1940
McCoy, 1932
Miller, 1946
Ortenburger and Ortenbur-
    ger, 1927
Ruthven, 1907
Schmidt, 1924a
Stejneger, 1902a
Stone, 1903
Van Denburgh, 1912
—— and Slevin, 1913
Yarrow, 1875

ARKANSAS
Burger, Smith, and Smith,
    1949
Burt and Hoyle, 1934
Dellinger and Black, 1938
Hurter and Strecker, 1909
McLain, 1899b
Ortenburger, 1929b
Schmidt, 1940, 1944
Strecker, 1908a, 1924
Taylor, 1935

CALIFORNIA
Bogert, 1930, 1937
Camp, 1913
Cooper, J. G., 1863, 1870
Cooper, M. D., 1870
Cowles and Bogert, 1936
Cronise, 1868
Gilmore, 1937
Girard, 1858
Grant, Chaffee, 1936
Grant, Chapman, 1936b
Grinnell and Camp, 1917
—— and Grinnell, 1907
—— and Storer, 1924
Hall and Grinnell, 1919
Hallowell, 1859
Klauber, 1928, 1930, 1934
Linsdale and Gressit, 1937
Lockley, 1948
McLain, 1899a
Mearns, 1907
Meek, 1905
Miller, L., 1932, 1942
Miller, R., 1946
Myers, 1933
Rivers, 1889
Rüthling, 1915
Seeliger, 1945
Shaw, 1947
Slevin, 1927, 1931
Stejneger, 1893
Stephens, 1921
Taft, 1944
Townsend, 1887
Van Denburgh, 1897, 1905,
    1912, 1924a
Von Bloeker, 1942

COLORADO
Burt, 1935
Cockerell, 1910, 1927
Ellis and Henderson, 1913,
    1915
Rivers, 1889

Rodeck, 1949
Schmidt, 1945a

CONNECTICUT
Babbitt, 1932
Babcock, 1928
Finneran, 1947, 1948a, b
Fisher, 1945
Lamson, 1935
Linsley, 1844
Storer, 1851

DELAWARE
Conant, 1945, 1947b
—— and Downs, 1940
Fowler, 1925a
Stone, 1906

DISTRICT OF COLUMBIA
Brady, 1924
Clark, 1930
Fowler, 1945
Hay, W. P., 1902
McAtee, 1918
McCauley, 1945

FLORIDA
Allen, E. Ross, 1938, [n.d.]
—— and Slatten, 1945
Audubon, 1926
Barbour and Stetson, 1931
Bartram, 1794, 1943
Baur, 1890b, 1893c
Beck, 1938
Bogert and Cowles, 1947
Brice, 1897
Brimley, 1907
Carr, 1935, 1937a, b, 1938a,
    b, 1940, 1942b, 1946
—— and Marchand, 1942
Conant, 1930
Deckert, 1918
DeSola, 1935
Engelhardt, 1912
Fisher, 1917

530

Fletcher, 1900
Fowler, 1906a
Gilmore, 1927
Goin, 1943, 1948
—— and Goff, 1941
Grant, 1911, 1946
Hallinan, 1923
Hamilton, 1947
Harper, 1940
Hay, O. P., 1916
Hubbard, 1893, 1894, 1896
Knight, 1871
Lönnberg, 1894, 1896
Marchand, 1942, 1944,
  1945b
Mowbray, 1922
Munroe, 1898
Neill, 1951
Roosevelt, 1917
Schroeder, 1924, 1931
Stejneger, 1918
Van Hyning, 1931, 1933
Wright, 1935
Young and Goff, 1939

GEORGIA
Barkalow, 1948
Brimley, 1907, 1910
DeSola and Abrams, 1933
Harper, 1926, 1940
Holbrook, 1849
Neill, 1948a, b, c, d, 1951
Wright, 1926
—— and Funkhouser, 1915

IDAHO
Seeliger, 1945
Slater, 1941
Van Denburgh, 1897

ILLINOIS
Blanchard, 1924
Burger, Smith, and Smith,
  1949
Burt and Hoyle, 1934
Cagle, 1941, 1942a, b,
  1944a, b, c, d, 1945, 1946,
  1948a, b
Cahn, 1931, 1933, 1937
Davis and Rice, 1883a
Edgren and Stille, 1948
Gaige, 1914
Garman, H., 1889, 1890a,
  1892
Hankinson, 1915, 1917
Hay, O. P., 1892b
Hurter, 1893, 1911

Kennicott, 1855
Kerr, 1895
Kirsch, 1895a
McLain, 1899b
Necker, 1938
Owens, 1941
Peters, 1942
Pope, 1944
Schmidt, 1938
—— and Necker, 1935
Smith, P., 1947, 1948
Stille, 1947
—— and Edgren, 1948
Weed, 1922, 1923
Wied, 1865

INDIANA
Blanchard, 1925
Blatchley, 1891, 1899, 1900,
  1906
Butler, 1892
Clark, 1935
Culbertson, 1907
Dury, 1932b
Edgren and Stille, 1948
Eigenmann, 1896
Evermann and Clark, 1916,
  1920
Fichter, 1947
Grant, Chapman, 1935a, b,
  1936a
Hahn, 1908
Hammann, 1939
Hay, O. P., 1887a, b, 1892b
Hughes, 1886
Kirsch, 1895a, b
Le Sueur, 1827
McAtee, 1907
Myers, 1926, 1927
Necker, 1935, 1938, 1939
Ortenburger, 1921
Pope, 1944
Ramsey, 1901
Schmidt and Necker, 1935
Shockley, 1949
Springer, 1928
Stille and Edgren, 1948
Swanson, 1939
Weed, 1922

IOWA
Bailey, 1941
Blanchard, 1923
Loomis, 1948
Muller, 1921
Ruthven, 1910a, 1912b.
  1919

Somes, 1911
Stromsten, 1923

KANSAS
Brenkelman, 1936
Brennan, 1934, 1937
Brumwell, 1940
Burt, 1927, 1931, 1933,
  1935, 1937
—— and Hoyle, 1934
Cragin, 1880a, b, 1885a, b,
  1886, 1894
Dice, 1923
Gloyd, 1928, 1932
Grant, 1927a
Hall and Smith, 1947
Hallowell, 1857a
Householder, 1916
Linsdale, 1927
Smith and Leonard, 1934
Taylor, 1929, 1943
Tihen, 1937
—— and Sprague, 1939
Williston, 1899

KENTUCKY
Bailey, 1933
Blanchard, 1925
Chenoweth, 1949
Dury and Gessing, 1940
—— and Williams, 1933
Funkhouser, 1925
Garman, H., 1894
Hibbard, 1936
Welter and Carr, 1939

LOUISIANA
Baur, 1890b, c, 1893c
Beyer, 1900
Burt, 1944
Carr, 1937a, 1949
Neill, 1951
Penn, 1943
—— and Pottharst, 1940
Strecker and Frierson, 1926
Viosca, 1923, 1927, 1931

MAINE
Brown, 1863
Fogg, 1862a, b
Fowler, 1942
Manville, 1939
Norton, 1929
Verrill, 1863
Williamson, 1832

MARYLAND
Brady, 1937

Mearns, 1897
Mertens, 1928
Mittleman, 1945b
Murphy, 1916
Nichols, 1917, 1939a, b, 1947
Noble, 1929
Otto, 1914
Raney and Lachner, 1942
Reed and Wright, 1909
Schoonhoven, 1911
Skinner, 1909
Smith, 1899
Stewart, 1947
Thompson, 1889
Townsend, 1926
Weber, 1928
Wright, 1918b, 1919

NORTH CAROLINA
Breder and Breder, 1923
Brimley, 1907, 1909, 1915,
    1917, 1918a, b, 1920a,
    1922, 1926, 1927, 1928,
    1942–1943
—— and Mabee, 1925
—— and Sherman, 1908
Coker, 1906a, b, 1920
Coues, 1871
—— and Yarrow, 1879
Dunn, 1917
Engels, 1942
Gray, 1941
Hay, O. P., 1923
Hildebrand and Hatsel,
    1926
Schmidt, K. P., 1916
Weller, 1930
Wright, 1918b

NORTH DAKOTA
Allen, J. A., 1874
Coues and Yarrow, 1878
Hankinson, 1929
Hayden, 1875
Telford and Stevens, 1942
Wheeler, 1947
Young, 1924

OHIO
Butler, 1887
Conant, 1932, 1934, 1938a,
    b
Dexter, 1948
Duellman, 1947
Fichter, 1947
Grant, 1935b
James, 1887
Kirtland, 1838

McLain, 1899b
Morse, 1901, 1903, 1904
Smith, W. H., 1882
Trautman, 1931
Wood and Duellman, 1947

OKLAHOMA
Anonymous, 1917, 1946a, b,
    c, 1947a, b
Burger, Smith, and Smith,
    1949
Burt, 1935
—— and Hoyle, 1934
Cope, 1893
Force, 1925, 1928, 1930
Glass, 1949
Hurter and Strecker, 1909
Moore and Ridney, 1941
Ortenburger, 1926a, b,
    1927a, b, 1929a, b
—— and Freeman, 1930
Self, 1938
Smith and Leonard, 1934
Trowbridge, 1937
Wickham, 1922

OREGON
Anderson and Slater, 1941
Brode, 1932
Evenden, 1948
Fitch, 1936
Gordon, 1931
Graf, Jewett and Gordon,
    1939
Harlan, 1837
Seeliger, 1945
Van Denburgh, 1897

PENNSYLVANIA
Atkinson, 1901
Conant and Downs, 1940
Cope, 1890
Cramer, 1935
Dunn, 1915a
Evermann, 1918
Fowler, 1916
Mattern and Mattern, 1917
Mittleman, 1945a, b
Netting, 1927a, 1932, 1935,
    1936a, b, c, 1939, 1944,
    1946
Pawling, 1939
Roddy, 1928
Rosenberger, 1936
Stewart, 1928
Stone, 1906
Surface, 1908
Wied, 1839

RHODE ISLAND
Anonymous, 1918
Bumpus, 1884–1886
Drowne, 1905

SOUTH CAROLINA
Catesby, 1730–1748
Chamberlain, 1938
Conant and Downs, 1940
Harper, 1940
Jopson, 1940
Neill, 1951
Obrecht, 1946
Pickens, 1927
Schmidt, 1924b
True, 1883

SOUTH DAKOTA
Coues and Yarrow, 1878
Over, 1923
Visher, 1914
Wied, 1865

TENNESSEE
Blanchard, 1922
Burger, Smith, and Smith,
    1949
Cagle, 1937
Edney, 1949
Gentry, 1941, 1944
King, 1939
Necker, 1934
Parker, 1937, 1939, 1948
Rhoads, 1895
Schoffman, 1949
Shoup, Peyton, and Gentry,
    1941
Smith and Glass, 1947
Stejneger, 1923
Troost, 1844
Wood, 1946

TEXAS
Brown, 1903
Burger, Smith, and Smith,
    1949
Burt, 1935
Cope, 1880, 1888, 1892,
    1893
Cragin, 1894
Gadow, 1905a
Garman, S., 1887, 1892
Gunter, 1945
Hallowell, 1857c
Hamilton, R., 1944, 1947
Harwood, 1932
Hurter and Strecker, 1909
Jones, Mounts, and Wolcott,
    1945

Mearns, 1907
Mittleman and Brown, 1947
Murray, 1938
Parks, Archibald, and Cald-
well, 1939
—— and Cory, 1936
Roemier, 1849
Schmidt and Smith, 1944
Shannon and Smith, 1949
Smith, H., 1946
—— and Brown, 1946
Stone, 1903
Strecker, 1902, 1908b, c,
1909a, b, c, 1910a, b,
1915, 1922, 1926a, b, c, d,
1927a, b, c, 1928a, b,
1929a, b, c, d, e, f, 1935
—— and Williams, 1927,
1928
Test, 1894

UTAH
Ruthven, 1907
Tanner, 1927
Woodbury, 1931
—— and Hardy, 1940, 1948

VERMONT
Babbitt, 1936
Loveridge, 1931
Thompson, 1842

VIRGINIA
Brady, 1924, 1925, 1927
Brimley, 1918a
Conant, 1945
Dunn, 1915b, c, 1918a,
1920c, 1936
Fowler, 1925c
Lynn, 1936
Richmond, 1945
—— and Goin, 1938

WASHINGTON
Baird and Girard, 1852
Blanchard, 1921
Cooper, 1860
Dice, 1916
Evenden, 1948
Johnson, 1942
Owen, 1940
Seeliger, 1945
Slater, 1939

Svihla and Svihla, 1933
Van Denburgh, 1897

WEST VIRGINIA
Bond, 1931
Frum, 1947
Green, 1937a, b, 1941
Llewellyn, 1940
Netting, 1940
Strader, 1936

WISCONSIN
Cahn, 1929
Edgren and Stille, 1948
Higley, 1889
Hoy, 1883
Lapham, 1852
Necker, 1939
Pearse, 1923a, b
Pope, 1928, 1931
—— and Dickinson, 1928
Schmidt, 1926
Stille and Edgren, 1948
Wagner, 1922

Canada

GENERAL
Mills, 1948
Provancher, 1874–1875

ALBERTA
Fowler, 1934

BRITISH COLUMBIA
Anonymous, 1932
Carl, 1944
Cowan, 1936, 1938
Evenden, 1948
Kermode, 1909, 1932
Logier, 1932
Lord, 1866

Newcombe, 1932
Thacker, 1924

MANITOBA
Criddle, 1919
Jackson, 1934
Norris-Elye, 1949
Seton, 1918

NOVA SCOTIA
Jones, 1865
Mackay, 1898
Piers, 1897

ONTARIO
Allin, 1940

Brown, 1928
Garnier, 1881
Logier, 1925, 1928, 1930,
1931, 1939, 1941
Nash, 1905, 1908
Patch, 1919
Snyder, Logier, and Kurata,
1942
Toner, 1936
—— and Edwards, 1938
Wright and Simpson, 1920

QUEBEC
Trapido and Clausen, 1938
Patch, 1925

Baja California

Belding, 1887
Bocourt, 1870–1909
Cowles and Bogert, 1936
Hallowell, 1854
Linsdale, 1932
Meek, 1905

Miller, 1946
Mocquard, 1899
Mosauer, 1935
Nelson, 1921
Schmidt, 1922

Seeliger, 1945
Shaw, 1946, 1947
Slevin, 1926
Terron, 1919
Van Denburgh, 1895, 1921

INDEX

[Where more than one page number is given, the principal reference appears in boldface. Under each family and genus in the text there are a brief description and a key to the forms included (except where there is only one). The species accounts are organized on this outline: common and technical names, range, distinguishing features, description, habits, breeding, feeding, economic importance. References are not given in the Index for items involved in this standard treatment.]